Sedimentology of Aqueous Systems

Sedimentology of Aqueous Systems

Edited by Cristiano Poleto and
Susanne Charlesworth

WILEY-BLACKWELL
A John Wiley & Sons, Ltd., Publication

Blackwell Publishing was acquired by John Wiley & Sons in February 2007. Blackwell's publishing program has been merged with Wiley's global Scientific, Technical and Medical business to form Wiley-Blackwell.

Registered office: John Wiley & Sons Ltd, The Atrium, Southern Gate, Chichester, West Sussex, PO19 8SQ, UK

Editorial offices: 9600 Garsington Road, Oxford, OX4 2DQ, UK
The Atrium, Southern Gate, Chichester, West Sussex, PO19 8SQ, UK
111 River Street, Hoboken, NJ 07030-5774, USA

For details of our global editorial offices, for customer services and for information about how to apply for permission to reuse the copyright material in this book please see our website at www.wiley.com/wiley-blackwell

Library of Congress Cataloguing-in-Publication Data

Sedimentology of aqueous systems / edited by Cristiano Poleto and Susanne Charleworth.
p. cm.
Includes bibliographical references and index.
ISBN 978-1-4443-3290-2 (hardcover : alk. paper) 1. Alluvium. 2. Sediment transport.
I. Poleto, Cristiano. II. Charlesworth, Susanne.
QE581S394 2010
551.3′53—dc22

2009042576

A catalogue record for this book is available from the British Library.

Set in 9 on 11.5 pt Sabon by Toppan Best-set Premedia Limited
Printed and bound in Singapore by Fabulous Printers Pte Ltd

1 2010

Contents

Contributors

Almudena Ordóñez
Oviedo School of Mines, University of Oviedo, Spain

Chauncey W. Anderson
United States Geological Survey, USA

Jonathan S. Barton
National Aeronautics and Space Administration, USA

Edson Campanhola Bortoluzzi
Passo Fundo University, Brazil

Susanne Charlesworth
Department of Geography, Environment and Disaster Management, Coventry University, UK

Eduardo De Miguel
Environmental Geochemistry Group, Madrid School of Mines, Spain

Gregory G. Fisk
United States Geological Survey, USA

Jeffrey W. Gartner
United States Geological Survey, USA

Janet Gaskin
University of Ottawa, Canada

G. Douglas Glysson
United States Geological Survey, USA

Daniel J. Gooding
United States Geological Survey, USA

John R. Gray
United States Geological Survey, USA

Mark G.J. Hartl
Centre for Marine Biodiversity and Biotechnology, School of Life Sciences, Heriot-Watt University, Edinburgh, UK

Nancy J. Hornewer
United States Geological Survey, USA

Christopher G. Ingersoll
United States Geological Survey, USA

Matthew C. Larsen
United States Geological Survey, USA

Ariane Laurenti
Department of Pathology, Federal University of Santa Catarina, Brazil

Donald D. MacDonald
MacDonald Environmental Sciences Ltd., Canada

Jamie P. Macy
United States Geological Survey, USA

Smokey A. Pittman
Graham Matthews and Associates, USA

Cristiano Poleto
Hydraulic Research Institute, Federal University of Rio Grande do Sul, Brazil

Patrick P. Rasmussen
United States Geological Survey, USA

Colin D. Rennie
University of Ottawa, Canada

Maria Alice Santanna dos Santos
Federal University of Santa Maria, Brazil

Marcos Antonio Villetti
Federal University of Santa Maria, Brazil

Scott A. Wright
United States Geological Survey, USA

Andrew C. Ziegler
United States Geological Survey, USA

Introduction: the sedimentology of aqueous systems

Cristiano Poleto[1] & Susanne Charlesworth[2]

[1]*Hydraulic Research Institute, Federal University of Rio Grande do Sul, Brazil*
[2]*Department of Geography, Environment and Disaster Management, Coventry University, UK*

Studies of sediments have been increasingly highlighted internationally because of their negative influence on the quality of the aquatic environment. The problems caused by these sediments may be physical, whereby they can lead to the aggradation of water bodies and the subsequent obstruction of engineered structures, or they may be chemical upon the release of pollutants previously transported in association with them. To mitigate these impacts, sedimentological studies need to be structured carefully to meet specific needs. This volume therefore aims to explore sediments logically, from their origins, impacts on the aquatic environment to their collection in the field and subsequent analytical methodologies, in order to provide a thorough examination of one the symptoms of anthropogenic impact on the environment.

There are therefore various themes that run through the book:

- establishment of the sources of sediments, and how they can be characterized and traced;
- transportation of sediments, their pollutant burden in the environment and their final destination;
- their potential to impact negatively on receiving waters and biota;
- the production of useful and accurate data for use in models and management strategies to mitigate such impacts.

The chapters are written by international experts in their fields and provide in-depth considerations of current and topical research in the field of aqueous sedimentology.

Sedimentology of Aqueous Systems, 1st edition.
Edited by Cristiano Poleto and Susanne Charlesworth.

Thus, Chapters 1 and 2 critically explore the limitations of approaches used in the collection of suspended and bed sediment samples in order to address the perennial problem of ensuring quality data are produced based on the collection of quality samples. They cover traditional techniques of extracting physical samples and advances in the various means of remotely monitoring sediment concentration.

Chapter 3 leads on from the collection of sediment samples in Chapters 1 and 2 by an in-depth consideration of sediment characteristics and characterisation. It therefore details the behaviour of particles in water, the forces involved in both holding them together and forcing them apart as well as the energy required in such processes. Techniques such as light scattering, X-ray diffraction and electron microscopy are critically examined and nuclear magnetic resonance and infrared spectroscopy bring the subject right up to date allowing the reactivity of particulates to be assessed and their ability to interact with the environment to be better understood.

Chapters 4 and 5 address the environment where 86% of us will be living by 2050 (UN Economic and Social Affairs 2008): urban areas. These two chapters give an overview of the sediments themselves, their sources, physicochemical characteristics and contribution to environmental degradation, and in particular the urban aquatic environment where the behavior of water is constrained into pipes and channels: "out of sight, out of mind". The concept of sustainable drainage (SUDS) is introduced here to tackle end-of-pipe solutions to urban aquatic degradation, which tend to treat the symptoms rather than the cause. Here the properties of the sediments given in Chapter 4 are applied as a management strategy in which people modify their behavior to take account of water, rather than the other way around.

Chapter 6 investigates the impacts contaminated sediments can have on biota by assessing toxicity using a variety of testing models. Environmental impact assessments are examined using an integrated approach, having built upon the individual tests.

Chapter 7 puts all of the previous chapters into context by addressing the tools available for assessing sediments, for use in evaluation of contaminated areas and as a first step in planning strategies for safe remediation and disposal. This chapter clarifies the guidelines for various environments, which tend to be used interchangeably, giving proper attention to the criteria used to tackle what is an important issue in managing contaminated sediment.

It is our hope that this volume can contribute to the advancement of knowledge, certainly, but can also assist in the preservation of the aquatic environment by providing some of the information needed in the study and management of these dynamic and vital areas.

Reference

UN Department of Economic and Social Affairs (2008) *World Urbanization Prospects: The 2007 Revision*. New York: United Nations. http://www.un.org/esa/population/publications/wup2007/2007WUP_ExecSum_web.pdf.

1 Surrogate technologies for monitoring suspended-sediment transport in rivers

John R. Gray[1] & Jeffrey W. Gartner[1] (editors)

Chauncey W. Anderson[1], Gregory G. Fisk[1], Jeffrey W. Gartner[1], G. Douglas Glysson[1], Daniel J. Gooding[1], John R. Gray[1], Nancy J. Hornewer[1], Matthew C. Larsen[1], Jamie P. Macy[1], Patrick P. Rasmussen[1], Scott A. Wright[1] & Andrew C. Ziegler[1]

[1]*United States Geological Survey, USA*

Advances in technologies for suspended-sediment transport monitoring programs in rivers show varying degrees of promise toward supplanting traditional data-collection methods based on routine collection of physical samples and subsequent laboratory analyses. Mostly commercially available technologies operating on bulk-, laser-, and digital-optic, pressure-difference, and acoustic principles have been or are the foci of field or laboratory tests by the US Geological Survey (USGS) and other organizations. Advantages and limitations associated with each suspended-sediment-surrogate technology, considered with deployment-site sedimentological characteristics and monitoring objectives, can be factored into the design of program networks using the most appropriate technology. Examples of factors that can limit or enhance the efficacy of a surrogate technology include cost (purchase, installation, operation, and data analysis), reliability, robustness, accuracy, measurement volume, susceptibility to biological fouling, volumetric- versus mass-concentration determinations, and suitability to the range of in-stream mass concentrations and particle-size distributions (PSDs). All of the *in situ* technologies require periodic site-specific calibrations to infer the sedimentary characteristics representative of the entire channel cross section.

In March 2009, the USGS endorsed bulk optics (turbidity) for use in operational suspended-sediment monitoring programs, the first sediment-surrogate technology to receive USGS endorsement. Other technologies are likewise being considered for USGS acceptance.

Nevertheless, hydroacoustic technologies show the most promise for use in operational suspended-sediment monitoring programs. A fixed-mounted, self-contained single-frequency acoustic backscatter instrument supported by appropriate deployment, calibration, and data-analyses protocols presents the prospect for automated collection of continuous time-series suspended-sediment-concentration data in selected river reaches. The anticipated adaption of a multi-frequency acoustic Doppler current profiler in fixed-mounted mode portends the potential for even more accurate monitoring of suspended-sediment concentration (SSC) and transport, possibly by particle-size classes. Laser-optic instruments deployed *in situ* or manually that provide PSDs and concentrations also show considerable promise.

Endorsement and broad-scale deployment of certifiably reliable sediment surrogate technologies supported by operational and analytical protocols are revolutionary concepts in fluvial sedimentology. The benefits could be enormous, providing for safer, more frequent and consistent, arguably more accurate, and ultimately less expensive fluvial-sediment data collection for use in managing the world's sedimentary resources.

Sedimentology of Aqueous Systems, 1st edition.
Edited by Cristiano Poleto and Susanne Charlesworth.

1.1 Introduction

Fluvial sediment and sorbed materials are the most widespread pollutants affecting US rivers and streams (US Environmental Protection Agency 2008). The need for reliable, comparable, cost-effective, spatially and temporally consistent data to quantify the clarity and sediment content of waters of the USA has never been greater. Yet resources dedicated to this need have been in decline for more than two decades. For instance, the number of sites at which the USGS collected nationally consistent daily sediment data in 2006 was about a quarter of the number operated in 1981 (David W. Stewart, USGS, personal communication 2008) (the USA has never had a federally funded, national sediment monitoring and assessment program analogous to the National Streamflow Information Program (USGS 2008a) for flow monitoring). This precipitous decrease in sediment monitoring over a quarter century by the USGS – the Federal agency tasked by the US Department of the Interior to collect, archive, and disseminate US water data, including fluvial sediment (Glysson & Gray 1997; USGS 2008b) – is due to several factors, principally cost (Gray *et al.* 2003). The decrease in monitoring is of particular concern, given that the physical, chemical, and biological damages attributable to fluvial sediment in North America alone are estimated to range from US$20 billion to US$50 billion annually (Pimental *et al.* 1995; Osterkamp *et al.* 1998, 2004; Gray & Osterkamp 2007). The relative dearth of adequate, consistent, and reliable data describing fluvial-sediment fluxes hinders development of technically supportable management and remedial plans around the world.

Historically, suspended-sediment flux data in the US have been produced by gravimetric analyses performed on physical samples collected by manual or automatic samplers (see Edwards & Glysson 1999; Bent *et al.* 2003; Davis 2005; Nolan *et al.* 2005; Gray *et al.* 2008). These traditional data-collection methods tend to be expensive, labor intensive, time-consuming, difficult, and under some conditions, hazardous. Specialized instruments and considerable training in their proper use are prerequisites for obtaining reliable samples. The characteristic paucity of the derived data – particularly at the higher flows that are most influential in mass transport of sediment – can lead to inadequate definition of the temporal variability in SSCs and suspended-sediment discharges, or loads (SSLs). Consequently, temporal interpolations and spatial corrections are commonly required to develop the requisite time series that is used with an associated time series of water-discharge data to produce sub-daily and daily records of SSL (Porterfield 1972; Koltun *et al.* 2006).

Sediment-surrogate technologies are defined as instruments coupled with operational and analytical methodologies that enable acquisition of temporally and (or) spatially dense fluvial-sediment data sets without the need for routine collection and analysis of physical samples other than for periodic calibration purposes. Selected sediment-surrogate technologies show varying degrees of promise toward providing the types, quality, and density of fluvial-sediment data needed to improve SSL computations. Potentially useful instruments and methods for inferring the physical characteristics of fluvial sediments (Bogen *et al.* 2003; Gartner *et al.* 2003; Gray *et al.* 2003a,b; Gray 2005; Topping *et al.* 2007; Gray & Gartner 2009) are being developed and tested worldwide. For example, through the informal USGS Sediment Monitoring Instrument and Analysis Research Program (Gray 2003; Gray & Simões 2008), the USGS and collaborators in other government agencies, academia, and the private sector are testing several instruments for measuring SSCs and, in some cases, PSDs. These instruments, operating on bulk-, laser-, and digital-optic, pressure-difference, and acoustic principles are being evaluated in North American rivers and laboratories. To make the transition from research to operational monitoring applications, these new technologies must be rigorously tested with respect to accuracy and reliability in different physiographic and (or) laboratory settings as appropriate, and their performances must be compared with data obtained by the aforementioned traditional methods and to available quality-control data. In most cases, performance comparisons should include concurrent collection of data by traditional and new techniques for a sufficient period – probably years – and in a variety of river types and flow conditions to identify potential bias and minimize differences in precision between the old and new technologies.

The *in situ* technologies presented herein require periodic site-specific calibrations to infer the sedimentary characteristics representative of the entire channel cross section or reach segment. This requirement is anticipated to be substantial for new river-monitoring applications, but may diminish as comparative data accumulate.

None of the technologies represents a panacea for sediment monitoring in all rivers under all flow and sediment-transport conditions. However, with careful matching of surrogate-monitoring technologies to selected river reaches and objectives, it is becoming possible to remotely, continuously, and accurately monitor SSCs and SSLs (and in some cases, PSDs) in a variety of river types, flow conditions, and sedimentological regimes. In some cases, the computed SSC values and perhaps other data types may be qualified with estimates of uncertainty (USGS 2005).

These are revolutionary concepts in the discipline of sedimentology when considered from an operational perspective. The benefits of such applied capability could be enormous, providing for safer, more frequent and consistent, arguably more accurate, and ultimately less expensive fluvial-data collection for use in managing the world's sedimentary resources.

This chapter describes five suspended-sediment-surrogate technologies evaluated in field or laboratory settings by the USGS for monitoring fluvial sediment with varying degrees of potential toward providing continuous, largely automated time-series data used for computing SSLs in rivers. All five of the *in situ* technological applications provide continuous SSC data, and at least two of those may provide PSD data.

The chapter starts with an overview of traditional instruments and techniques for suspended-sediment sampling, against which the surrogate technologies are evaluated. Descriptions of the theory, applications, some advantages, limitations, and costs of each surrogate technology are presented and compared. A subjective evaluation of the efficacy of each technology concludes this chapter. Use of firm, brand, or trade names are for identification purposes only and do not constitute endorsement by the US Government.

1.1.1 Background: traditional suspended-sediment-sampling techniques

Suspended sediment is that part of the total-sediment load (Fig. 1.1) carried in suspension by the turbulent components of the fluid or by Brownian movement (ASTM International 1998). Instruments and methods for collecting suspended-sediment data in the USA have evolved considerably since 1838 when the US Army Corps of Engineers' Captain Andrew Talcott first sampled the Mississippi River (Federal Interagency Sedimentation Project 1940). The earliest suspended-sediment samples were collected by use of instantaneous samplers such as an open container or pail. By 1939, at least nine different types of sediment sampler were being used by US agencies. Most of the samplers had been developed by independent investigators, lacked calibrations, and were deployed using a variety of methods. A 1930s survey of sediment-sampling equipment used in the US indicated that the 30 instantaneous samplers studied had limited usefulness either because of poor intake-velocity characteristics or because of the short filament of water–sediment mixture sampled (Federal Interagency Sedimentation Project 1940; Nelson & Benedict 1950; Glysson 1989).

In 1939, six US Federal agencies and the Iowa Institute of Hydraulic Research organized a committee to consider the development of sediment samplers, sampling techniques, and laboratory procedures, and to coordinate such work among the Federal agencies "actively concerned with the

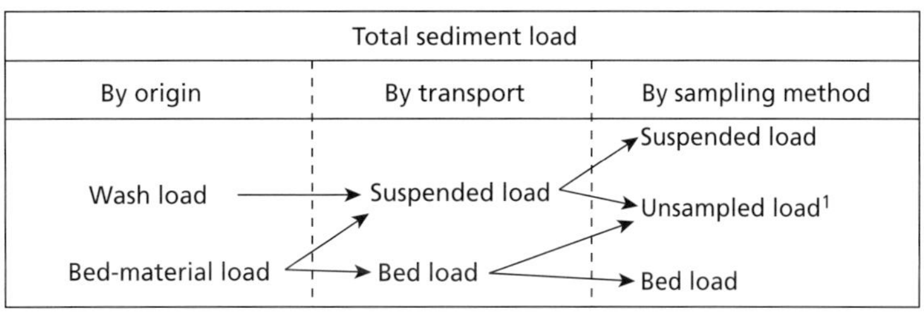

[1]That part of the sediment load that is not collected by the depth-integrating suspended-sediment and pressure-difference bedload samplers used, depending on the type and size of the sampler(s). Unsampled-load sediment can occur in one or more of the following categories: (a) sediment that passes under the nozzle of the suspended-sediment sampler when the sampler is touching the streambed and no bedload sampler is used; (b) sediment small enough to pass through the bedload sampler's mesh bag; (c) sediment in transport above the bedload sampler that is too large to be sampled reliably by the suspended-sediment sampler; and (d) material too large to enter the bedload-sampler nozzle.

Fig. 1.1 Components of total-sediment load considered by origin, by transport, and by sampling method.
From Diplas *et al.* (2008).

sedimentation problem" (US Department of Agriculture 1965). This committee has evolved into three entities: the present-day Subcommittee on Sedimentation of the Advisory Committee on Water Information; Technical Committee; and Federal Interagency Sedimentation Project (FISP) (Sedimentation Committee of the Water Resources Council 1976; Skinner 1989; Glysson & Gray 1997; Federal Interagency Sedimentation Project 2008; Subcommittee on Sedimentation 2008). The purpose of the FISP is to study methods and equipment used in measuring the sediment discharge of streams and to improve and standardize equipment and methods where practicable. Through the FISP, an integrated system of sediment samplers, sampling procedures, and analytical methods was developed and is codified in US Federal sediment-monitoring standards (Federal Interagency Sedimentation Project 2008; Edwards & Glysson 1999) and incorporated to a large degree into international standards (ISO 1992a,b, 1997, 2002, 2005). Today, FISP products and techniques form the framework for collection of consistent, reliable, quality-assured fluvial-sediment data in the USA and many other countries.

The bulk of suspended-sediment data collected by US agencies are acquired using manually deployed FISP isokinetic samplers (Davis 2005), and traditional sampling methods described by Edwards & Glysson (1999), Nolan *et al.* (2005), and Gray *et al.* (2008). These include rigid-bottle samplers (bottle samplers), and flexible bag samplers (bag samplers) that fill at a rate determined by the product of the ambient stream velocity at the sampler nozzle and the nozzle's area. These samplers are designed to collect a representative velocity-weighted sample of the water–sediment mixture. FISP isokinetic samplers are designed to ensure that the water velocity entering the intake nozzle is within about 10% of the stream velocity incident on the nozzle throughout the samplers' operable velocity range. If the velocity of water entering the nozzle differs substantially from the ambient velocity, a bias in the SSC and PSD values computed for the sample may result (Federal Interagency Sedimentation Project 1941; Gray *et al.* 2008) (Fig. 1.2). This bias is a result of differing momentums between water and the entrained sediment, and can be particularly pronounced when sand-size material constitutes a substantial fraction of the material in suspension.

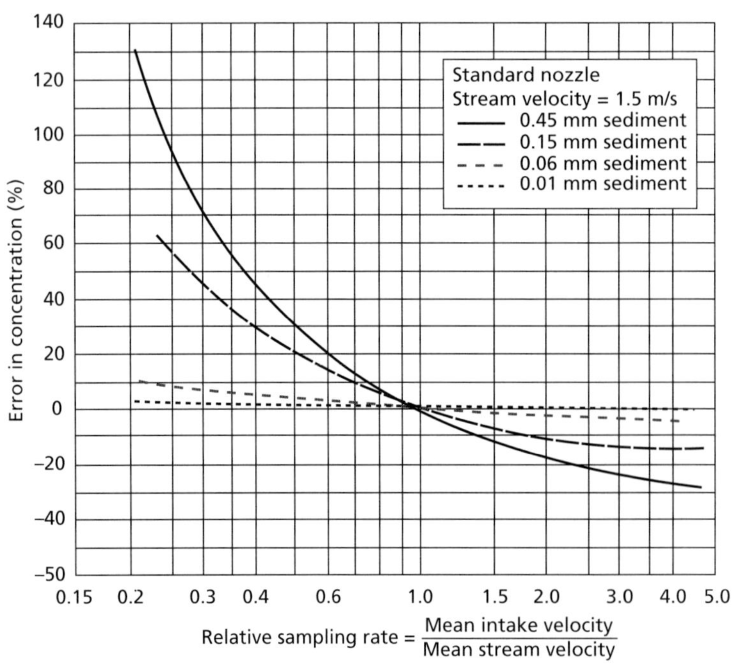

Fig. 1.2 Effect of sampling rates on measured SSCs for four sediment-size distributions.
From Gray *et al.* (2008); adapted from the Federal Interagency Sedimentation Project (1941).

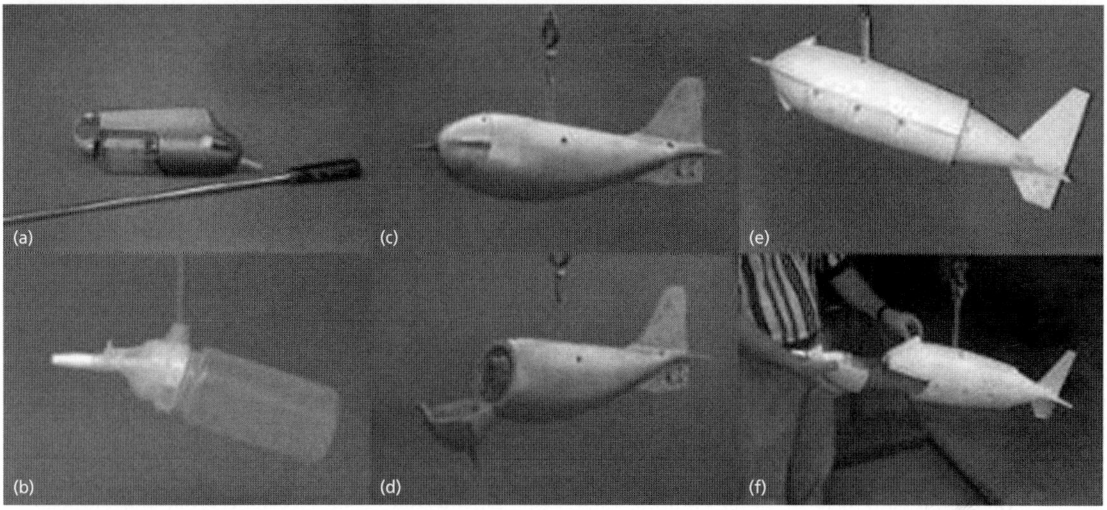

Fig. 1.3 Examples of Federal Interagency Sedimentation Project suspended-sediment samplers. (a) A US DH-48 rigid-bottle sampler; (b) a US DH-81 rigid-bottle sampler; (c) a US D-74 rigid-bottle sampler closed, and (d) open; (e) a US D-96 flexible-bag sampler closed, and (f) open.

A list of FISP suspended-sediment samplers and selected attributes is provided by Davis (2005) and Gray *et al.* (2008). Examples of FISP rigid-bottle- and flexible-bag-type samplers are shown in Fig. 1.3.

A depth-integrating sampler collects and accumulates a velocity- or discharge-weighted sample as it descends and ascends through the water column provided that an appropriate constant transit rate is not exceeded in either transit direction, and the sample container does not overfill. A point-integrating sampler uses an electrically activated valve, enabling the operator to sample points isokinetically either in parts of, or throughout, the water column. Both types of samplers integrate the water column from the water surface to within about 0.1 meters (m) of the bed.

When properly deployed in a single vertical (or, in the case of the point-integrating sampler, at multiple points in a vertical), FISP isokinetic samplers provide representative samples for the parts of the stream sampled. When deployed using either the equal-discharge-increment (EDI) or equal-width-increment (EWI) sampling method (Edwards & Glysson 1999; Nolan *et al.* 2005), an isokinetic sampler integrates a sample proportionally by velocity and area, resulting in a discharge-weighted sample that contains an SSC and PSD representative of the suspended material in transport throughout the cross section at the time that of sample collection.

Although the aforementioned manual samplers have considerable benefits – most notably the acquisition of demonstrably reliable suspended-sediment data from rivers – they have inherent drawbacks. For example, total costs associated with the manual deployment of isokinetic samplers and subsequent sample analytical costs can be substantial or even prohibitive with respect to available resources. Several safety considerations must be addressed any time a hydrographer works in, over, or near a watercourse. The sparse temporal distribution of the derivative data – often but a single observation per day – requires that daily SSL computations be based on estimated SSC values and (or) indexed to another more plentiful if imperfect predictive data source such as river discharge by a sediment-transport curve (Glysson 1987; Gray *et al.* 2008).

1.1.2 Performance criteria for concentrations and particle-size distributions produced by suspended-sediment-surrogate technologies

The reliability and efficacy of data produced by a sediment-surrogate technology are predicated on the adequacy of its calibrations. Two general types of calibration are used: instrument calibrations and cross-section calibrations. Instrument calibration refers in a statistical sense to the precision and variance of data derived from the surrogate measurement in the sampled region (the instrument-measurement realm) to an actual value in the corresponding realm ascertained by independent measurement. Cross-section calibration refers to correlation of the derived data to the mean constituent value occurring in the full stream cross section or stream segment at the time of the measurement, typically using FISP samplers and sampling techniques. Although the instrument-measurement realm generally corresponds to a volume, it is referred to herein in practical terms with respect to the instrument sensor as a point for a local, minute-volume measurement; a water column; or a beam (or average of multiple beams).

Derivations of true mean cross-section constituent values are unlikely from consistently false instrument-measurement-realm values, similar to the axiomatic "garbage in, garbage out" concept in computer science. On the other hand, inferences of false mean cross-section constituent values from true instrument-measurement-realm values can and often do occur. False inferences from true surrogate data can result from heterogeneity typically associated with the occurrence and transport of suspended sediment in the cross section, and is the reason for the need for cross-section calibrations. Therefore, the most meaningful measure of a surrogate technology's reliability is derived from calibrations performed within the instrument-measurement realm. Hence, criteria to evaluate sediment-surrogate technologies should be based solely on instrument calibrations in the instrument-measurement realm, if possible. However, the ultimate measure of the efficacy of a surrogate technology to monitor suspended sediments in rivers is its ability to quantify adequately the sedimentary characteristics of interest over the entire cross section.

Validation of a suspended-sediment-surrogate technology requires evaluation criteria and a well-conceived and -administered testing program (Gray *et al.* 2002; Gray & Glysson 2005). The following are some qualitative criteria for selecting and deploying a surrogate technology:

- capital and operating costs should be affordable with respect to the objectives of the monitoring program in which the surrogate instrument is deployed;
- the technology should be able to measure SSCs, and in some cases, PSDs, throughout the range of interest (but not necessarily throughout the entire potential environmental range);
- the equipment should be robust and reliable, that is, prone to neither failure nor signal drift;
- the method should be sufficiently simple to deploy and operate by a field technician with a reasonable amount of appropriate training;
- the derived data should be relatively simple and straightforward to use in subsequent computations and (or) accompanied by standard analytical procedures as computational routines for processing the data.

Quantitative criteria for acceptable accuracies of the derived data are difficult to develop for all potential applications, in part because of substantial differences in river sedimentary and flow regimes. For example, accuracy criteria for rivers transporting mostly silt and clay should be set more stringently (intolerant of larger-magnitude uncertainties) than those for rivers that transport comparatively large fractions of sand. However, there is a clear need for consistency in PSD and SSC criteria on the part of instrument developers, marketers, and users.

To this end, quantitative acceptance criteria developed for PSD and SSC data produced by a laser-diffraction instrument (Gray *et al.* 2002) have been generalized for evaluating data from other suspended-sediment surrogate instruments. At least 90% of PSD values between 0.002 and 0.5 mm median diameter are required to be ±25% of true median diameters. In the absence of a more rigorous evaluation, this criterion has been applied to all particle sizes in suspension.

SSC acceptance criteria range from ±50% uncertainty at lowest SSCs to ±15% uncertainty for SSC's exceeding 1 gram per liter (g/L). The criteria presented in Table 1.1 are adapted from Gray *et al.* (2002).

These criteria pertain solely to the performance of a surrogate technology within its physical realm of measurement. Routine calibrations to correlate instrument signals to mean cross-sectional SSC values are required for all of the *in situ* instruments presented herein.

Table 1.1 Acceptance criteria for SSC data. The data are considered acceptable when they meet these criteria 95% of the time.

Suspended-sediment concentration		Acceptable uncertainty
Minimum (g/L)	Maximum (g/L)	± Percent
0	<0.01	50
0.01	<0.1	50-25 computed linearly
0.1	<1.0	25-15 computed linearly
1.0	—	15

Adapted from Gray *et al.*(2002).

1.1.3 Ranges in US suspended-sediment concentrations and suspended-sediment discharges

Because of the spatial and temporal variability in river sedimentological regimes, only generalities regarding the expected range of SSCs and PSDs in rivers can be made in the absence of site-specific data. Rainwater (1962) produced an empirically derived map of the 48 conterminous United States showing mean SSC ranges for rivers, generalized over the entire land area, for seven logarithmically based SSC ranges. The SSC ranges were computed and delineated as average annual discharge-weighted mean SSCs, derived from annual measured SSL values divided by their paired annual streamflow values at streamgages. Computed SSC values in the largest range exceeded about 48 g/L.

Meade & Parker (1985) simplified the Rainwater (1962) map into four SSC ranges: less than 0.3 g/L; 0.3–2 g/L; 2–6 g/L; and more than 6 g/L (Fig. 1.4). They also produced a similar-type map for Alaska, USA, using other information sources (Robert Meade, personal communication 1985). These maps (Fig. 1.4) also portray mean annual SSLs from selected river basins to the coastal zone depicted by half circles at river mouths. The area of each half circle is proportional to the average annual sediment mass discharged to the coastal zone. The maps can serve as initial, general indicators of the suitability of a selected sediment-surrogate technology in a river reach of interest.

Additional information on the range of SSCs in US rivers is available from Smith *et al.* (1987), who computed percentile values for SSC data collected at 267 streamgages in medium and large river basins as part of the original USGS National Stream Quality Accounting Network (NASQAN) (USGS 2008c). The 25th, 50th, and 75th SSC percentiles were 0.02, 0.07, and 0.19 g/L, respectively. In 1995, the NASQAN network was redesigned to focus on the nation's largest river basins – the Mississippi (including the Missouri and Ohio), Columbia, and Colorado Rivers, and the Rio Grande. Horowitz (USGS, personal communication 2008) calculated the 10th, 25th, 50th, 75th, and 90th SSC percentiles for the 41 NASQAN streamgages in these large river basins for the period 1994–2006 as 0.01, 0.03, 0.12, 0.32, and 0.74 g/L, respectively.

Many streams transport near-zero SSCs at various times. At the other extreme, SSCs measured during surface runoff from 1989 to 1991 in the Little Colorado River Basin, Arizona and New Mexico, USA, commonly exceeded 100 g/L (Graf *et al.* 1996). SSC values at the Paria River at Lees Ferry streamgage, Arizona, USA, exceeding 1000 g/L have been reported (Beverage & Culbertson 1964).

In general, most of a river's annual sediment budget is transported during infrequent high-flow periods concomitant with relatively large SSCs. Any proposed suspended-sediment surrogate technology deployment should consider not only the statistics quoted above, but also the potential maximum SSC and, where appropriate, maximum particle sizes that might be transported in the period of interest.

1.1.4 Information germane to suspended-sediment-surrogate technology costs

After surrogate-technology efficacy is resolved, cost considerations are often of penultimate interest. The cost of producing reliable, quality-assured suspended-sediment data can be separated into four categories:

- the purchase price of the instrument;
- other capital costs associated with installation, and initial operation of the instrument;
- operational costs to maintain and calibrate the instrument;
- analytical costs to evaluate, reduce, compute, review, store, and disseminate the derived data.

Of these four categories, only the purchase price is straightforward to quantify. The others are dependent on several factors, including site location and physical characteristics, hydrological and

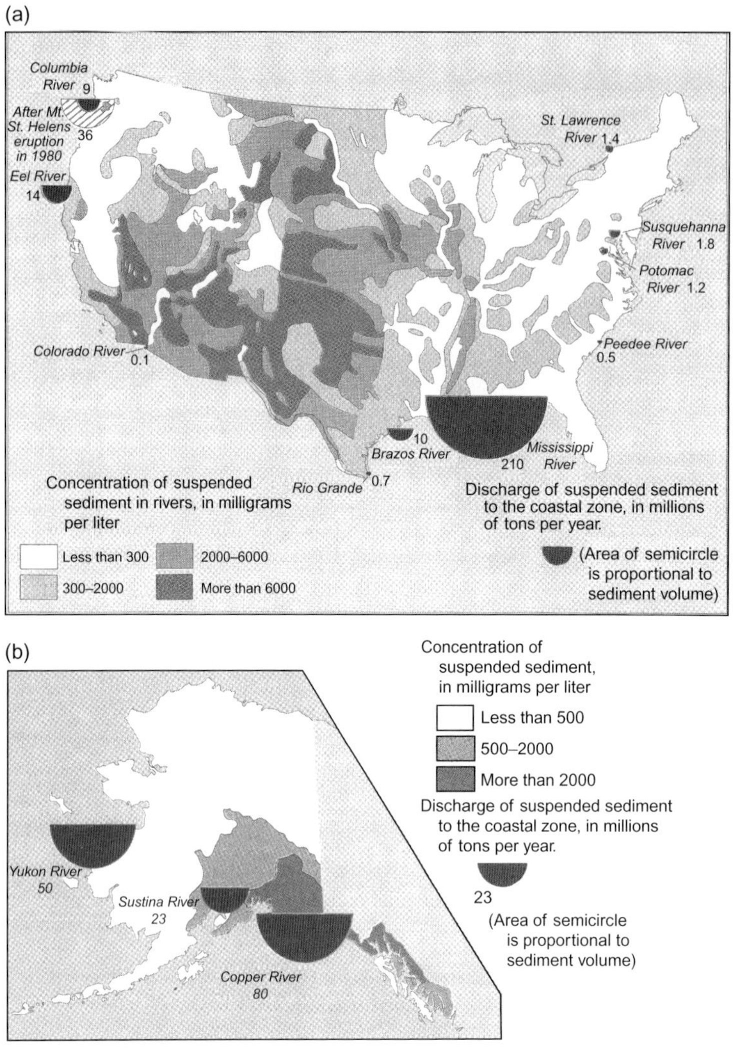

Fig. 1.4 Discharge of suspended sediment to the coastal zone, in millions of metric tonnes per year.

sedimentological regime, availability of electrical power, limitations associated with accessibility, safety considerations, and the time and complexity associated with data analysis. Additionally, any such information inevitably becomes obsolete due, in part, to technological advances, marketing competition, and changes in currency valuation. Hence, relative purchase prices are proffered for the surrogate instruments described herein compared with the actual (summer 2008) purchase price for the most common of the instruments, an *in situ* fully equipped turbidimeter. In some instances, other relevant cost information for a given technology that is considered reliable is provided. That information may be considered in light of the fact that the cost to compute, store, and provide a year's worth of daily SSL data at a USGS streamgaging station in 2001 (adjusted for inflation in 2008 dollars) is estimated to range from US$24,000 to US$78,000 (Gray 2003).

1.2 Technological advances in suspended-sediment-surrogate monitoring

The need for more affordable daily and more frequent time-series data, and for data collected with less risk to field personnel, coupled with advanced technological capabilities, is leading to a new era in fluvial-sediment monitoring. The following sections describe theoretical principles (Gray & Gartner 2004), selected examples of field applications, and advantages and limitations of five suspended-sediment-surrogate technologies that cover a range of transport conditions and are considered to be acceptable or promising by the USGS.

1.2.1 Turbidity (bulk optics)

Patrick P. Rasmussen, John R. Gray, Andrew C. Ziegler, G. Douglas Glysson, & Chauncey W. Anderson

1.2.1.1 Background and theory

Turbidity is an expression of the optical properties of a sample that cause light rays to be scattered and absorbed rather than transmitted in straight lines through the sample (Ziegler 2003; Anderson 2005). According to the USGS (2004), "Turbidity itself is not an inherent physical property of water (as is, for example, temperature), but rather is a measure of light scattering through a liquid as measured by detectors with known geometry," and hence is operationally defined. Measurements of turbidity are the most common means of determining water clarity and computing SSC in US rivers (Pruitt 2003). The instrument-measurement realm of a turbidimeter is usually a point in a stream (Secchi disk measurements being a notable exception). Both instrument and cross-section calibrations are normally performed.

The configuration of detectors and the source of light are important factors in the response of the turbidity instrument. Although comparisons among instruments with differing designs are often robust, they can also vary according to the character of the sample's matrix and particulates. Results from an interagency workshop held in 2002 demonstrated that turbidity data from different sources and instrumentation can be highly variable and are often in disagreement with each other, even when instrument-calibration methods are similar (Gray & Glysson 2003). In effect, instruments with different detector geometries and light sources often do not make equivalent measurements.

To reduce the variability among instruments measuring identical in-stream turbidity conditions, a USGS protocol (Anderson 2005) requires that turbidity data be reported based on instrument design in one of ten units, comprising eight new reporting units in addition to the two established reporting units, the nephelometric turbidity unit and the formazin nephelometric unit (USGS 2008d). These ten reporting units provide a systematic method by which to characterize the type of turbidimeter used and are intended to improve the comparability of turbidity data.

Commercially available optical instruments operate on one of two bulk-optic principles. Transmissometers use a light source beamed directly at the sensor. The instrument measures the fraction of light from a collimated light source (typically within the visible range at about 660 nm) that reaches a light detector. The fraction of light reaching the detector is converted to a beam attenuation coefficient, which is related to SSC. Few turbidimeters operate on the transmissometry principle. Nephelometers measure visible or infrared (IR) light scattered by suspended particles (rather than light transmitted through particles). They measure scattering in a (SSC-dependent) volume less than a few cubic centimeters. Most turbidimeters measure 90° scattering. Optical backscatterance instruments (OBS) (Downing *et al.* 1981; Downing 1983) are a type of nephelometer designed to measure less than 180° backscattered IR light in a volume on the order of a few cubic centimeters. Figure 1.5 shows examples of nephelometry and optical-backscatter sensors.

Two instruments widely used for *in situ* applications are the YSI Model 6136 turbidimeter (manufactured by YSI, Inc.), which measures IR scatter at 90°, and OBS-3+ (manufactured by Campbell Scientific, Inc.), which measures IR backscattered at about 140–160°. Transmittance and scatterance are functions of the density, size, color, index of refraction, and shape of suspended particles (Conner & De Visser 1992; Sutherland *et al.* 2000).

In summer 2008, the purchase price of an *in situ* nephelometric turbidimeter with sonde, wiper, and controller was about US$5000. The cost of an OBS and cable without a wiper was about equal to the average cost of a fully equipped *in situ* nephelometric turbidimeter.

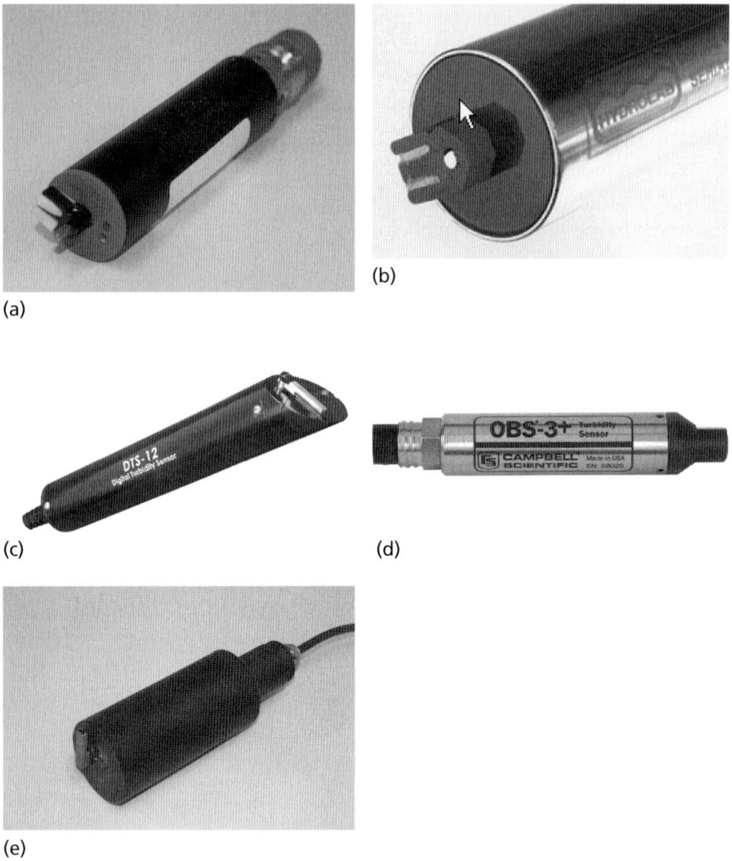

Fig. 1.5 Photographs showing nephelometry sensors. (a) YSI model 6136; (b) Hydrolab turbidity sensor with wiper; (c) Forrest Technology Systems model DTS-12; (d) Campbell Scientific Inc. model OBS 3+; (e) Hach Solitax with wiper. All photographs reproduced with permission.

Bulk-optical instruments lack moving parts (unless outfitted with optical wipers), can be deployed *in situ* to collect time-series data, and provide rapid-sampling capability. The technology is relatively mature, and has been shown to provide reliable data at several USGS streamgages (Uhrich 2002; Melis *et al.* 2003; Schoellhamer & Wright 2003; Uhrich & Bragg 2003; Rasmussen *et al.* 2005) and other sites (Lewis 2002; Pratt & Parchure 2003).

The validity of data produced by bulk-optic instruments can be compromised by at least two in-stream conditions. Biological fouling ("biofouling") of the optical windows of sensors, which results in the tendency for the output to shift from the calibration curve to spuriously larger values over timescales of days or more, remains a problem, particularly in warmer, microbiologically active waters. Commercially available mechanical wiper systems for some sensors may alleviate this problem.

Additionally, turbidity levels exceeding the instrument's maximum measurement limit results in sensor saturation. When saturation occurs, constant values equal to the turbidimeter's upper measurement limit are output, creating a turbidity trace with a "plateau" comprising erroneously low turbidity data. This phenomenon tends to occur at the higher flows and higher SSCs that are most influential in sediment transport. Figure 1.6 shows a hydrograph and turbidity trace for the USGS streamgage on the Kansas River near DeSoto, Kansas, USA, for the period April 12 to May 24, 2002. The turbidity trace for periods encompassing April 22 and May 14 (Fig. 1.6) show the characteristic "saturation plateau" when the in-stream turbidity level exceeded the turbidimeter's maximum recording level.

Maximum SSC limits for turbidimeters depend in part on instrument specifications and the ambient PSD. The OBS instrument has a generally linear

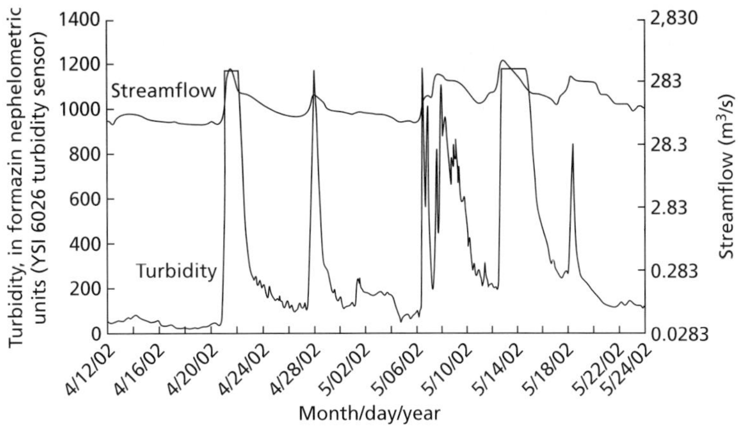

Fig. 1.6 Comparison of continuous measurements of streamflow and turbidity, April 12–May 24, 2002, for USGS streamgage on the Kansas River at DeSoto, Kansas, USA. Turbidimeter saturation occurs around April 22 and May 14.
Adapted from Rasmussen *et al.* (2005).

response at SSCs less than about 2 g/L for clay and silt, and 10 g/L for sand (Ludwig & Hanes 1990), although Kineke & Sternberg (1992) describe the capability to measure SSCs up to about 320 g/L (in the nonlinear region of the OBS response curve). Specifications for an OBS instrument marketed by Campbell Scientific, Inc. (2008) lists an applicable range of 50–500 g/L; however this should be verified by the user for local sediment characteristics. The upper SSC limit for transmissometers depends on optical path length, but may be as low as about 0.05 g/L (D & A Instrument Co. 1991). Thus, transmissometers are more sensitive at low SSCs whereas OBS sensors have superior linearity in highly turbid water (Downing 1996) and are less prone to signal saturation.

Because of the relation between turbidity and PSD, inferences of SSCs from turbidity measurements (like all single-frequency optical and acoustical instruments) are best suited for application at sites with relatively stable PSDs. OBS signal gain is inversely related to grain size (Sutherland *et al.* 2000). Laboratory investigations of Conner & De Visser (1992) indicate OBS signal gain is minimally affected by changes in PSD in the range 200–400 μm but greatly affected by changes when particles are smaller than about 100 μm. They caution against using OBS when changes in the PSD occur and the suspended material is less than 100 μm. Additionally, the OBS signal can vary as a function of particle color. Sutherland *et al.* (2000) found a strong correlation between observed and predicted OBS measurements of varying SSCs and ratios of black and white suspended sediment. They found the smallest OBS signal-gain response for black sediment and the largest for white sediment, with responses from other colors falling between. They suggest that the level of blackness of particles acts to absorb the near-infrared signal of the OBS, thus modifying its output. Hence, caution should be exercised in deployments under varying PSD and particle-color conditions, unless the instrument is recalibrated for ambient conditions.

Turbidity is often proportional to SSC in the water column within the measuring range of the sensor. Empirical relations between turbidity and SSC have been modeled using linear regression analysis (Walling 1977; Gilvear & Petts 1985; Buchanan & Schoellhamer 1995; Lewis 1996; Christensen *et al.* 2000; Uhrich & Bragg 2003; Lietz & Debiak 2005; Rasmussen *et al.* 2005). If continuously monitored water-discharge and turbidity data are available on the same time interval for a site, the derived unit-value SSCs can be multiplied by their paired water-discharge data to compute continuous SSL without the need for interpolation or estimation. When the turbidity-SSC model is considered adequate as described below, continuous turbidity data calibrated with SSC data from samples collected over a range of flows can provide a more reliable and reproducible SSC time series. When the turbidity-SSC model is considered inadequate, use of water discharge and turbidity may improve model performance sufficiently to justify use of the bivariate model to produce an SSC time series. Upon derivation of an acceptable SSC time series, SSL can be computed from these data and their paired water-discharge time series without the need for interpolation or estimation. Guidelines based on this approach for computing SSC values from continuous turbidity data (or, when appropriate, continuous turbidity and streamflow data) have been produced by Rasmussen *et al.* (2009) and endorsed for collecting and storing SSC and SSL data by the USGS.

The turbidity-based computational scheme has several benefits:

- no subjective interpolation or estimation is required, although the hydrologic judgment and statistical prowess of the analyst may be important in the derivation of the equation used to convert turbidity, or turbidity and water discharge, to SSCs;
- the computational procedure is precisely reproducible;
- the scheme takes full advantage of the available data and computational resources, hence, substantially reduces the time and effort to compute SSL records;
- estimates of uncertainty can be computed for the SSC time series.

An adequate model calibration dataset consists of an appropriate number of instantaneous SSC samples and concurrent turbidity and streamflow measurements made over most of the observed range of hydrologic conditions for the period of record. Another factor that should be considered when determining the adequacy of the number of samples in a calibration dataset is the amount of variability in the relation between turbidity and SSC. The larger the variability in the relation between turbidity and SSC at a site, the greater the need to collect more calibration data.

The key factor for computing time series of SSC data from periodic instantaneous SSC, time series of turbidity, and streamflow data is the type and goodness-of-fit of the regression model used in the computation. A simple linear regression model relating turbidity to SSC is often sufficient for reliable computations of SSC. A multiple linear regression model relating both turbidity and streamflow to SSC may significantly improve the usefulness of the simple turbidity linear regression model. Typically, addition of a streamflow variable is more likely to improve the turbidity-SSC regression if more than about 20% of the suspended-sediment mass is sand-size material (between 62 and 2000 μm median diameter), as inferred from research by Gray *et al.* (2000) on differences between SSCs and total suspended solids measurements.

Prediction intervals are determined to evaluate the uncertainty of SSC regression-computed values (Helsel & Hirsch 2002). Prediction intervals define a range of values for the regression estimate associated with a known level of uncertainty. For a given turbidity value, the 90% prediction interval represents a range of values within which there is a 90% certainty that the true SSC value lies.

Once an acceptable regression model is developed, it can be used to compute SSC within and outside of the period of record used in model development. Maintaining a long-term SSC record requires ongoing collection of turbidity and streamflow time-series data and sample collection for reanalysis and verification of the current SSC regression model. The method for validating the regression model is affected by the frequency of sample collection and the purpose of the study. Regression models can be validated annually (or at some other frequency as needed based on the nature of the monitored hydrologic system and its watershed), after new data have been collected, or on the basis of other valid criteria. Owing to variability in hydrology and other factors, one such period may experience an extreme condition compared with another, such as in floods or droughts, urbanization, wildfire, or implementation of best-management practices. Ergo, a regression model to compute SSC should never be considered static, but rather to represent a set period in a dynamic system in which additional data will help verify changes in the SSC regression relation.

1.2.1.2 Example field evaluations

Continuous turbidity measurements have been shown to provide reliable continuous SSC values with a quantifiable uncertainty at the USGS streamgage on the Little Arkansas River at Sedgwick, Kansas, USA. The adequacy of the calibration dataset was evaluated using duration curves of turbidity and streamflow (Fig. 1.7). The number of samples is often cited as the primary criterion for determining if a dataset is adequate. Although the sample total is important, their broad distribution over the range of

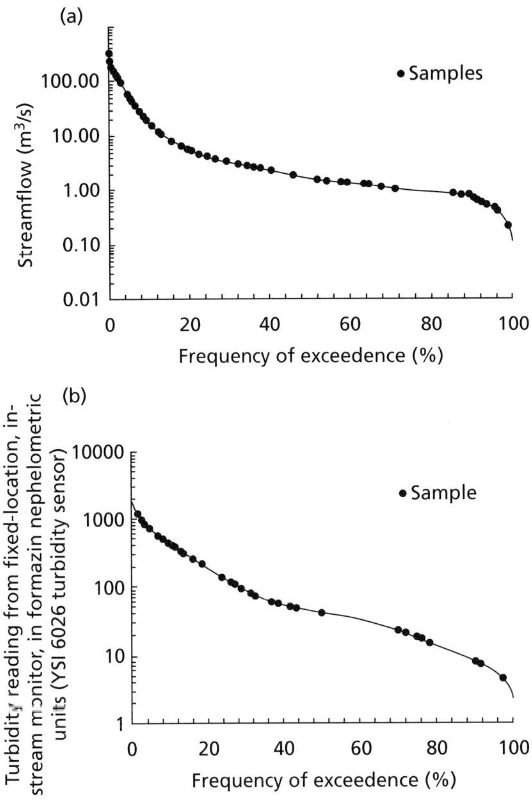

Fig. 1.7 Duration curves for (a) streamflow, and (b) turbidity measured in samples collected in the Little Arkansas River at Sedgwick, Kansas, USA, 1999–2005.

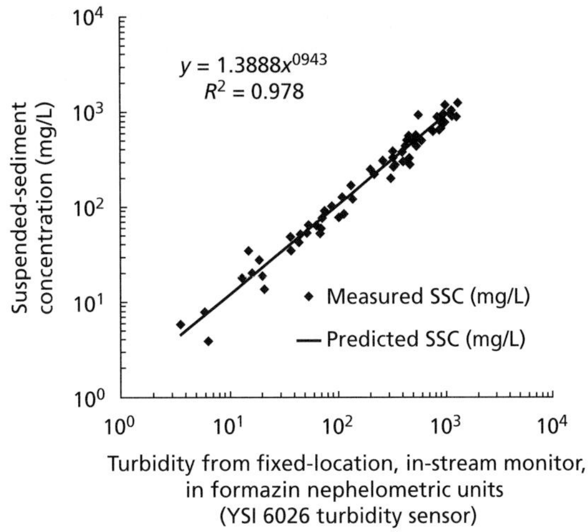

Fig. 1.8 Comparison of field turbidity in formazin nephelometric turbidity units and SSC for the Little Arkansas River at Sedgwick, Kansas, USA, 1999–2006.

observed turbidity, SSC, and streamflow values for the site is paramount in developing a reliable model.

Simple linear regression analysis explained in Rasmussen *et al.* (2009) was used to develop a site-specific univariate model using turbidity to compute time-series SSC (Fig. 1.8). The model explains about 98% of the variance in SSC. Continuous SSLs computed from the model and paired water discharge–SSC time-series datasets are available online (USGS 2005).

Base-10 logarithmic transformation is one of several mathematical functions that can be used to transform datasets to meet the assumptions for linear regression analysis. Other considerations should include the ease of retransforming the results from the model and the bias associated with the retransformation. The computed SSC values must be retransformed to their original units, a step that introduces a bias (usually negative) in computed SSC values (Miller 1951; Koch & Smillie 1986) unless the data are perfectly and positively correlated. To correct for retransformation bias, Duan (1983) introduced a nonparametric bias-correction factor called the "smearing" estimator. Duan's (1983) smearing estimator is insensitive to non-normality in the distribution of regression residuals about a logarithmically transformed model. A method proposed by Cohn *et al.* (1989) assumes normally distributed residuals about the logarithmic model and results in an exact minimum variance unbiased estimator and its variance.

Schoellhamer *et al.* (2002) describe a successful multi-station, multi-year field investigation in California's San Francisco Bay and Delta system. OBS sensors at each station are calibrated with SSC from water samples collected at each site. San Francisco Bay OBS sensors are calibrated to point samples (described in Section 1.1) and San Francisco Delta OBS sensors are calibrated to discharge-weighted, cross-sectionally averaged SSC values. SSL is determined by multiplying the discharge-weighted, cross-sectionally averaged SSC by water discharge, accounting for tide-driven bi-directional flow (Schoellhamer *et al.* 2002).

1.2.1.3 Summary: turbidity (bulk optics) as a suspended-sediment-surrogate technology

Two types of bulk-optic instruments – turbidimeters and optical-backscatter sensors – have been shown to provide reliable data at several field sites at which the limitations of the instrument have not been exceeded. Owing in part to the fact that bulk-optic instruments are the most common and among the most reasonably priced of the suspended-sediment-surrogate technologies, results from a considerable amount of research and evaluation associated with the technology are available to improve and better qualify the derived SSC data. One such outcome was the USGSs development and endorsement of guidelines for converting continuous turbidity time-series data (or continuous turbidity and water-discharge time-series data) to SSC and SSL time-series data (Rasmussen *et al.* 2009).

The primary advantage of regression-based estimates using continuous turbidity measurements over discrete sample collection is typified by the SSC time series for the Little Arkansas River near Sedgewick, Kansas, USA. Regardless of flow conditions, SSC and SSL values are obtained continuously at the interval in which turbidity and water discharges are recorded (Fig. 1.9).

Turbidity as an SSC surrogate, however, has drawbacks. For example, turbidity time-series data derived from a single point in the stream at the sensor location may not be representative of the sedimentary conditions of the river cross section. Biofouling of optical windows may require frequent site visits to clean and recalibrate the instrument (many sensors offer an integrated wiper, considerably reducing biofouling). A lack of consistency in measurement characteristics among commercially available instruments impinges on the comparability of turbidity measurements (Landers 2003; Ziegler 2003). Instrument response to grain size, composition, color, shape, and coating can be variable, and hence, can reduce the accuracy of derived SSC values. Perhaps most importantly, saturation of the turbidimeter signal can occur, resulting in constant, erroneous SSC values above the saturation limit. Saturation often occurs at high SSCs that tend to occur concomitant with high flows, which are the most influential in suspended-sediment-flux magnitudes. Hence, some knowledge of the turbidimeter measurement range and site sedimentological characteristics is desirable before deploying a continuous turbidimeter for calculating SSC and sediment transport.

1.2.2 Laser diffraction

Jeffrey W. Gartner & John R. Gray

1.2.2.1 Background and theory

Laser diffraction instruments exploit the principle of small-angle forward light scattering to infer PSDs and volume SSCs. These instruments measure scattering over a sufficiently wide range of small forward scattering angles to allow determination of PSD information over a wide range (typically 1 : 100 or

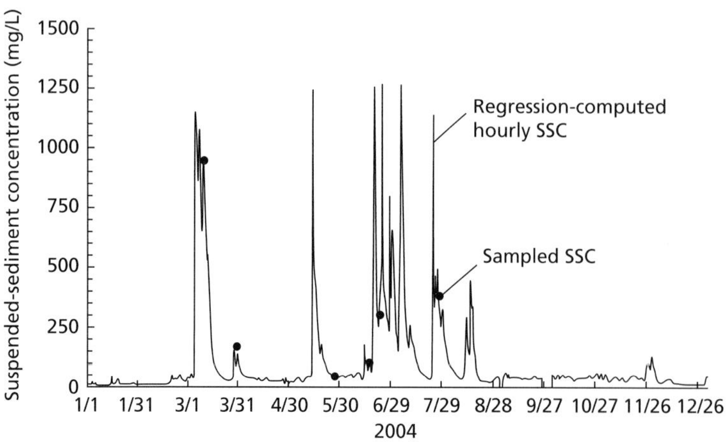

Fig. 1.9 Hourly regression-computed and sampled SSCs, Little Arkansas River near Sedgwick, Kansas, USA, 2004.

(a) (b)

Fig. 1.10 Laser *in situ* scattering and transmissometers. (a) a LISST-100 *in situ* instrument; (b) an in-development LISST-SL (streamlined) manually deployable instrument (photographs courtesy of Sequoia Scientific, Inc.).

1:200) of particle sizes. Scattering by spheres (larger than the wavelength of light) at small angles is equal to diffraction by apertures of the same diameter (Swithenbank *et al.* 1977; Agrawal *et al.* 1991; Agrawal & Pottsmith 1994). In addition, scattering is determined almost completely by light diffracted by the particle; any light transmitted through the particle does not affect the small angle measurement, thus, this method of determining size distributions is mostly insensitive to changes in particle color or composition (Agrawal & Pottsmith 2000). However, departure from spherical shape produces changes in estimated PSDs and SSCs; laser diffraction instruments provide the equivalent sphere-size distribution (Agrawal *et al.* 2008).

Commercially available instruments to measure PSD using laser diffraction have been available for laboratory use since the early 1980s, for example instruments made by Malvern Instruments and Coulter Corporation to name two manufacturers. The first attempt to apply the technology for *in situ* application used a commercial laboratory instrument adapted for ocean use (Bale & Morris 1987). A self-contained version of a laser diffraction instrument that could be deployed in an autonomous mode and determined PSD in eight size classes is described by Agrawal & Pottsmith (1994).

A more advanced and commercially available version of the instrument (Agrawal *et al.* 1996; Agrawal & Pottsmith 2000) capable of providing time series of PSDs and volume SSC values is the Laser In Situ Scattering and Transmissometry (LISST)-100 (Sequoia Scientific, Inc. 2008). The LISST-100 (Fig. 1.10a), with an overall length (minus cable) of 87 cm and diameter of 13 cm, measures optical transmission, water temperature, and hydrostatic pressure in addition to PSD and volume SSC. The LISST uses a 670-nm wavelength solid-state laser. The standard sample path of this device is a cylindrical volume with a diameter of approximately 6 mm and a length of 50 mm, although versions with shorter laser-path lengths are available for highly turbid environments. The instrument uses a 32-ring detector with logarithmically increasing radii to measure scattering intensity at 32 small forward angles that correspond to 1.25–250 μm (LISST-100B), 2.5–500 μm (LISST-100C), or 7.5–1500 μm (LISST_FLOC). The inner radius (smallest-scattering angle) of the ring detector corresponds with the largest measured particles and the outer radius (largest-scattering angle) corresponds with the smallest measured particles. The measured scattering intensity distribution is also referred to as the volume scattering function (VSF) (Pottsmith and Bhogal 1995; Agrawal and Pottsmith 2000). In practice, to determine PSDs and volume SSCs, the measured VSF is first corrected with a background scattering distribution. The corrected VSF is mathematically inverted to determine a PSD that would produce the multi-angle scattering that fits the measured observation in the 32-ring detector. Details of the inversion process can be found in Agrawal & Pottsmith (2000). Volume SSC is calculated from the inverse of the corrected scattering distribution divided by the volume conversion constant, an empirical calibration constant supplied by the manufacturer. Although laboratory versions of laser diffraction instruments are available from several manufacturers, the authors are aware of only one (Sequoia Scientific, Inc. 2008) that produces commercially available instruments designed for *in situ* applications and manual deployment.

The purchase price of one of the laser instruments (*in situ* and manually deployed) described in this section ranges from about two to six times that for a fully equipped turbidimeter, depending on the instrument of interest. The instrument-measurement realm of the *in situ* instruments described herein is a point in a stream. When used for measurement of PSD or volume SSC, they do not require routine instrument calibrations.

The LISST-100, which has been field and laboratory tested, has been shown to successfully determine PSDs of natural materials and the size of mono-sized particle suspensions within about a 10% accuracy (Traykovski *et al.* 1999; Gartner *et al.* 2001; Meral 2008). It can also be used to determine mass SSC from volume SSC if particle density is known (Traykovski *et al.* 1999; Gartner *et al.* 2001; Melis *et al.* 2003). Unlike single-frequency optical backscatter instruments, these instruments are not subject to potential inaccuracies associated with changes in PSDs if the particle sizes fall within the range of instrument sensitivity (Agrawal & Pottsmith 2000). Onboard memory and power allow high temporal resolution sampling at intervals up to 5 Hz during field studies that range in time scales from days (or tidal cycles) to months. In addition to analysis of PSDs and concentrations of inorganic material, LISST instruments are now being used increasingly for analysis of size distribution and population concentration and mixing dynamics of organic material such as phytoplankton (see, for example, Serra *et al.* 2001, 2003; Karp-Boss *et al.* 2007).

There are limitations associated with the use of LISST instruments for determining size distribution of suspended sediment. The scattering model (Mie theory) requires absence of multiple light scattering; thus, there is an upper SSC limit because of the presence of multiple scattering from particles at high SSC. Agrawal & Pottsmith (2000) found multiple scattering effects occurred when optical transmission was less than 30%. The limiting SSC is a function of particle-size distribution, laser-path length, and SSC; it ranges from tenths of a gram per liter (for small particle sizes) to a few grams per liter (for larger particle sizes). In addition, as is the case with all types of *in situ* optical instruments, biofouling can degrade measurements.

These problems can be addressed with anti-fouling shutters or optical blocks that reduce the laser path length (Sequoia Scientific, Inc. 2008). For example, reducing the optical path in water from the standard 5 cm to 3 mm has been effective in extending measurement limits to 2–3 g/L of fine material. For still higher SSCs, a LISST-Infinite was developed as part of a research-and-development project with the USGS. The LISST-Infinite, a prototype of which was tested by the USGS (Konrad *et al.* 2006), pumps a water–sediment sample to the instrument, and then uses automated multi-stage dilution (as necessary) before measuring PSDs and SSCs with a built-in LISST-100. Thus, the measurable SSC limit is, in theory, extended to the highest SSCs of material that can be pumped to the LISST-100 (Yogesh Agrawal, Sequoia Scientific, Inc., personal communication 2008). However, the process of pumping the water-sediment sample from a point in the channel may alter the original size distribution. Still another version of the LISST-100, the LISST-FLOC, is designed to measure larger particles such as flocculated estuarine marine particles.

As previously presented, laser diffraction techniques historically have interpreted the light scattered by natural particles as 'equivalent spheres', i.e. an ensemble of spheres with identical angular scattering properties. However, spherical particles are rarities in nature. Angular scattering from irregularly shaped particles is different to that from spheres. An irregular particle scatters light similarly to that of a spherical particle that is ¼- to ½-phi larger than the irregular particle's median diameter (Agrawal *et al.* 2008). For example, a natural particle of diameter 10 µm may be inferred as a 12- to 14-µm particle using laser diffraction. Agrawal *et al.* (2008) quantified the multi-angle laser scattering characteristics of natural particles. They interpreted the measured laser light scattering as random shaped particles rather than spheres, an interpretation that produced results consistent with sieved samples.

An instrument somewhat similar to the LISST-100, the LISST-25, measures mean SSC and Sauter mean particle size (the diameter of a sphere that has the same volume/surface area ratio as the particle of interest) in two size classes (2.5–63 μm and 63–500 μm) (Sequoia Scientific, Inc. 2008). The LISST-25 is based on the same principles as the LISST-100, but, unlike the LISST-100, it determines SSC through a weighted summation of the output of ring detectors rather than the inversion of intensity distribution to obtain size distribution. The weighted sum can be affected by use of comet-like shaped focal plane detectors (Yogesh Agrawal, Sequoia Scientific, Inc., personal communication 2008).

A cable-suspended, streamlined, isokinetic version of the LISST-100, the LISST-SL (Fig. 1.10b), is being developed for manual river deployment. The LISST-SL is designed to address the potential problem of flow disturbance associated with the size and shape of the conventional LISST-100 instruments. The LISST-SL features the capability of real-time velocity measurement that is in turn used to control a pump to withdraw a filament of water and route it through the laser beam at the ambient current velocity (Gray *et al.* 2004; Agrawal & Pottsmith 2006). This isokinetic flow-through capability is a prerequisite for reliably ascertaining the suspended-sediment properties in all but the shallowest or most sluggish rivers. The performance of the LISST-SL is being evaluated by the FISP (2008).

1.2.2.2 Example field evaluation

Laser diffraction sensors are being investigated as an alternative monitoring protocol for tracking reach-scale suspended-sediment supply at a USGS streamgage on the Colorado River at Grand Canyon, Arizona, USA, located 164 km downstream from Glen Canyon Dam (Melis *et al.* 2003; Topping *et al.* 2004). A canyon wall-mounted LISST-100 provides continuous PSD and SSC data for computing suspended-sediment transport that may reduce uncertainty in estimates of the transport of sand and finer material.

An example of data collected by the LISST-100B at the Colorado River at the Grand Canyon streamgage is shown in Fig. 1.11. Data were obtained by averaging 16 measurements at 2-minute intervals during a 24-hour deployment in July 2001. The time series of 720 LISST-100B measurements obtained from a single point in the river compare favorably with cross-sectional data obtained concurrent with some of the LISST-100B measurements using an isokinetic bag sampler and techniques described by Nolan *et al.* (2005). In addition, the LISST-100B also recorded the increase of variance in the SSC of sand-size particles expected with increasing flows (Melis *et al.* 2003); peak SSC values ranged between 0.06 and 0.14 g/L (60–140 mg/L).

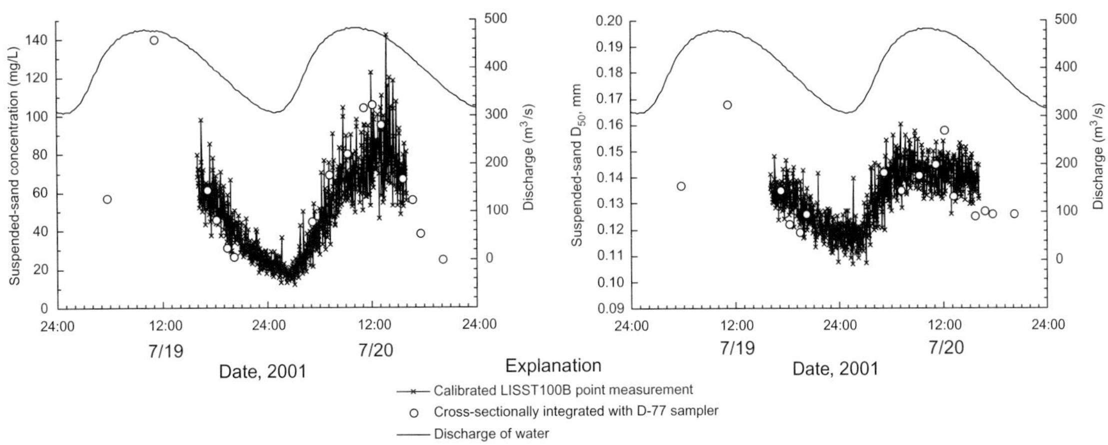

Fig. 1.11 Comparison of SSC (left) and median grain sizes (right) measured at the USGS streamgage, Colorado River at Grand Canyon, Arizona, USA, using a LISST-100-B and a US D-77 bag sampler.
From Topping *et al.* (2004).

1.2.2.3 Summary: laser diffraction as a suspended-sediment surrogate technology

A major advantage of the LISST technology is real-time measurement of PSD in 32 ¼-phi-diameter size classes, a capability shared by no other currently available sediment-surrogate monitoring instrument. LISST instruments do not require instrument calibration when used for PSD or volume SSC.

Nevertheless, the technology has some limitations. The measurement is a point sample. In addition, SSC measurements are in volume units, thus requiring estimates or measurements of sediment density to convert to mass SSC units. When deployed *in situ*, the LISST is susceptible to biofouling unless anti-fouling shutters are used. Reductions in data accuracy due to the presence of non-spherical particles and loss of data from signal saturation can occur. Finally, the cost of a LISST instrument is two to six times that of a fully equipped *in situ* turbidimeter. However, for applications that require long-term repetitions of at-a-point or spatially dense measurements, especially if PSD data are required, the LISST suite of instruments may represent the most cost-effective approach for suspended-sediment data acquisition.

1.2.3 Digital Optical Imaging

Daniel J. Gooding

1.2.3.1 Background and theory

A digital optic-image analysis and pattern recognition system that does not require routine calibration has been developed and is being adapted to quantifying SSCs and selected size and shape characteristics of suspended sediment in water samples. The technology, commercially promoted by the medical industry in the 1990s to quantify cells in a blood sample, computes size statistics based on automated measurements of individual particles. Volumetric SSC is inferred from the size statistics.

The technology, in development and testing at the USGS Cascades Volcano Observatory, Vancouver, Washington, USA, was conceptualized for application in the laboratory. However, a field version is planned for testing as part of a stream-side pumping system. The technology may eventually be adapted for use in manually deployed isokinetic sediment samplers. The cost for a complete unit without environmental packaging is similar to that for a fully equipped turbidimeter. The instrument-measurement realm of a digital-optic measurement is a point. Like the LISST technology, routine instrument calibrations are unnecessary.

The principal components of the system are up to three charged-coupled-device progressive scan cameras (each with a selected lens) and a multi-port flow-through cell. Each lens is affixed to the flow-through cell using extension tubes, keeping a precise optical alignment between the cameras, lenses, targeted area, and backlighting (Fig. 1.12a). All components other than the flow-through cell, for which a patent is pending, and extension tubes are commercially available.

The key component of the system, and the only part developed explicitly for this application, is the multi-port flow-through cell (Fig. 1.12b). The flow-through cell serves two purposes: to separate particles into fractions smaller and larger than 75 µm, thus enabling a relatively unobstructed analysis of the smaller particles; and to disjoin and isolate particles to create a more robust digital image of each particle. If imaged particles are separated, or can be digitally separated, they easily can be identified, measured, and counted by the software.

Computing SSC is based on four attributes derived from the images: particle population, particle shape, grayscale relation to turbidity, and the amount of light passing through the entire image. The amount of light (average image brightness) and average image grayscale are measured over a sequence of several images from the flow-through cell taken within 2–6 seconds. The net changes for brightness and grayscale are relative to a reference image using clear water contrast against the sample images. Particle volumes are estimated by calculating a "z" axis length (the third unmeasured axis in the two-dimensional image) based on the particle shape, texture, chord length, and the particle center of gravity from the two-dimensional image.

A multi-camera configuration measures PSDs in the range of 4–4000 µm. This three-order-of-magnitude range cannot be accomplished using a single magnification, hence the use of multiple cameras and lenses is required. The software is designed to integrate images from up to three cameras depending on the particle-size range required by the

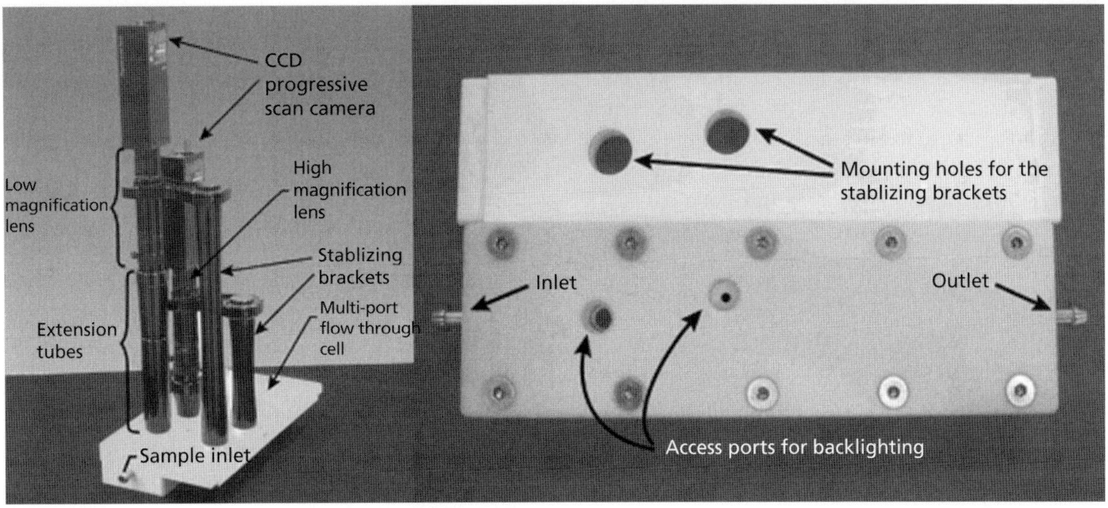

(a) (b)

Fig. 1.12 Suspended-sediment digital optic-imaging components. (a) Cameras atop encased lenses with extension tubes and encased flow-through cell (fiber-optic cable not shown). (b) Multi-port flow-through cell (patent pending).

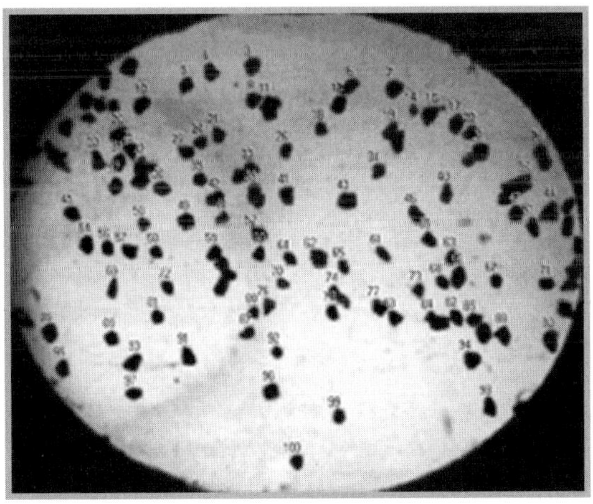

Fig. 1.13 A morphologically transformed image of a water-sediment mixture illuminated by cross-polarization. Each sediment particle and a possible aggregate appearing as a single particle are numbered.

application. Once an image of the water–sediment mixture in the flow-through cell is captured, morphological transformations (successions of pixel-level image processing) are conducted. The final image is used to extract discrete particle information such as maximum and minimum lengths, shape and area (Kindratenko 1997) (Fig. 1.13). Although there may be an upper SSC-measurement limit, any such value is still to be determined.

Inherent complexities involved with imaging individual particles in a liquid medium can create impediments to extracting usable information from the binary images, which usually contain fewer textural details than appear in the original image. Despite some loss of detail in the image, the derived solid-phase images, referred to as "blobs," are better suited for analysis by the imaging software – particularly for conducting discrete analyses such as particle-edge detection and for computing the size and shape characteristics of individual sediment particles in the final analysis.

The flow-through cell design results in effective dispersion of most particles to render most particle boundaries distinguishable. In the event of incomplete particle dispersion and (or) large SSCs that increase the incidence of imaged-particle overlap, the software uses interpretations based on image normalization, segmentation, and other imaging analysis tools to aid in identifying individual particles.

Balance in contrast is essential for obtaining useful images of sediment particles. As part of the prototype lens assembly, two in-line polarized filters are oriented 50–70° from cross-polarization between the illumination source and target area. This assembly helps darken bright areas created by translucent particles and reduces scattered light caused by refraction and reflectance of the material being imaged. A suitable diffuser is needed for the backlighting to assure balanced lighting throughout the image.

Turbidity, caused by organic and colloidal material, is another hindering factor in obtaining an assessable particle image for analysis. The use of a near-ultraviolet wavelength of 0.45–0.5 μm produces a sharper image of particles. Also, with the shorter wavelength, there is less light scatter due to reflectance and refraction as occurs when using the full visible light spectrum. Figure 1.14 shows suspended material finer than 62 μm at an SSC of 10 g/L (10,000 mg/L) in a sample that was seeded with a small number of 125- to 250-μm particles that were digitally enhanced by the software.

In some cases, the binary image could still be degraded by turbidity, depending on the nature of the factors causing the turbidity. If the spatial correlation of the background cannot be automatically resolved, automatic detection of particle boundaries becomes less precise or unattainable. More analysis and development is required in this regard.

Perhaps the most difficult task in the automatic calculation of size characteristics of imaged blobs deals with connected, aggregated, and overlapping particles that appear as a single blob on the image. The software's segmentation algorithm works well in identifying discrete particles within aggregates by detecting disparities within clusters. Because of this and other possible hindering factors, it is desirable to analyze several images from the same water–sediment sample to better characterize the actual volume SSC computed from poor-quality images. The software is designed to analyze selected layers of the image starting with well-delineated and easily identifiable particles, leaving characterization of those particles that are obscured or that otherwise present definitional problems for the final and most computationally intensive analyses. Research on the photo-imaging technology continues to focus on refining the software to maximize automatic interpretation of aggregates.

For example, the software is able to distinguish a blob as two discrete particles, labeled as 100 and 102 (numbers appear above respective blobs) (Fig. 1.15). Although the blob labeled as 99 may be two connected or overlapping particles, the software interpreted the blob as a single particle. Very fine sand composes the sample material used in this image. Using a microscope, it was observed that some of the sand grains are indeed made up of two naturally fused minerals that gave some of the single particles a barbell-shape appearance.

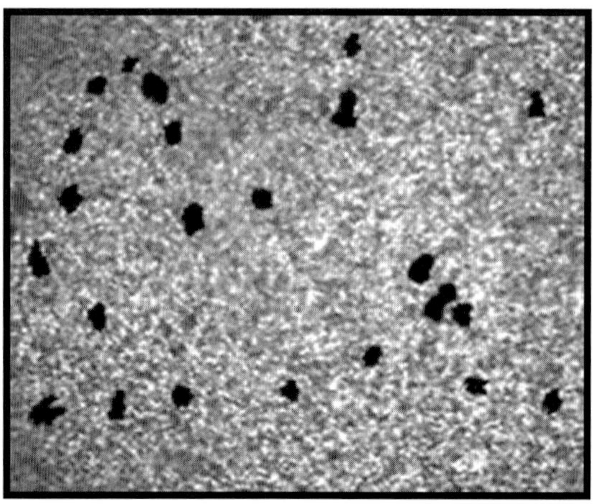

Fig. 1.14 A morphologically transformed image of a water-sediment mixture composed of 10 g/L of material finer than 62 μm, seeded with 125- to 250-μm particles that appear as dark blobs.

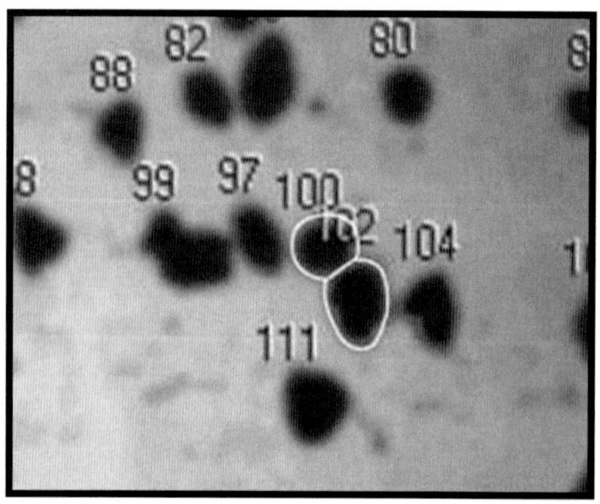

Fig. 1.15 A morphologically transformed image of a water–sediment mixture composed of 62–125 μm particles showing potentially inconsistent interpretation of overlapping or connected particles.

1.2.3.2 Status of laboratory evaluation

Research in quantitative digital-optic analysis for suspended-sediment particles has so far been limited to laboratory conditions at the USGS Cascades Volcano Observatory, Vancouver, Washington, USA. The technology calculates, enumerates, and sums volumes of individual moving particles photographed in a flow-through cell. There are no routine requirements for validation of the technology, although cross-section calibrations will be required if deployed in the field in the future.

Several challenges remain in rendering this laboratory-based technology acceptable for laboratory or riverine deployment. Partly hidden particles, aggregates, and other anomalies can result in less-accurate measurements, as can higher turbidity levels. The multi-port flow-through cell design reduces these problems; however, imaging bias can still occur, such as at very large SSC of clay-size particles. Analytical results are expressed in volume/volume units and not in more commonly used mass/volume units, requiring assumptions on the value of particle density or collection and analysis of samples for SSC and (or) particle density. Reliable PSD and SSC estimates can be difficult to obtain when the image becomes "noisy" because of several factors. Aggregates, organics, air bubbles, and stagnant material within the viewing area can cause the image to become corrupted and numerically unstable. Special safeguards incorporated into the software help overcome these obstacles.

If the source of the imaging problems is identified, then there may be geometric and statistical solutions to the problem. For example, image-to-image comparisons can be used to check for stationary particles that have adhered to the flow-through cell windows viewing area. This particular group of pixels becomes useless for analytical purposes until the area has cleared. The software recognizes the recurring blob and will not use the occupied pixels in sequential calculations until the area clears or changes. Air bubbles could be counted as particles, but with their distinctive geometric attributes the software can easily identify them as such and remove them from subsequent SSC calculations.

There are inherent difficulties for digital-imaging systems to perform well in real-world environments. However, if the problems can be identified and quantified and the number of complicating environmental variables minimized, it may be feasible to achieve practical quantitative results for measuring SSC and PSDs in riverine environments.

1.2.3.3 Summary: digital optical imaging as a suspended sediment surrogate technology

Digital-optic imaging technology remains in the research and development phase and has yet to be deployed for testing beyond the laboratory. Other than the flow-through cell and lens extensions, the technology is composed of off-the-shelf parts available at a cost similar to that of a fully equipped turbidimeter. Routine instrument calibrations are unnecessary.

Pending completion of testing and development, several inferences on limitations based on its attributes can be made:

- The technology can be affected by some of the same drawbacks as those for the bulk-optic and laser technologies. These drawbacks include issues associated with samples drawn from a single point, biofouling of the optic lenses, and upper measurement limits;
- Assumptions or measurements of mean particle density are required to convert volume SSC values to mass SSC values;
- Because the flow-through cell system is designed to separate aggregated sediments, it is not suitable for ascertaining SSCs of flocculents.

1.2.4 Pressure difference

John R. Gray, Nancy J. Hornewer, Matthew C. Larsen, Gregory G. Fisk, & Jamie P. Macy

1.2.4.1 Background and theory

The pressure-difference technique for monitoring SSC relies on measurements from two precision pressure-transducer sensors arrayed at different, fixed elevations in a water column. The difference in pressure readings is converted to a fluid-density value, from which SSC is inferred after correcting for water temperature (dissolved-solids concentrations in fresh-water systems are rarely large enough to be of consequence in the density computation). One of the first uses of the pressure-difference technique for measuring fluid density was applied to crude oil in pipes (William Fletcher, Design Analysis Associates, Inc., personal communication 1999).

The specific weight of the water–sediment mixture from measured pressure differences in a water column between two pressure-transducer orifices anchored at different depths can be calculated by the following equation:

$$\gamma = (p_1 - p_2)/(z_2 - z_1) \quad (1)$$

where: γ is the specific weight of the fluid; p_1 and p_2 are the simultaneous pressure measurements at orifices 1 and 2, respectively; and z_1 and z_2 are the simultaneous measurements of the distances to the water surface from orifices 1 and 2, respectively.

The difference in the distances from the fixed orifices to the water surface is a constant value. SSC is calculated as the difference in the specific weights of the water–sediment mixture and that of pure water at the same temperature as the ambient streamflow. Implicit assumptions in the method are that the simultaneous pressure measurements represent the same water surface, and that the density of the water–sediment mixture above the lower sensor is more or less equal to that above the higher sensor. Exceptionally sensitive pressure transducers are required. The technology has both laboratory and field applications (Lewis & Rasmussen 1999). The purchase price of the technology is similar to that for a fully equipped turbidimeter. In theory, the installation should require a minimum of maintenance other than removal of debris from the in-stream sensor assembly. The instrument-measurement realm is a water column. Instrument calibrations can be accomplished by sampling in or near the instrumented water column with a suspended-sediment sampler, although they are often supplanted by cross-section calibrations.

The technique has been applied in the laboratory with promising results of better than 3% accuracy (0.543 ± 0.014 g/L) for determining mass concentration of suspensions of glass microspheres (Lewis & Rasmussen 1999). However, application of this technique in the field can be complicated by a low signal-to-noise ratio associated with low-to-moderate SSC, turbulence, large dissolved-solids concentrations, and large water-temperature variations. Additionally, analyses may be complicated by density variations in the suspended material. William Fletcher (Design Analysis Associates, Inc., personal communication 2005) indicated that calculations based on a moving average of the pressure-difference data tended to provide a smoother time series of SSC that was more comparable to SSC data derived from water–sediment samples obtained by methods described by Nolan *et al.* (2005).

1.2.4.2 Example field evaluations

Information on the field performance of the pressure-difference technology is available from USGS streamgages on the lower Río Caguitas in Puerto Rico (Larsen *et al.* 2001) and near the mouth of the Paria River in Arizona, USA. Continuous pressure-difference data were collected during October–December 1999 at the Río Caguitas streamgage using a Double Bubbler Pressure Differential instrument developed by Design Analysis Associates, Inc. (2008) (Figs 1.16 and 1.17). Most of the annual sediment discharge in the lower Río Caguitas occurs in runoff from a few storms when SSC exceeds about 0.5 g/L. The maximum SSC measured at the streamgage during the Double Bubbler tests based on water samples collected by an automatic pumping sampler was 17.7 g/L.

The analytical procedure involved data smoothing and removal of outliers. To calculate the weight density of suspended sediment and dissolved solids the weight density of pure water at 27 °C was subtracted from the smoothed data values. Even with these manipulations, this test of the Double Bubbler instrument in Puerto Rico showed relatively poor agreement among discharge, SSC, and the manipulated water-density data measured by the Double Bubbler (Fig. 1.18). The Double Bubbler data contained a large amount of signal noise, making interpretation difficult. Lacking a thermistor for temperature compensation, 12 of 15 base-flow instrument measurements inferred negative SSC values (an impossibility) concomitant with in-stream measured SSC values of 0.01–0.1 g/L (10–100 mg/L). However, all but two of the samples collected during seven high-flow periods showed concomitant increases in inferred positive SSC values.

A complicating factor in the pressure-difference method is in-stream turbulence, which introduces noise about equal to the magnitude of the signal of

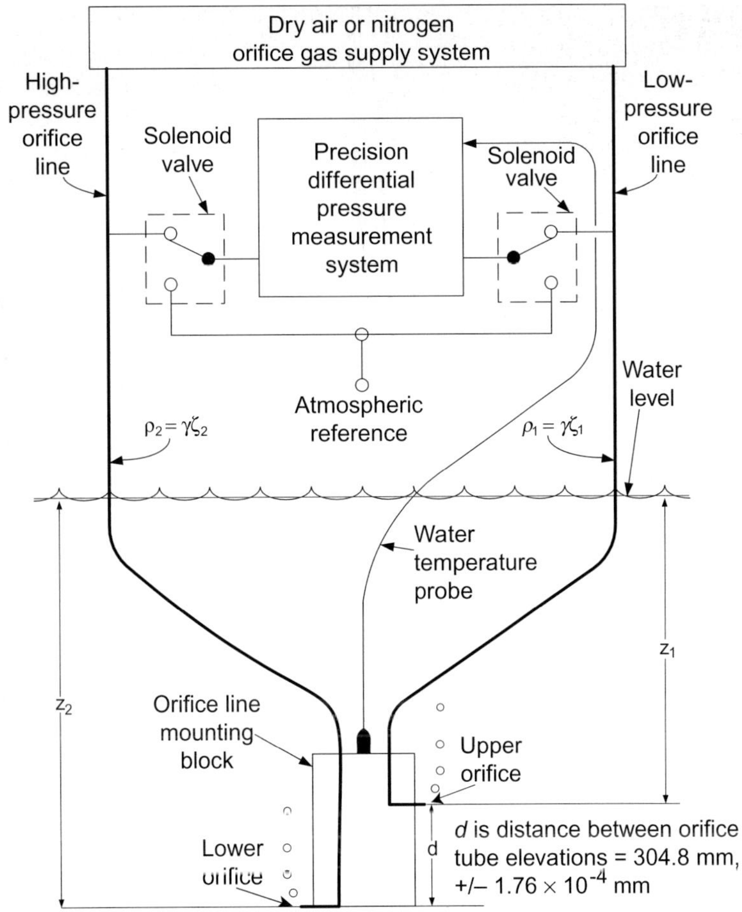

Fig. 1.16 Schematic of the Double Bubbler Pressure Differential instrument. Adapted from Larsen *et al.* (2001).

interest, particularly during high discharges that occur more or less concomitant with the largest SSC levels. Additionally, diel and storm-related fluctuations in water temperatures must be accounted for by using a continuously logging temperature sensor (the daily range in water temperatures at the Río Caguitas streamgage is as much as 10 °C). The high relative humidity characteristic of this humid-tropical site can also complicate the use of the Double Bubbler because of the sensitivity of the narrow-diameter bubbler gas lines to moisture, unless the gas lines are equipped with dryer tubes. This test of the Double Bubbler instrument showed the need for temperature compensation, and possibly the need to deploy the instrument at a site where weight densities of higher flows might be substantially larger than those measured at the Río Caguitas streamgage during the Double Bubbler tests.

In 2004, the Puerto Rico Double Bubbler system was transferred to the USGS streamgage on the Paria River at Lees Ferry, Arizona, USA, and augmented with a continuous water-temperature sensor. SSCs near 10^3 g/L have been measured during storm runoff at this streamgage. Deployment of the Double

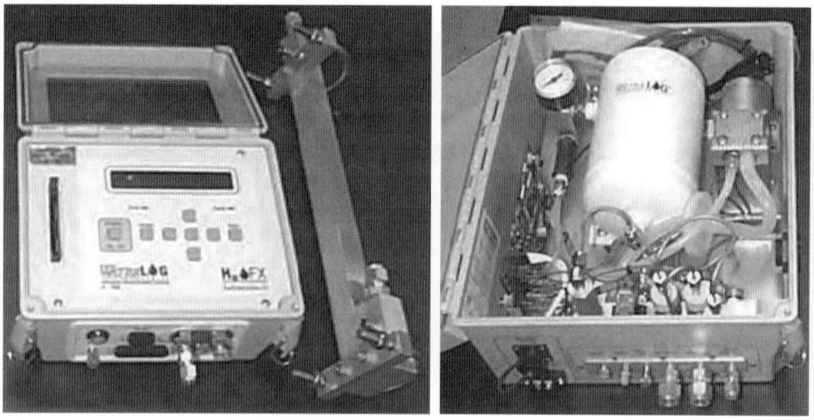

(a) (b)

(c)

Fig. 1.17 Double Bubbler Pressure Differential Instrument. (a) controller and orifice bar, (b) air compressor and tank assembly, and (c) in-stream components before installation. Photographs a and b courtesy of Design Analysis Associates, Inc. (2008).

Bubbler in the Paria River was predicated on the hypothesis that the expected large weight densities, ranging up to about double that of pure water under hyperconcentrated streamflow conditions (Beverage & Culbertson 1964), would prove to be within the Double Bubbler's operating range.

Double Bubbler data were collected, at 5-minute intervals, during periods of elevated flow at the Paria River streamgage from July 2004 through September 2006. Data collected from 14 periods of storm runoff were examined and compared with results from suspended-sediment samples collected during the storm runoff. The samples were collected using a combination of automated-pump samplers, depth-integrating samplers in a single vertical and deployed in the cross section, and dip samples (Nolan *et al.* 2005; Edwards & Glysson 1999). The elevated flows had peaks ranging from about 7–90 m^3/s; the maximum SSC measured was 382 g/L in water from an automated-pump sampler. A total of 261 suspended-sediment samples were collected during the 14 storm-runoff periods, and 86% of those samples had SSC values larger than 50 g/L. Double-Bubbler data were collected only during periods when water levels immersed both pressure sensors (the instrument was not fully submerged during normal shallow flows).

Double Bubbler data were filtered to remove outliers but not smoothed, because smoothing appeared to have little effect on reducing signal noise for data collected at this site. Water-temperature data were continuously recorded near the Double-Bubbler orifices. The weight density of suspended sediment and dissolved solids was calculated by subtracting the weight density of pure water, corrected for temperature, from the filtered data.

Similar to data collected at the Río Caguitas in Puerto Rico, the Double Bubbler data collected at the USGS streamgage on the Paria River at Lees Ferry, Arizona, USA, had a large amount of signal noise, also making interpretation difficult. Relations between measured SSC and SSC calculated from

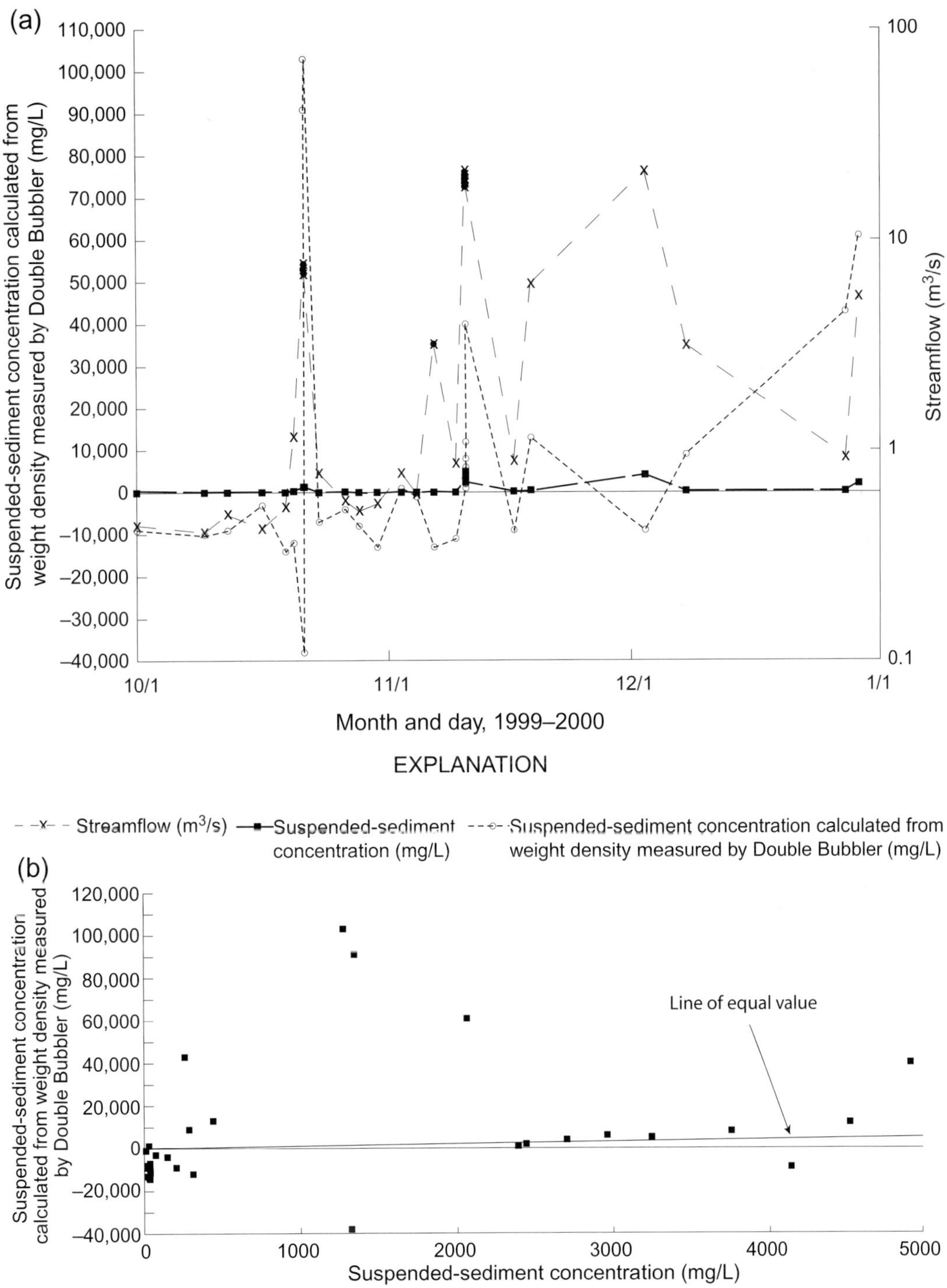

Fig. 1.18 Data for the USGS streamgage on the Rio Caguitas, Puerto Rico, October 1999 to January 2000. (a) Time series of streamflow, SSCs from samples, and SSCs calculated from weight densities of suspended sediments and dissolved solids measured using the Double Bubbler; symbols denote measured values, dashed interpolation lines are included for viewing purposes only; (b) scatter plot of measured SSCs from samples and those calculated from the Double Bubbler. Streamflow and sediment data are instantaneous samples, and each Double Bubbler SSC value, calculated from weight density, is a 30-minute mean of measurements made at 5-minute intervals.

Double Bubbler data lacked consistency, as illustrated by Fig. 1.19. Although parts of the record more or less show agreement between Double Bubbler-derived SSC data and those from analyses of physical samples, none of the sampled SSC values on January 10–12, 2005, was among the dozens of Double Bubbler values exceeding about 220 g/L. However, the veracity of the larger Double Bubbler measurements cannot be dismissed out-of-hand as measurement artifacts; essentially all of the physical-sample SSC values plot among Double Bubbler data, and all but the largest Double Bubbler SSC value are less than the historical maximum SSC of 1,080 g/L reported by Beverage & Culbertson (1964) for the Paria River streamgage.

It has been surmised that bed movement during Paria River Double Bubbler tests caused the lower orifice to become partly or fully blocked at times, contributing to erroneous data. In their tests of an *in situ* densimeter (pressure-difference monitoring system), Tollner *et al.* (2005) identified the passage of bed forms between the densimeter's orifices and fluid turbulence as potential complicating factors in SSC computations. They conclude that densimeter measurements, although feasible under laboratory conditions, are unreliable in general field conditions.

The USGS experience with the Double Bubbler cannot unequivocally support or refute Tollner *et al.*'s (2005) conclusion. However, because of its strong theoretical underpinnings, continuous monitoring capability, and – not unimportantly – a lack of any other proven surrogate technology for providing SSC time-series data in highly concentrated and hyperconcentrated streamflow conditions, the pressure-difference technique continues to be evaluated.

1.2.4.3 Summary: pressure difference as a suspended sediment surrogate technology

The pressure-difference technology was tested to ascertain if it could fulfill what may be a unique niche in suspended-sediment monitoring because, at least in theory, its performance improves as SSCs increase. The technology is relatively robust, being prone to neither signal drift nor biofouling, and is comparatively inexpensive. The technology doubles as a redundant stage sensor for the site. The theoretical underpinnings of the technology are relatively simple and straightforward. Given a valid set of temperature-compensated measurements at higher SSC values that are adequately filtered and smoothed to reduce the effects of turbulence, the technology may provide a time series of SSC that is ultimately superior to the periodic datasets obtained by traditional methods. The instrument can be calibrated using single-vertical samples. The water-column measurements are theoretically more representative of the mean cross-section SSC than point measurements.

In spite of its sound theoretical underpinnings, the field performance of the Double Bubbler in Puerto Rico and northern Arizona, USA, has yet to be fully resolved. Research is continuing into whether development and use of empirical relations from calibration data in lieu of the theoretical considerations are warranted. The required computational scheme presupposes that the SSC in the vertical profile between the sensors is more or less equal to that above the higher sensor. This assumption is difficult to verify and may not be valid. The technology is unreliable for measuring SSC at less than about 10 g/L, and the actual lower measurement threshold may be at a somewhat larger SSC. The technology is incapable of measuring SSC when the top orifice is out of water. Spurious data are numerous and are believed to be associated with flow turbulence or orifice blockage by bedforms. Continuous pressure-difference measurements may be useful in developing a continuous SSC trace under some circumstances but are not yet considered sufficiently reliable to replace traditional suspended-sediment-monitoring techniques.

1.2.5 Acoustic backscatter

Jeffrey W. Gartner & Scott A. Wright

1.2.5.1 Background and theory

Attempts to characterize SSC from *in situ* acoustic backscatter sensors (ABS) have increased in recent years. In contrast to traditional methods using analyses of water samples utilizing gravimetric or other techniques, use of ABS to estimate SSC is non-intrusive, far less labor intensive for the derived data density, more or less unaffected by biofouling, and results in a continuous time series of SSC. Use of ABS is appealing because SSC profiles can be obtained in

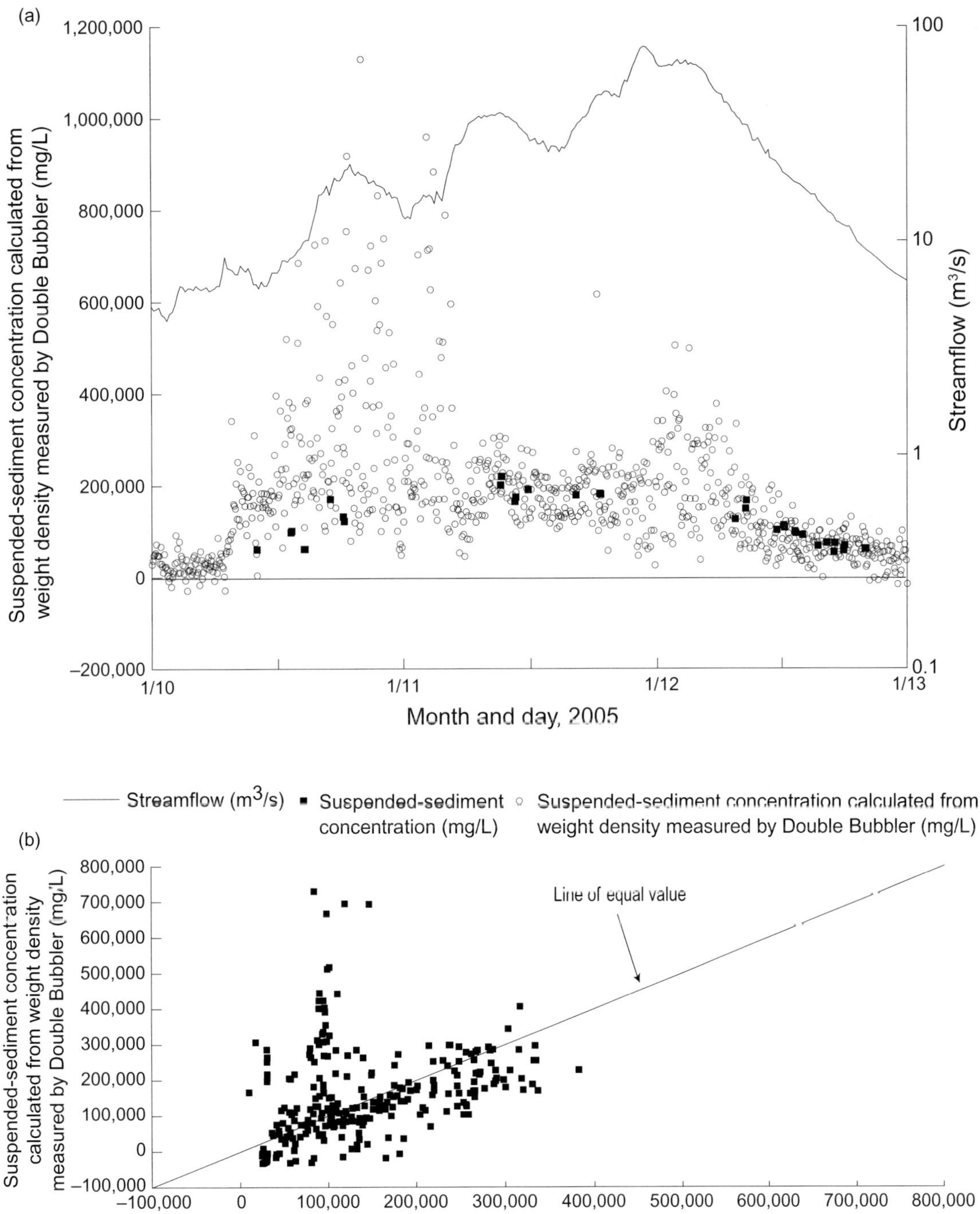

Fig. 1.19 Data for the USGS streamgage on the Paria River at Lees Ferry, Arizona, USA, July 2004 through September 2006. (a) Time series of streamflow, SSCs from samples, and SSCs calculated from weight densities of suspended sediments and dissolved solids measured using the Double Bubbler for a storm in January 2005; (b) scatter plot of measured SSCs from samples and those calculated from the Double Bubbler. Streamflow and sediment data are instantaneous samples, and the Double Bubbler SSC values, calculated from weight densities, are from measurements made at 5-minute intervals.

the acoustic beam, which typically characterize the sedimentary content of multiple orders of magnitude more water than point samplers. Like bulk-optic techniques, empirical calibrations are required to convert the ABS measurements to SSC. Complex post-processing requires compensations for physical properties of ambient water such as temperature, salinity, and pressure, and, in some cases, suspended materials. Additional compensations are needed for instrument characteristics such as frequency, power, and transducer design.

The purchase price of a commercially available single-frequency Doppler *in situ* instrument is about two to four times that of a fully equipped turbidimeter. Because biofouling has little if any effect on the performance of the sensor, field-maintenance costs are probably less than that for a turbidimeter. The instrument-measurement realm is multiple conic beams. Instrument calibrations can be performed using physical samples collected within the volume of the beam; however, they are often supplanted by cross-section calibrations.

The development and application of the ABS technology can be broadly grouped into two approaches, based primarily on the instrumentation type and target application (the underlying theory is equivalent for the two approaches). The first approach uses specially designed acoustic instrumentation often using multiple frequencies to compute SSCs and grain sizes over relatively short ranges (1–2 m). This approach has primarily been applied using fixed deployments to study near-bed sediment transport processes in the marine environment. There are ample publications describing the development and application of this approach (see, for example, Hanes *et al.* 1988; Sheng & Hay 1988; Hay 1991; Thorne *et al.* 1991, 1993, 1995, 1996; Hay & Sheng 1992; Thorne & Campbell 1992; Crawford & Hay 1993; Richards *et al.* 1996; Schaafsma & Hay 1997; Thorne & Hardcastle 1997; Thorne & Buckingham 2004; Thorne & Meral 2008). A review paper by Thorne & Hanes (2002) provides a good overview of the technique. This approach requires calibration of a "system constant" for each instrument, which is typically accomplished in the laboratory (Thorne & Hanes 2002). At least one commercially available instrument that uses this technique but lacks Doppler capability is available (Aquatec Group 2008).

The second approach uses commercially available *in situ* acoustic Doppler current profiles (ADCPs; the term ADCP is used generically and does not imply a particular manufacturer unless specified.) This approach is particularly suited to monitoring suspended-sediment flux because ADCPs provide three-dimensional velocity profiles as well as acoustic backscatter information. As stated above, the underlying theory is the same, though for the ADCP approach the sonar equations are typically formulated in logarithmic form (i.e. in decibels (dB); see next section) whereas for the first approach the linear form of the equations are used (i.e. in terms of pressure or voltage). The increasing popularity of ADCPs for characterizing hydrodynamics in fluvial, estuarine, and coastal environments has facilitated the concurrent estimation of suspended-sediment properties in these environments as well.

Theoretical aspects of the ADCP approach have been well documented (see, for example, Thevenot *et al.* 1992; Reichel & Nachtnebel 1994; Deines 1999; Gartner 2004). Applications have been documented for a wide range of environments (see, for example, Schott & Johns 1987; Thevenot *et al.* 1992; Thevenot & Kraus 1993; Jay *et al.* 1999; Klein 2003; Gartner 2004; Topping *et al.* 2004, 2006, 2007; Hoitink & Hoestra 2005; Hortness 2006; Wall *et al.* 2006; Tessier *et al.* 2008; among many others). At least one commercial software product is available to convert backscatter to SSC (Land & Jones 2001). Comparisons of SSC computed from acoustic backscatter with SSC values determined from water samples have been found to agree within about 10–20% (Thevenot *et al.* 1992; Thorne *et al.* 1991; Hay & Sheng 1992).

The theoretical development presented below is constructed in terms of the logarithmic form of the sonar equations, which is the typical form used for the ADCP approach. This form is particularly suited to this approach because commercially available ADCPs typically provide the conversion factor from raw backscatter counts to decibels (see below), which facilitates accounting for transmission losses and empirical calibration of backscatter to SSC. The logarithmic form of the sonar equations can be inverted to obtain an expression for SSC:

$$SSC_{computed} = 10^{(A+(B*RB))} \tag{2}$$

The exponent of eqn. 2 contains a term for the relative acoustic backscatter, *RB*, measured by an instrument such as an ADCP as well as terms for an intercept, *A*, and slope, *B*, determined by regression of concurrent ABS with known mass SSC measurements ($SSC_{measured}$) on a semi-log plane in the form of $\log(SSC_{measured}) = A + (B*RB)$. The relative backscatter is the sum of the echo level measured at the transducer plus the two-way transmission losses (Thevenot *et al.* 1992) as defined below.

In its simplified form, the sonar equation (Urick 1975) can be written as:

$$RL = SL - 2TL + TS \quad (3)$$

where: *RL* is the reverberation level; *SL* is the source level, which is the intensity of emitted signal that is known or measurable; *2TL* is the two-way transmission loss; and *TS* is the target strength, which is dependent on the ratio of wavelength to particle diameter.

All variables in eqn. 3 are measured in decibels. In terms of ADCP parameters, $RL = K_c(E - Er)$, where *E* is ADCP echo intensity recorded in counts, *Er* is ADCP received signal strength indicator (RSSI) reference level (the echo baseline when no signal is present), in counts, and K_c is the RSSI scale factor used to convert counts to decibels. K_c varies among instruments and transducers and has a value of 0.35–0.55 (Deines 1999). The two-way transmission loss is defined as:

$$2TL = 2(\alpha_w + \alpha_s)R + 20\log R \quad (4)$$

where: *R* is the range to the ensonified volume, in meters; α_w is an absorption coefficient for water; α_s is an attenuation coefficient accounting for viscous and scattering losses due to suspended sediment (see below), both in decibels per meter; $2(\alpha_w+\alpha_s)R$ is the combined transmission loss due to water absorption and sediment attenuation; and $20\log R$ is the loss due to spreading.

The absorption coefficient for water is a function of acoustic frequency, salinity, temperature, and pressure (Schulkin & Marsh 1962). Because of non-spherical spreading in the transducer near field, the spreading loss is different in near and far transducer fields. The transition between near and far transducer fields is called the critical range, $R_{critical}$. $R_{critical} = \pi a_t / \lambda$ where a_t is the transducer radius, in centimeters, and λ is acoustic wavelength. The near-field correction, ψ, for spreading loss can be calculated from the formula in Downing *et al.* (1995) as:

$$\psi = \left[1 + 1.35Z + (2.5Z)^{3.2}\right] / \left[1.35Z + (2.5Z)^{3.2}\right] \quad (5)$$

where: *Z* is $R/R_{critical}$.

As an example, $R_{critical}$ is 167 cm for a 1200-kHz ADCP with a 5.1-cm diameter transducer.

For the particle-size range and acoustic frequencies of interest here, attenuation from suspended sediment consists of a viscous loss component and a scattering loss component (Flammer 1962; Richards *et al.* 1996). In the presence of suspended sediments that are generally less than 100–200 μm, the viscous and scattering components of attenuation change in opposing ways to changes in size (for typical ADCP transducer frequencies). Attenuation from viscous losses increases inversely with sediment size. Attenuation from scattering losses increases directly with sediment size. Scattering characteristics are a function of λ to particle circumference $2\pi a_p$, where a_p is particle radius. When $\lambda >> 2\pi a_p$, most of the scattering pattern propagates backward; however, as λ approaches $2\pi a_p$, the scattering pattern increases in complexity, and when $\lambda << 2\pi a_p$ half the scattered pattern propagates forward and the remainder is scattered through all directions (Flammer 1962). In the case of 1200-kHz acoustic sources, $\lambda = 2\pi a_p$ for 400-μm diameter particle size. Taken together, scattering- and viscous-loss terms account for little attenuation with 1200-kHz frequency unless particle size is very small or SSCs are very high, in which case corrections for attenuation are needed. However, in the case of higher frequencies, total attenuation may need to be accounted for even at lower SSC if particles are very small (viscous losses) or larger than about 100- to 150-μm diameter (scattering losses). The result is a nonlinear (backscatter intensity) response at high SSC (Hamilton *et al.* 1998). Although a function of frequency, attenuation from sediment may need to be accounted for in the presence of as little as 0.1 g/L (Libicki *et al.* 1989; Thorne *et al.* 1991); multiple scattering produces nonlinear response when SSC is on the order of 10 g/L (Sheng & Hay 1988; Hay 1991). Thorne *et al.* (1991) found that, in the case of 3.0- and 5.65-MHz acoustic frequencies, attenuation from fine sands may become significant at ranges on the order of a meter when

SSC levels approach 0.1 g/L. Attenuation due to presence of sediment can be accounted for following Flammer (1962). A coefficient, ζ, is defined as:

$$\zeta = K(\gamma-1)^2\{S/[S^2+(\gamma+\tau)^2]\}+(K^4a_p^3)/6 \quad (6)$$

where: $K = 2\pi/\lambda$; γ is the particle or aggregate wet density divided by the fluid density; $\tau = 0.5 + 9/(4\beta a_p)$; $S = [9/(4\beta a_p)][1 + 1/(\beta a_p)]$; $\beta = [\omega/2\nu)]^{0.5}$; $\omega = 2\pi f$, f is frequency in Hz; and ν is the kinematic viscosity of water, in stokes. The two-way attenuation from suspended particles, $2\alpha_s$ in decibels per centimeter, is equal to (8.68)(ζ)(SSC), where SSC is dimensionless (1000 ppm = 0.001) and 8.68 is the conversion from nepers to decibels. The first term in eqn. 6 is the attenuation from viscous losses and the second term is the attenuation from scattering losses. An alternative form for the scattering loss component can be found in Richards *et al.* (1996).

From a practical standpoint, it is not necessary to know the source level, nor is it typically feasible to measure all the characteristics of suspended material required to directly model target strength (Thevenot *et al.* 1992; Reichel & Nachtnebel 1994). Therefore, following the derivation of Thevenot *et al.* (1992), eqn. 3 is cast in terms of relative backscatter, $RB = RL + 2TL$. After appropriate substitutions, the sonar equation can be written in the desired form in terms of SSC and relative backscatter as:

$$SSC = 10^{(-0.1K_2+0.1RB)} \quad (7)$$

where: K_2 is a parameter that includes terms for source level, target strength, ensonified volume, and mass of suspended material.

The theoretical parameters $A = -0.1K_2$ and $B = 0.1$ are appropriate for an SSC of uniform particles of the same mass and other properties. For a distribution of particles in the field, agreement with the theoretical values is experimentally checked by regression of RB with measured estimate of SSCs at the same location. Thevenot *et al.* (1992) determined the coefficient $-0.1K_2$ to be equal to 0.97 and 1.43 for laboratory and field calibrations, respectively. They determined values for the coefficient multiplying RB to be 0.077 (laboratory) and 0.042 (field). Thus eqn. 7 can be used to compute a time series of SSC from ADCP ABS at any distance from the acoustic transducer where valid backscatter data are available once appropriate transmission losses and slope and intercept values are determined. An alternative approach is to assume the theoretical value for the slope, B, equal to 0.1 and determine an appropriate value of intercept, $A = \log_{10}(SSC_{measured}) - 0.1RB$.

Limitations of the acoustic technique are well described in the literature (e.g. Reichel & Nachtnebel 1994; Hamilton *et al.* 1998). One critical limitation is the fact that it is not possible to differentiate between concurrent changes in SSC and PSD (without sufficient calibrations) when using a single-frequency instrument, as changes in both SSCs and PSDs can result in a change in the backscatter signal strength. In addition, there is an appropriate acoustic frequency for a given PSD. Errors in estimates of SSC will increase if a substantial fraction of the suspended material includes particles that are too large or too small for a response by a given frequency. For these reasons, techniques or instruments that utilize more than one acoustic frequency are preferable to single frequency methods. Several applications of multi-frequency instrumentation have successfully characterized both SSC and mean particle size (Hay & Sheng 1992; Crawford & Hay 1993; Thorne *et al.* 1996; Topping *et al.* 2007).

Finally, an alternative approach for segregating size fractions using a single acoustic frequency has been developed by Topping *et al.* (2006, 2007) on the Colorado River at Grand Canyon, Arizona, USA. This approach segregates the silt-clay and sand components of the suspension by taking advantage of the fact that silt-clay tends to dominate acoustic attenuation whereas sand tends to dominate backscatter. Side-looking ADCPs are mounted on the river bank that profile across the river width; after removing the two-way transmission losses, the slope of the backscatter profile yields the attenuation coefficient, which is strongly correlated with silt-clay SSC, while the acoustic backscatter is strongly correlated with sand SSC. The potential to segregate "wash load" from "bed material suspended load" in sand-bedded rivers warrants future testing of this methodology in a wider range of environments.

1.2.5.2 Example field application

A multi-instrument, multi-frequency system has been established at the USGS streamgage on the Colorado River at Grand Canyon, Arizona, USA, to produce data from which continuous SSCs and SSLs can be computed (Topping *et al.* 2007). The system uses

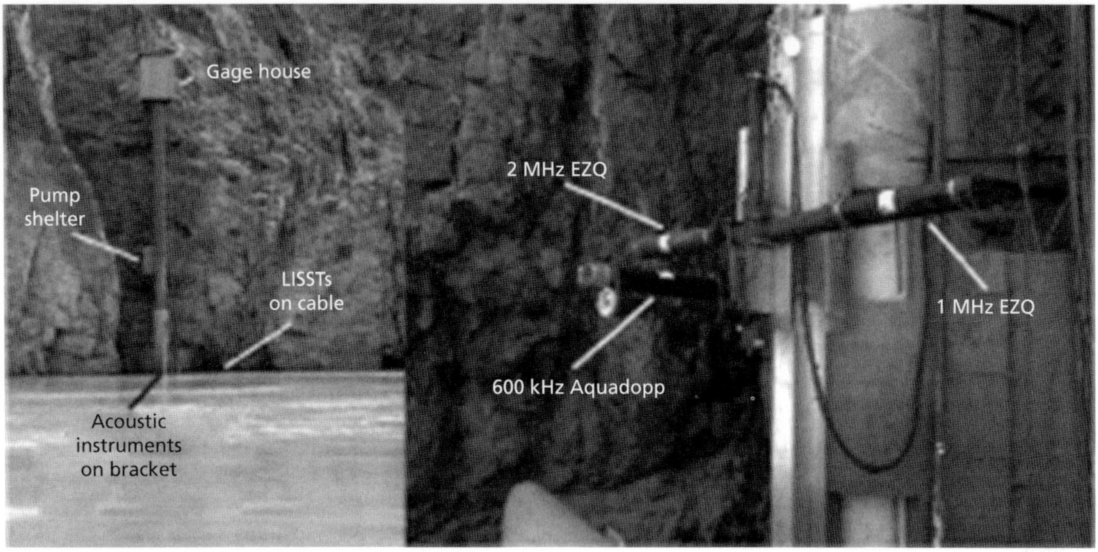

Fig. 1.20 Photograph of an array of the three acoustic Doppler current profilers used to estimate SSCs and PSDs in the Colorado River in Grand Canyon, Arizona, USA.
From: Topping *et al.* (2007).

three single-frequency (1.0 and 2.0 MHz, and 600 kHz) side-looking ADCPs (Fig. 1.20). A post-processing technique is applied to analyze (1) acoustic attenuation to compute the suspended silt-clay size fraction, and (2) acoustic backscatter to compute the suspended-sand fraction in a size range applicable for each frequency. Topping *et al.* (2007) indicate that the approach is applicable for monitoring SSC over the ranges of 0.01–20 g/L (silt-clay) and 0.01–3 g/L (sand); results are within 5% of those computed by conventional methods. In addition, the method calculates median grain size within 10% of that measured by conventional means. Topping *et al.* (2007) infer a greater accuracy with this technique than with a conventional sampling regime largely due to the substantially greater sample frequency and volume. Figure 1.21 shows comparisons of SSC from three-frequency acoustic backscatter, calibrated pump, and LISST measurements.

1.2.5.3 Summary: acoustic backscatter as suspended sediment surrogate technology

As a surrogate for SSC, acoustic backscatter holds several advantages over other suspended-sediment-surrogate technologies. Unlike point measurements, profiles of acoustic backscatter measurements from Doppler velocity instruments can cover a substantial part of the water depth or river cross section; they can integrate orders of magnitude more flow than other methods that rely on at-a-point or single-vertical measurements. Sediment fluxes in the beam can be computed and empirically indexed to the mean cross-sectional SSC value. These data in turn can be used with continuous water-discharge data to compute unit- and daily-value sediment fluxes at the monitoring site. Unlike optic-based surrogate instruments, biological fouling is not a problem.

In addition to some major advantages over other surrogate techniques, the acoustic backscatter method has some limitations. Similar to optical surrogate techniques, a single-frequency source cannot differentiate between change in PSD and change in SSC without calibration and there is an appropriate frequency for a given particle size and a somewhat narrow frequency range for which the method is appropriate for a given size distribution. A series of calculations are required for the reduction and analysis of the acoustic signals; thus until standard operating procedures are developed and adopted for this

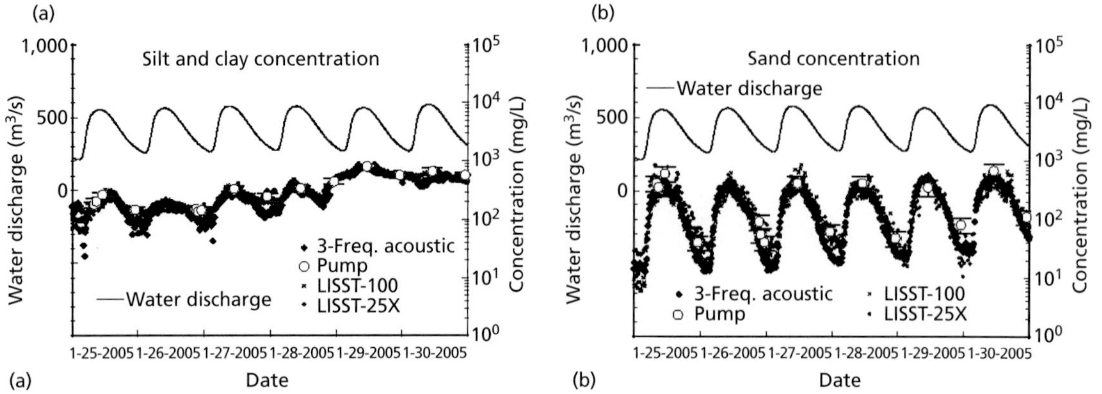

Fig. 1.21 Comparisons of SSCs from three-frequency acoustic backscatter, calibrated pump, and LISST measurements (a) suspended-silt and -clay concentration and (b) suspended-sand concentration.
From Topping *et al.* (2007).

technique, considerable time and effort for a user to compute a time series of SSC from ABS may be required. The cost of a single-frequency *in situ* instrument is about double that for a fully equipped turbidimeter, but the field maintenance cost is expected to be less than that for a turbidimeter.

1.3 Summary and conclusions

Five surrogate technologies for monitoring suspended-sediment-transport characteristics have been or are being tested and evaluated by the USGS toward deployment in operational sediment-transport monitoring programs. The five technologies are bulk optics (turbidity), laser optics, digital optics, pressure difference, and acoustic backscatter. None of the *in situ* technologies measures the surrogate constituent of interest over the entire cross section. Hence, most if not all of the technologies require cross-section calibration. Although most of the *in situ* instruments are routinely calibrated, this step is sometimes bypassed in favor of cross-section calibration.

Table 1.2 summarizes selected attributes of the five suspended-sediment-surrogate technologies presented herein. All of the technologies, with suitable calibration, provide time series of computed SSC at sub-daily sampling frequencies at-a-point (three optical technologies), in a single vertical (pressure-difference technology), or along one more cone-shaped beams (acoustic technology) in streamflow. The capability for providing computed time series of SSC is a major advantage over the relatively sparse data produced by traditional methods for collecting and computing records by conventional methods described by Porterfield (1972), Edwards & Glysson (1999), and Nolan *et al.* (2005). The routine need to estimate SSC values for periods lacking sample data and to interpolate between known or estimated SSC values interjects an unquantifiable degree of uncertainty in traditionally derived sediment-discharge values. The reduction in uncertainty associated with the availability of continuous surrogate data likely will result in a more accurate computation of sediment discharges even considering uncertainties associated with instrument-measurement realm or cross-section calibration of surrogate measurements.

Spatial correlations between any surrogate measurement and its respective mean value in the cross section are still required. However, because of the relatively large ensonified volume associated with acoustic surrogate techniques, correlations associated with the acoustic-backscatter technology are at least theoretically less variable than those for the single-vertical pressure-difference technology, which in turn are theoretically less variable than those for the at-a-point measurements obtained by bulk, laser, or digital-optics technologies.

The most common surrogate technology is turbidity (bulk-optics). Turbidity has been shown to provide sufficiently reliable data for computing SSC

Table 1.2 Summary of selected attributes of five suspended-sediment surrogate technologies.

Technology	Turbidity (bulk optics)		Laser		Digital optic imaging	Pressure difference	Hydroacoustic	
Instrument or type	*In situ* turbidimeter	*In situ* OBS	*In situ* LISST-100	Manually deployed LISST-SL	Multi-camera stream-side pumping system	*In situ* Double Bubbler	*In situ* single-frequency acoustic Doppler profiler	*In situ* multiple-frequency acoustic Doppler profiler
Price relative to *in situ* turbidimeter	*ca*. \$5,000 (summer 2008)	About 1×	About 5×	About 6×	About 1×–2×	About 1×	About 2×–4×	Unknown
Approximate concentration measurement range	Standard 0–2 g/L. Available at larger ranges	Standard 0–5 g/L. Available at larger ranges	Depending on versions: 0–2 g/L (particle size dependent)	About 0–2 g/L (particle size dependent)	0–10 g/l; future testing may elucidate a larger upper limit	Larger than about 10 g/L, but needs more research; theoretically no upper limit	Signal attenuation limited as function of PSD and frequency	Signal attenuation limited as function of PSD and frequency
Approximate measurement range, PSD (mm)	Does not measure PSD	Does not measure PSD	0.0025–0.5 or 0.00125–0.25	0.0025–0.5 or 0.00125–0.25	0.004 – 4.0	Does not measure PSD	Does not measure PSD Particle size dependent. Ratio circumference to wavelength <1	May measure sand versus silt/clay content. Particle size dependent. Ratio circumference to wavelength <1

Table 1.2 *Continued*

Technology	Turbidity (bulk optics)		Laser		Digital optic imaging	Pressure difference	Hydroacoustic	
Measurement metric basis to routinely compute mean cross-sectional values	Calibrated to SSC from physical samples in mass units	Calibrated to SSC from physical samples in mass units	Calibrated to SSC from physical samples in mass units; PSD in 32 size classes; volume SSC, converted to mass SSC if density known	Calibration may be unnecessary; PSD in 32 size classes; volume SSC, converted to mass SSC if density known	Calibrated to SSC from physical samples in mass units	Calibrated to SSC from physical samples in mass units	Calibrated to SSC from physical samples in mass units	Calibrated to SSC from physical samples in mass units. If PSD, by variable response to selected frequencies
Ancillary measurements	None	None	Depth and water temperature	Depth, ambient velocity, water temperature	None	Stage	Index velocity. Depth if oriented down	Index velocity. Depth if oriented down
Reliability and robustness	Optical window may foul, causing signal to drift with time. Sensor may saturate at larger SSC	Optical window may foul, causing signal to drift with time. Sensor may saturate at larger SSC	Requires anti-fouling device or bioblock. Sensor may saturate at larger SSC; PSD larger than 0.5 mm not included in calculations.	PSD larger than 0.5 mm not included in calculations.	Accuracy may decrease with window fouling; software will correct for this within yet-undefined limits	Low SSC data unreliable; veracity of higher SSCs unresolved	More or less unaffected by fouling. Responds almost solely to entrained sediment	More or less unaffected by fouling. Responds almost solely to entrained sediment

Region of measurement	Fixed point	Fixed point	Fixed point; device may be used in profiling mode by cable suspension	Point, vertical, or multiple verticals by cable suspension	Fixed point	Single fixed vertical, mean SSC value	Conic beam with data available at selected distances from the sensor	Conic beam with data available at selected distances from the sensor
Accuracy for derivation of suspended-sediment data	When within measurement range has been used to develop reliable SSC-turbidity regression relations	When within measurement range has been used to develop reliable SSC-turbidity regression relations	Deemed reliable in some field applications	Lab sedimento-logical tests completed 2008; field sedimento-logical and isokinetic tests in 2009	Unresolved. Preliminary tests show accurate PSD results for silt and fine sand, additional testing is planned	Unresolved based on two field tests; additional evaluation required	Shown useful in field applications where PSD does not change dramatically	Shown to provide accurate silt-clay versus sand-size fractions in one field deployment
Potential for meeting U.S. Geological Survey accuracy criteria	High for mass SSC depending on nature of the turbidity-SSC relation	High for mass SSC depending on nature of the turbidity-SSC relation	High for volume SSC and for PSD	High for volume SSC and for PSD	High for PSD; accuracy is still not determined for SSC	Low for mass SSC	Moderate for mass SSC	High for mass SSC; moderate for silt-clay versus sand-size fractions
Potential for application in suspended-sediment monitoring programs	Endorsed by the USGS, given appropriate in-stream sedimentological conditions, calibrations, and ability to maintain instruments	Endorsed by the USGS, given appropriate in-stream sedimentological conditions, calibrations, and ability to maintain instruments	High (given appropriate in-stream sedimentological conditions, known density, and ability to maintain instruments)	High (among potential uses, perform calibrations for *in situ* instruments)	High for laboratory SSC and PSD; moderate for field applications	Unknown pending additional testing using modifications of the physical system and algorithms	Moderate (given appropriate in-stream sediment-ological conditions, and calibration)	High for SSC; moderate for silt-clay versus sand-size fractions

in several varied field settings so as to warrant USGS endorsement for use in operational sediment-monitoring programs. However, instrument-sensor saturation can result in failure to record usable data during periods of high SSCs associated with higher streamflows, which tend to be the most influential in sediment-transport calculations. SSC computed from at-a-point turbidity data may not be representative of the mean cross-sectional SSC, particularly when sand-size material composes an appreciable fraction of total suspended-sediment transport. The presence of biofouling can cause bias in signal accuracy or render the data unusable if the optical surface is not kept clean manually or by using a mechanical wiper. Two fully equipped turbidimeters and one optical backscatterance meter purchased in the summer 2008 each cost about US$5000. This cost can be a small fraction of the annual cost associated with monitoring suspended-sediment transport using traditional techniques. However, the potential for additional site visits for maintenance, cleaning, or the collection of calibration samples can result in increased operating costs.

Similar to bulk-optical sensors, laser-optic instruments also are prone to biofouling and signal saturation at high SSC. However, these instruments have the major advantage in providing continuous PSDs from which volumetric SSC can be calculated, as well as mass SSC if particle density is known or can be confidently estimated. The cost of the LISST suite of instruments (the only commercially available *in situ* instruments using forward (multi-angle) laser light scattering measurements) ranges from two to six times that of a fully equipped turbidimeter.

The digital-optic surrogate technique determines volume SSC by enumerating and summing the volumetric characteristics of individual sediment particles from a digital image of a filament of sample in a flow-through cell. Real-time measurements of particles between 4 and 4000 μm are possible and the system requires no routine calibration. The technology's performance is currently limited to laboratory analyses, although it may have applications for bank-operated pumping systems or for manual deployment in rivers. Similar to the LISST instrument, results are expressed in volume/volume relations and not the more common mass/volume units. Indistinct particle boundaries can reduce measurement accuracy, as can high turbidity from organic or colloidal material. The cost of off-the-shelf instrument parts is one to two times that for a fully equipped turbidimeter.

Research on the pressure-difference technology (Double Bubbler) implies that its use should be limited to SSCs exceeding at least 10 g/L, which is generally larger than the suitable SSC range for the other surrogate techniques examined herein (with the exception of the LISST-Infinity laser instrument). This relatively robust technology, the cost of which is similar to that of a fully equipped turbidimeter, measures SSC in a fixed water column. The theoretical underpinnings of this technology are straightforward and its field application is relatively simple. However, performance of the pressure-difference technology has been marginal at best in field tests in Puerto Rico (maximum SSCs approaching 20 g/L) and Arizona, USA (maximum SSCs 10^2–10^3 g/L). Nevertheless, potential remains for use of this technology because it may provide time series of very high SSC that cannot be resolved using other surrogate techniques.

The acoustic backscatter technology shows the most promise for meeting the needs of suspended-sediment monitoring programs. Mounted *in situ* in a side-looking (or, less often, upward-looking) orientation, the technology is relatively robust and can integrate several orders of magnitude more flow than those technologies that make point measurements. Results using a three-frequency instrument array at the USGS streamgage on the Colorado River at Grand Canyon, Arizona, USA, have compared well with manually collected calibration data for sand-size material in the range 0.01–3 g/L and for finer material in the range 0.01–20 g/L. At present, the cost of using a three-frequency Doppler array (three separate instruments such as used at the USGS streamgage on the Colorado River at Grand Canyon) is about sixfold that for a fully equipped turbidimeter. Although at least one multi-frequency ABS is commercially available, it lacks Doppler (velocity) capability. Research and development efforts toward production of a reasonably priced multi-frequency hydroacoustic instrument are underway.

1.4 Prospects for operational surrogate monitoring of suspended-sediment transport in rivers

This chapter has described five surrogate technologies for monitoring characteristics important to understanding properties of sediment transport in rivers. Some characteristics common to these five technologies include the following:

- all address measurement of fluvial-sediment characteristics that are difficult, expensive, and (or) dangerous to directly measure with sufficient frequency to adequately define their spatial and temporal variability;
- all are generally affordable – ranging from about the cost of a fully equipped turbidimeter (about US$5000 in 2008) to about sixfold that cost for the more expensive laser-diffraction technologies;
- all (with the possible exception of the laser-diffraction and digital-optic technologies) require site-specific calibrations, although the need for calibration is expected to diminish over time;
- all require derivation of coefficients equating values recorded by the surrogate instrument to the mean cross-section constituent value;
- all but turbidity, which is endorsed by the USGS for use in operational sediment-monitoring programs, require additional testing and evaluation.

The USGS endorsement of SSC and SSL computations from turbidity measurements notwithstanding, none of the technologies is suitable for monitoring all the suspended-sediment characteristics in all rivers under all flow and sediment-transport conditions. Nevertheless, if care is exercised in matching surrogate technologies to appropriate river and sediment conditions, it is becoming possible to monitor SSC and SSL remotely and continuously in a variety of rivers over a range of flow and sedimentary conditions within generally acceptable accuracy limits. Endorsement and broad-scale deployment of certifiably reliable sediment-surrogate technologies supported by operational and analytical protocols are revolutionary concepts in fluvial sedimentology. The benefits could be enormous, providing for safer, more frequent and consistent, arguably more accurate, and ultimately less expensive fluvial-sediment data collection for use in managing the world's sedimentary resources.

Acknowledgments

This chapter benefited from the contributions and efforts of several individuals other than the authors. The manuscript was improved by the reviews provided by Michael Singer, University of St Andrews, UK, and James D. Fallon and Broderick E. Davis, USGS, Minneapolis, Minnesota, and Vicksburg, Mississippi, respectively. Annette L. Ledford, USGS, Reston, Virginia, devoted considerable effort in the development of the chapter's figures and tables. Arthur J. Horowitz's (Atlanta, Georgia, USA) research on the sedimentary properties of selected US rivers was excerpted. The laser-optic and hydroacoustic sections benefited from research led by David A. Topping, USGS, Flagstaff, Arizona.

References

Agrawal, Y. C. & Pottsmith, H. C. (1994) Laser diffraction particle sizing in STRESS. *Continental Shelf Research*, **14**, 1101–21.

Agrawal, Y. C. & Pottsmith, H. C. (2000) Instruments for particle size and settling velocity observations in sediment transport. *Marine Geology* **168**(1–4), 89–114.

Agrawal, Y. C. & Pottsmith, H. C. (2006) The isokinetic streamlined suspended-sediment profiling LISST-SL – status and field results. In: Proceedings of the 8th Federal Interagency Sedimentation Conference, Reno, Nevada, April 2–6 2006, 288–295, http://pubs.usgs.gov/misc_reports/FISC_1947-2006/pdf/1st-7thFISCs-CD/8thFISC/Session%202C-1_Agrawal.pdf.

Agrawal, Y. C., McCave, I. N. & Riley, J. B. (1991) Laser diffraction size analysis. In *Principles, Methods, and Application of Particle Size Analysis*, J. P. M. Syvitski (ed.), 119–28. New York: Cambridge University Press.

Agrawal, Y. C., Pottsmith, H. C., Lynch, J. & Irish, J. (1996) Laser instruments for particle size and settling velocity measurements in the coastal zone. In Proceedings Oceans 1996, Institute of Electrical and Electronic Engineers, Fort Lauderdale, Florida, 23–26 September, 1135–42.

Agrawal, Y. C., Whitemore, A., Mikkelsen, O. A. & Pottsmith, H. C. (2008) Light scattering by random shaped particles and consequences on measuring suspended sediments by laser diffraction. *Journal of Geophysical Research* **113**, C04023, 11pp.

Anderson, C. W. (2005) Turbidity (version 2). US Geological Survey Techniques of Water Resources Investigations Book 9, chapter A6, section 6.7, http://water.usgs.gov/owq/FieldManual/Chapter6/6.7_contents.html.

Aquatec Group. (2008) http://www.aquatecgroup.com/aquascat.html.

ASTM International. (1998) Terminology for fluvial sediment. Designation D 4410-98, 6p.

Bale, A. J. & Morris, A. W. (1987). In-situ measurements of particle size in estuarine waters. *Estuarine, Coastal and Shelf Science* **24**, 253–63.

Bent, G. C., Gray, J. R., Smith, K. P. & Glysson, G. D. (2003) A synopsis of technical issues for monitoring sediments in highway and urban runoff. In *The National*

Highway Runoff Data and Methodology Synthesis, G. E. Granato, C. Zenone, & P. E. Cazenas (eds), 113–63. US Department of Transportation Publication No. FHWA-EP-03-054, 1, http://ma.water.usgs.gov/fhwa/products/ofr00497.pdf.

Beverage J. P. & Culbertson, J. K. (1964) Hyperconcentrations of suspended sediment. *Journal of the Hydraulics Division*, Proceedings of the American Society of Civil Engineers, **90**, 117–28.

Bogen, J., Fergus, T. & Walling, D. E. (2003) Erosion and sediment transport in rivers, technological and methodological advances. International Association of Hydrological Sciences, Publication 283, 238pp.

Buchanan, P. A. & Schoellhamer, D. H. (1995). Summary of suspended-solids concentration data, Central and South San Francisco Bays, California, water years 1992 and 1993. US Geological Survey Open-File Report 94-543, 15pp.

Campbell Scientific, Inc. (2008) http://www.d-a-instruments.com/.

Christensen, V. G., Xiaodong, J. & Ziegler, A. C. (2000) Regression analysis and real-time water-quality monitoring to estimate constituent concentrations, loads, and yields in the Little Arkansas River, South-Central Kansas 1995–99. US Geological Survey Water-Resources Investigations Report 00-4126, 36pp., http://ks.water.usgs.gov/Kansas/pubs/reports/wrir.00-4126.html.

Cohn, T. A., DeLong, L. L., Gilroy, E. J., Hirsch, R. M. & Wells, D. K. (1989). Estimating constituent loads. *Water Resources Research*, **25**(5), 937–42.

Conner, C. S. & De Visser, A. M. (1992) A laboratory investigation of particle size effects on optical backscatterance sensor. *Marine Geology* **108**, 151–59.

Crawford, A. M. & Hay, A. E. (1993) Determining suspended sand size and concentration from multifrequency acoustic backscatter. *Journal of the Acoustical Society of America* **94**(6), 3312–24.

D & A Instrument Company. (1991) OBS-1 & 3 Instruction manual. D & A Instrument Company, Port Townsend, Washington, 41pp.

Davis, B. E. (2005) A guide to the proper selection and use of federally approved sediment and water-quality samplers: US Geological Survey Open-File Report 2005-1087, 20pp., http://pubs.usgs.gov/of/2005/1087/.

Deines, K. L. (1999) Backscatter estimation using broadband acoustic Doppler current profilers. Proceedings of the IEEE Sixth Working Conference on Current Measurement, San Diego, CA, March 11–13 1999, 249–253.

Design Analysis Associates. (2008) http://www.waterlog.com/.

Diplas, P., Kuhnle, R. A., Gray, J. R., Glysson, G. D. & Edwards, T. E. (2008) Sediment transport measurements. In *Sedimentation Engineering*, M. H. García (ed.), *American Society of Civil Engineering Manuals and Reports on Engineering Practice*, no.110, 307–53.

Downing, J. P. (1983). An optical instrument for monitoring suspended particles in ocean and laboratory. In *OCEANS 1983, San Francisco, California, August 29-September 1 1983, Proceedings: 199–202.*

Downing, J. P. (1996) Suspended sediment and turbidity measurements in streams – What they do and do not mean. In *Automatic Water Quality Monitoring Workshop*, Richmond, B.C., February 12–13 1996.

Downing, J. P., Sternberg, R. W. & Lister, C. R. B. (1981) New instrumentation for the investigation of sediment suspension processes in the shallow marine environment. *Marine Geology* **42**, 19–34.

Downing, A., Thorne, P. D. & Vincent, C. E. (1995) Backscattering from a suspension in the near field of a piston transducer. *Journal of the Acoustical Society of America*, **97**(3), 1614–20.

Duan, N. (1983) Smearing estimate – a nonparametric retransformation method. *Journal of the American Statistical Association*, **78**(383), 605–10.

Edwards, T. E. & Glysson, G. D. (1999) Field methods for collection of fluvial sediment. US Geological Survey Techniques of Water-Resources Investigations Book 3, Chapter C2, 89p. http://water.usgs.gov/osw/techniques/Edwards-TWRI.pdf.

Federal Interagency Sedimentation Project. (1940) *Field practice and equipment used in sampling suspended sediment.* Iowa City, Iowa University Hydraulics Laboratory, Interagency Report no.1, 175pp., http://fisp.wes.army.mil/Report%201.pdf.

Federal Interagency Sedimentation Project. (1941) *Laboratory investigation of suspended sediment samplers.* Iowa City, Iowa University Hydraulics Laboratory, Interagency Report no.5, 99pp., http://fisp.wes.army.mil/Report%205.pdf.

Federal Interagency Sedimentation Project. (2008) http://fisp.wes.army.mil/.

Flammer, G. H. (1962) Ultrasonic measurement of suspended sediment. US Geological Survey Bulletin 1141-A. Government Printing Office, Washington, DC, 48pp.

Gartner, J. W. (2004) Estimating suspended solids concentrations from backscatter intensity measured by acoustic Doppler current profiler in San Francisco Bay, California. *Marine Geology* **211**, 169–87.

Gartner, J. W., Cheng, R. T., Wang, P. F. & Richter, K. (2001) Laboratory and field evaluations of LISST-100 instrument for suspended particle size determinations. *Marine Geology* **175**(1–4), 199–219.

Gartner, J. W., Mueller, D. S., Wall, G. R. & Gray, J. R. (2003) Breakout session 4: other fluvial-sediment surrogates. In *Proceedings of the Federal Interagency Workshop on Turbidity and Other Sediment Surrogates, April 30–May 2 2002, Reno, Nevada, 21–27.* http://pubs.usgs.gov/circ/2003/circ1250/.

Gilvear, D. J. & Petts, G. E. (1985) Turbidity and suspended solids variations downstream of a regulating reservoir. *Earth Surface Processes Landforms*, **10**(4), 363–73.

Glysson, G. D. (1987). Sediment-transport curves. US Geological Survey Open-File Report 87-218, 47pp., http://pubs.er.usgs.gov/usgspubs/ofr/ofr87218.

Glysson, G. D. (1989). 100 years of sedimentation study by the USGS. In Proceedings of the International Symposium, Sediment Transport Modeling, Sam S.Y. Wang, ed. American Society of Civil Engineers, New Orleans, August 14–18 1989, 260–65.

Glysson, G. D. & Gray, J. R. (1997). Coordination and standardization of Federal sedimentation activities. Expanding Sediment Research Capabilities in Today's US Geological Survey. February 4–7, 1997, 2pp., http://water.usgs.gov/osw/techniques/workshop/.

Graf, J. B., Wirt, L., Swanson, E. K., Fisk, G. G. & Gray, J. R. (1996) Streamflow transport of radionuclides and other chemical constituents in the Puerco and the Little Colorado River Basins, Arizona and New Mexico: US Geological Survey Water-Supply Paper 2459, 89pp., http://pubs.er.usgs.gov/usgspubs/wsp/wsp2459.

Gray, J. R. (2003) US Geological Survey suspended-sediment surrogate research, Part I: Call for a sediment monitoring instrument and analysis research program. In Proceedings of the Virginia Water Research Symposium 2003, Water Resource Management for the Commonwealth, Virginia Polytechnic Institute and State University, Blacksburg, October 8-10 2003, 53–57.

Gray, J. R. (ed.). (2005) Proceedings of the Federal Interagency Sediment Monitoring Instrument and Analysis. Workshop, September 9–11 2003, Flagstaff, Arizona: US Geological Survey Circular 1276, 46pp., http://water.usgs.gov/pubs/circ/2005/circ1276/. Appendix 4 is available only online at: http://water.usgs.gov/osw/techniques/sediment/sedsurrogate2003workshop/listofpapers.html.

Gray, J. R., Agrawal, Y. C. & Pottsmith, H. C. (2004) The LISST-SL streamlined isokinetic suspended-sediment profiler. In *Proceedings of the 9th International Symposium on River Sedimentation, Yichang, China, October 18-21 2004*, vol. IV, Cheng Liu (ed.), 2549–555. Tsinghua University Press, http://water.usgs.gov/osw/techniques/Lisst_Gray_Agrawal_Pottsmith.pdf.

Gray, J. R. & Gartner, J. W. (2004) Surrogate technologies for continuous suspended-sediment monitoring in the United States. In Proceedings of the 9th International Symposium on River Sedimentation, October 18-21 2004, Yichang, China, vol. IV, 2515–22. Tsinghua University Press,.

Gray, J. R. & Gartner, J. W. (2009) Technological advances in suspended-sediment surrogate monitoring. *Water Resources Research*, **45**, W00D29, doi: 10.1029/2008WR007063, 20pp.

Gray, J. R. & Glysson, G. D. (eds) (2003) Proceedings of the Federal Interagency Workshop on Turbidity and Other Sediment Surrogates, April 30–May 2 2002, Reno, Nevada, US Geological Survey Circular 1250, 56pp., http://pubs.er.usgs.gov/usgspubs/cir/cir1250.

Gray, J. R. & Glysson, G. D. (2005) Attributes of a sediment monitoring instrument and analysis research program. In Proceedings of the Federal Interagency Sediment Monitoring Instrument and Analysis Research Workshop, September 9–11 2003, Flagstaff, Arizona, 6pp., http://water.usgs.gov/osw/techniques/sediment/sedsurrogate2003workshop/gray_glysson.pdf.

Gray, J. R., Glysson, G. D. & Edwards, T. E. (2008) Suspended-sediment samplers and sampling methods. In *Sediment Transport Measurements*, M. Garcia (ed.), Sedimentation Engineering – Processes, Measurements, Modeling, and Practice, American Society of Civil Engineers, Manual 110, Chapter 5.3, 320–39.

Gray, J. R., Glysson, G. D. & Mueller, D. S. (2002) Comparability and accuracy of fluvial-sediment data – a view from the US Geological Survey. In Proceedings of the ASCE Specialty Conference, Hydraulic Measurements & Experimental Methods Conference, Estes Park, Colorado, July 28-August 1 2002, 6pp. http://water.usgs.gov/osw/techniques/asce.pdf.

Gray, J. R., Glysson, G. D., Turcios, L. M. & Schwarz, G. E. (2000) Comparability of suspended-sediment concentration and total suspended solids data. US Geological Survey Water-Resources Investigations Report 00-4191, 14pp., http://water.usgs.gov/osw/techniques/sedimentpubs.html.

Gray, J. R., Gooding, M. T. S., Topping, D. J. & Rasmussen, P. P. (2003a) US Geological Survey suspended-sediment surrogate research, Part II: Optic technologies. In Proceedings of the Virginia Water Research Symposium 2003, Water Resource Management for the Commonwealth, Virginia Polytechnic Institute and State University, Blacksburg, October 8–10 2003, 58–64.

Gray, J. R. & Osterkamp, W. R. (2007) A vision for a worldwide fluvial-sediment information network. Keynote Paper, In Proceedings of the 10th International Symposium on River Sedimentation, Moscow, Russia, August 1-4 2007, Vol. I, 43–54, http://www.irtces.org/zt/10isrs/lunwen/Session%200/Symposium_0_5.htm.

Gray, J. R., Patiño, E. & Larsen, M. C. (2003b). US Geological Survey suspended-sediment surrogate research, Part III: Acoustic and pressure-differential technologies. In *Proceedings of the Virginia Water Research Symposium 2003, Water Resource Management for the Commonwealth, Virginia Polytechnic Institute and State University, Blacksburg, October 8–10 2003*, 65–69.

Gray, J. R. & Simões, F. J. M. (2008) Estimating sediment discharge. In M. Garcia (ed.), *Sedimentation Engineering – Processes, Measurements, Modeling, and Practice*, American Society of Civil Engineers, Manual 110, Appendix D, 1067–88.

Hamilton, L. J., Shi, Z. & Zhang, S. Y. (1998) Acoustic backscatter measurements of estuarine suspended cohesive sediment concentration profiles. *Journal of Coastal Research* **14**(4), 1213–24.

Hanes, D. M., Vincent, C. E., Huntley, D. A. & Clarke, T. L. (1988) Acoustic measurements of suspended sand concentration in the C2S2 experiment at Stanhope Lane, Prince Edward Island. *Marine Geology* Elsevier Science Publishers, Amsterdam, 185–96.

Hay, A. E. (1991) Sound scattering from a particle-laden, turbulent jet. *Journal of the Acoustical Society of America*, **90**(4) pt 1, 2055–74.

Hay, A. E. & Sheng, J. (1992) Vertical profiles of suspended sand concentration and size from multi-frequency acoustic backscatter. *Journal of Geophysical Research*, **97**(C10), **15**,661–77.

Helsel D. R. & Hirsch, R. M. (2002) Statistical methods in water resources – Hydrologic analysis and interpretation. US Geological Survey Techniques of Water-Resources Investigations, book 4, chap. A3, 510pp., http://pubs.usgs.gov/twri/twri4a3/

Hoitink, A. J. F. & Hoekstra, P. (2005) Observations of suspended sediment from ADCP and OBS measurements in a mud-dominated environment. *Coastal Engineering*, **52**(2), 103–18.

Hortness, J. E. (2006) Using acoustic backscatter technology to measure suspended sediment concentrations in Idaho streams. In Proceedings, 8th Federal Interagency Sedimentation Conference, April 2-6 2006, Reno, NV, CD-ROM, ISBN 0-9779007-1-1.

ISO. (1992a) Measurement of liquid flow in open channels – method for measurement of bedload discharge. Technical report 92-9212, 20pp.

ISO. (1992b) Liquid flow measurement in open channels – sampling and analysis of gravel-bed material. International Standard ISO 9195, 9pp.

ISO. (1997) Measurement of liquid flow in open channels –bed-material sampling. International Standard ISO 4364, 39pp.

ISO. (2002) Measurement of liquid flow in open channels –methods for measurement of characteristics of suspended sediment. International Standard ISO 4363, 27pp.

ISO. (2005) Liquid flow in open channels – sediment in streams and canals – Determination of concentration, particle size distribution and relative density. International Standard ISO 4365, 47pp.

Jay, D. A., Orton, P., Kay, D. J., Fain, A. & Baptisa, A. M. (1999) Acoustic determination of sediment concentrations, settling velocities, horizontal transports and vertical fluxes in estuaries. In Proceedings: 6th Working Conference on Current Measurement, 3/11–3/13 1999, San Diego, CA, 258–63.

Karp-Boss, L., Azevedo, L. & Boss, E. (2007) LISST-100 measurements of phytoplankton size distribution: evaluation of the effects of cell shape. *Limnology and Oceanography: Methods* **5**, 396–406.

Kindratenko, V. (1997). Development and application of image analysis techniques for identification and classification of microscopic particles. PhD thesis, University of Antwerp, Belgium 1997, Part 2, Shape Analysis.

Kineke, G. C. & Sternberg, R. W. (1992) Measurements of high concentration suspended sediments using the optical backscatterance sensor. *Marine Geology* **108**, 253–58.

Koch, R. W. & Smillie, G. M. (1986). Bias in hydrologic prediction using log-transformed regression models. *Water Resources Bulletin*, **22**(5), 717–32.

Klein, H. (2003) Investigating sediment re-mobilisation due to wave action by means of ADCP echo intensity data: field data from the Tromper Wiek, western Baltic Sea. *Estuarine, Coastal and Shelf Science*, **58**(3), 467–74.

Koltun, G. F., Eberle, M., Gray, J. R. & Glysson, G. D. (2006) User's manual for the Graphical Constituent Loading Analysis System (GCLAS). US Geological Survey Techniques and Methods Book 4, Chap. C1, 50pp. http://pubs.er.usgs.gov/usgspubs/tm/tm4C1.

Konrad, C. H., Pottsmith, H. C., Melis, T. S. & Rubin, D. M. (2006) Real-time analysis of concentrated fluvial suspended sediments. In Proceedings of the 8th Federal Interagency Sedimentation Conference, Reno, Nevada, April 2-6 2006, 585–91. http://pubs.usgs.gov/misc_reports/FISC_1947-2006/pdf/1st-7thFISCs-CD/8thFISC/Session%208C-2_Konrad.pdf.

Land, J. M. & Jones, P. D. (2001) Acoustic measurement of sediment flux in rivers and near-shore waters. In Proceedings of the 7th Federal Interagency Sedimentation Conference, March 25-29 2001, Reno, Nevada, III 127–34., http://pubs.usgs.gov/misc_reports/FISC_1947-2006/pdf/1st-7thFISCs-CD/7thFISC/7Fisc-V1/7FISC1-3.pdf.

Landers, M. N. (2003) Summary of blind sediment reference sample measurement session. In *Proceedings of the Federal Interagency Workshop on Turbidity and Other Sediment Surrogates, April 30-May 2 2002, Reno, Nevada*, J. R. Gray & G. D. Glysson (eds), 29–30. US Geological Survey Circular 1250, http://water.usgs.gov/osw/techniques/turbidity.html.

Larsen, M. C., Figueroa-Alamo, C., Gray, J. R. & Fletcher, W. (2001) Continuous automated sensing of streamflow density as a surrogate for suspended-sediment concentration sampling. In Proceedings of the 7th Federal Interagency Sedimentation Conference, Reno, Nevada, March 25–29, vol. I, III-102 to III-109, http://pubs.usgs.gov/misc_reports/FISC_1947-2006/pdf/1st-7thFISCs-CD/7thFISC/7Fisc-V1/7FISC1-3.pdf.

Lewis, J. (1996) Turbidity-controlled suspended sediment sampling for runoff-event load estimation. *Water Resources Research*, **32**(7), 2299–310.

Lewis, J. (2002) Estimation of suspended sediment flux in streams using continuous turbidity and flow data coupled with laboratory concentrations. In Proceedings of the Federal Interagency Workshop on Turbidity and Other Sediment Surrogates, April 30-May 2 2002, Reno, Nevada, J. R. Gray & G. Douglas Glysson (eds). US Geological Survey Circular 1250, 3pp., http://water.usgs.gov/osw/techniques/TSS/LewisTSS.pdf.

Lewis, A. J. & Rasmussen, T. C. (1999) Determination of suspended sediment concentrations and particle size distributions using pressure measurements. *Journal of Environmental Quality*, **28**, 1490–96.

Libicki, C., Bedford, K. W. & Lynch, J. F. (1989). The interpretation and evaluation of a 3-MHz acoustic back-scatter device for measuring benthic boundary layer sediment dynamics. *Journal of the Acoustical Society of America* **85**(4), 1501–11.

Lietz, A. C. & Debiak, E. A. (2005) Development of rating curve estimators for suspended-sediment concentration and transport in the C-51 canal based on surrogate technology, Palm Beach County, Florida 2004-05. US Geological Survey Open-File Report 2005-1394 19pp., accessed March 2008, at http://pubs.usgs.gov/of/2005/1394/.

Ludwig, K. A., Hanes, D. M. (1990). A laboratory evaluation of optical backscatterance suspended solids sensors exposed to sand-mud mixtures. *Marine Geology* **94**, 173–79.

Meade, R. H. & Parker, R. S. (1985). Sediment in rivers of the United States. US Geological Survey Water-Supply, Paper 2275, 49–60.

Melis, T. S., Topping, D. J. & Rubin, D. M. (2003) Testing laser-based sensors for continuous in situ monitoring of suspended sediment in the Colorado River, Arizona. In *Erosion and Sediment Transport Measurement in Rivers, Technological and Methodological Advances* J. Bogen, T. Fergus & D. E. Walling (eds), 21–27. International Association of Hydrological Sciences Publication 283.

Meral, R. (2008) Laboratory evaluation of acoustic back-scatter and LISST methods for measurements of suspended sediments. *Sensors*, **8**, 979–93.

Miller, C. R. (1951) Analysis of flow-duration sediment rating curve method of computing sediment yield: US Bureau of Reclamation, 55pp.

Nelson, M. E. & Benedict, P. C. (1950) Measurement and analysis of suspended sediment loads in streams. Transactions of the American Society of Civil Engineers, paper no. 2450, 891–918.

Nolan, K. M., Gray, J. R. & Glysson, G. D. (2005) Introduction to suspended-sediment sampling. US Geological Survey Scientific Investigations Report 2005-5077, http://pubs.er.usgs.gov/pubs/sir/sir20055077.

Osterkamp, W. R., Heilman, P., Gray, J. R., and 33 others. (2004) An invitation to participate in a North American sediment-monitoring network. American Geophysical Union, EOS, 84, no.40, October 5 2004, 386–89.

Osterkamp, W. R., Heilman, P. & Lane, L. J. (1998) Economic considerations of a continental sediment-monitoring program. *International Journal of Sediment Research* **13**(4), 12–24, http://water.usgs.gov/osw/techniques/Osterkamp.html.

Pimentel, D., Harvey, C., Resosudarmao, P., *et al.*. (1995) Environmental and economic costs of soil erosion and conservation benefits. *Science*, **267**, 1117–23.

Porterfield, G. (1972) Computation of fluvial-sediment discharge. US Geological Survey Techniques of Water-Resource Investigations Book 3, Chapter C3, 66pp., http://water.usgs.gov/pubs/twri/twri3-c3/.

Pottsmith, H. C. & Bhogal, V. K. (1995) In-situ particle size distribution in the aquatic environment. In *14th World Dredging Congress, Amsterdam, The Netherlands, Nov. 19, 1995*.

Pratt, T. & Parchure, T. (2003) OBS calibration and field measurements. In *Proceedings of the Federal Interagency Workshop on Turbidity and Other Sediment Surrogates, April 30-May 2 2002, Reno, Nevada*, J. R. Gray & G. D. Glysson (eds). US Geological Survey Circular 1250, 3pp., http://water.usgs.gov/pubs/circ/2003/circ1250/.

Pruitt, B. A. (2003) Uses of turbidity by states and tribes. In Proceedings of the Federal Interagency Workshop on Turbidity and Other Sediment Surrogates, April 30-May 2 2002, Reno, Nevada, J. R. Gray & G. D. Glysson (eds), 31–46. US Geological Survey Circular 1250, http://pubs.usgs.gov/circ/2003/circ1250/.

Rainwater, F. H. (1962). Stream composition of the conterminous United States: US Geological Survey Hydrologic Atlas 61, 3 maps, http://pubs.er.usgs.gov/usgspubs/ha/ha61.

Rasmussen, P. P., Gray, J, R., Glysson, G. D. & Ziegler, A. C. (2009) Guidelines and procedures for computing time-series suspended-sediment concentrations and loads from in-stream turbidity-sensor and streamflow data. US Geological Survey Techniques and Methods Report 3 C4, http://pubs.usgs.gov/tm/tm3c4/.

Rasmussen, T. J., Ziegler, A. C. & Rasmussen, P. P. (2005) Estimation of constituent concentrations, densities, loads, and yields on Lower Kansas River, Northeast Kansas, using regression models and continuous water-quality modeling, January 2000 through December 2003. US Geological Survey Scientific Investigations Report 2005-5165, 117pp., http://pubs.usgs.gov/sir/2005/5165/.

Reichel, G. & Nachtnebel, H. P. (1994) Suspended sediment monitoring in a fluvial environment: Advantages and limitations applying an acoustic Doppler current profiler. *Water Research*, **28**(4), 751–61.

Richards, S. D., Heathershaw, A. D. & Thorne, P. D. (1996) The effect of suspended particulate matter on sound attenuation in seawater. *Journal of the Acoustical Society of America* **100**(3), 1447–50.

Schaafsma, A. S. & Hay, A. E. (1997) Attenuation in suspensions of irregularly shaped sediment particles: A two-parameter equivalent spherical scatterer model. *Journal of the Acoustical Society of America* **102**(3), 1485–502.

Schoellhamer, D. H., Buchanan, P. A. & Ganju, N. K. (2002) Ten years of continuous suspended sediment concentration monitoring in San Francisco Bay and Delta. In Proceedings: Turbidity and Other Sediment Surrogates Workshop, April 30 – May 2 2002, Reno, Nevada, http://water.usgs.gov/osw/techniques/TSS/schoelhamer.pdf.

Schoellhamer, D. H. & Wright, S. A. (2003) Continuous measurement of suspended-sediment discharge in rivers by use of optical backscatterance sensors. In, J. Bogen, T. Fergus & D. E. Walling (eds), 28–36. International Association of Hydrological Sciences Publication 283.

Schott, F. & Johns, W. (1987) Half-year-long measurements with a buoy-mounted acoustic Doppler current profiler in the Somali Current. *Journal of Geophysical Research*, **92**(C5), 5169–76.

Schulkin, M. & Marsh, H. W. (1962) Sound absorption in sea water. *Journal of the Acoustical Society of America*, **34**(6), 864–65.

Sedimentation Committee of the Water Resources Council. (1976) Proceedings of the Third Federal Inter-Agency Sedimentation Conference: March 22-25 1976, Denver, Colorado, 7 chapters, variously paged, http://pubs.usgs.gov/misc_reports/FISC_1947-2006/pdf/1st-7thFISCs-CD/3rdFISC/3FiscTOC.PDF.

Sequoia Scientific, Inc. (2008) http://www.sequoiasci.com/default.aspx?SectionName=home.

Serra, T., Colomer, J., Cristina, X. P., Vila, X., Arellato, J. B. & Casamitjana, X. (2001) Evaluation of laser in situ scattering instrument for measuring concentration of phytoplankton, purple sulfur bacteria, and suspended inorganic sediments in lakes. *Journal of Environmental Engineering*, November, 1023–30.

Serra, T., Granata, T., Colomer, J., Stips, A., Mohlenberg, F. & Casamitjana, X. (2003) The role of advection and turbulent mixing in the vertical distribution of phytoplankton. *Estuarine, Coastal and Shelf Science*, **56**, 53–62.

Sheng, J. & Hay, A. E. (1988) An examination of the spherical scatter approximation in aqueous suspensions of sand. *Journal of the Acoustical Society of America* **83**(2), 598–610.

Skinner, J. V. (1989) History of the Federal Interagency Sedimentation Project. In Proceedings of the International Symposium, Sediment Transport Modeling, American Society of Civil Engineers, New Orleans, August 14–18 1989, S. S. Y. Wang (ed), 266–71.

Smith, R. A., Alexander, R. B. & Wolman, M. G. (1987) Water-quality trends in the nation's rivers: *Science*, **235**, 1607–15.

Subcommittee on Sedimentation. (2008) http://acwi.gov/sos/, accessed September 30, 2008.

Sutherland, T. F., Lane, P. M., Amos, C. L. & Downing, J. (2000) The calibration of optical backscatter sensors for suspended sediment of varying darkness levels: *Marine Geology* **162**, 587–97.

Swithenbank, J., Beer, J. M., Taylor, D. S., Abbot, D. & McCreath, C. C. (1977) A laser diagnostic technique for the measurement of droplet and particle size distribution. In *Experimental Diagnostics in Gas Phase Combustion Systems: Progress in Astronautics and Aeronautics*, vol. **53**, B. J. Zinn (ed.), 421–47.

Tessier, C., Le Hir, P., Lurton, X. & Castaing, P. (2008) Estimation of suspended sediment concentration from backscatter intensity of Acoustic Doppler Current Profiler. *Comptes Rendus – Geoscience*, **340**(1), 57.

Thevenot, M. M., Prickett, T. L. & Kraus, N. C. (1992). Tylers Beach, Virginia, dredged material plume monitoring project 27 September to 4 October 1991: Dredging Research Program Technical Report DRP-92-7, US Army Corps of Engineers, Washington, D.C, 204pp.

Thevenot, M. M. & Kraus, N. C. (1993) Comparison of acoustical and optical measurements of suspended material in the Chesapeake Estuary. *Journal of Marine Environmental Engineering*, **1**, 65–79.

Thorne, P. D. & Buckingham, M. J. (2004) Measurements of scattering by suspensions of irregularly shaped sand particles and comparison with a single parameter modified sphere model, *Journal of the Acoustical Society of America* **116**(5), 2876–89.

Thorne, P. D. & Campbell, S. C. (1992) Backscattering by a suspension of spheres. *Journal of the Acoustical Society of America* **92**(2), 978–86.

Thorne, P. D. & Hanes, D. M. (2002) A review of acoustic measurements of small-scale sediment processes. *Continental Shelf Research* **22**, 603–32.

Thorne, P. D. & Hardcastle, P. J. (1997) Acoustic measurement of suspended sediments in turbulent currents and comparison with in-situ samples. *Journal of the Acoustical Society of America* **101**(5), 2603–14.

Thorne, P. D., Hardcastle, P. J. & Hogg, A. (1996) Observations of near-bed suspended sediment turbulence structures using multifrequency acoustic backscatter. In *Coherent Flow Structures in Open Channels*, P. J. Ashworth, S. J. Bennet, J. L. Best & S. J. McLelland (eds). John Wiley & Sons Ltd.

Thorne, P. D., Hardcastle, P. J. & Souslby, R. L. (1993). Analysis of acoustic measurements of suspended sediment. *Journal of Geophysics Research* **98**(C1) 899–910.

Thorne, P. D., Vincent, C. E., Harcastle P. J., Rehman, S. & Pearson, N. (1991) Measuring suspended sediment concentrations using acoustic backscatter devices. *Marine Geology*, **98**, 7–16.

Thorne, P. D., Waters, K. R. & Brudner, T. J. (1995) Acoustic measurements of scattering by objects of irregular shape. *Journal of the Acoustical Society of America* **97**(1), 242–51.

Thorne, P. D. & Meral, R. (2008) Formulations for the scattering properties of suspended sandy sediments for use in the application of acoustics to sediment transport. *Continental Shelf Research*, **28**(2), 309–17.

Tollner, E. W., Rasmussen, T. C., Upchurch, B. & Sikes, J. (2005) Simulated moving bed form effects on real-time in-stream sediment concentration measurement with denitometry. *ASCE Journal of Hydraulic Engineering*, **132**(12), 1141–44.

Topping, D. J., Melis, T. S., Rubin, D. M. & Wright, S. A. (2004) High-resolution monitoring of suspended-sediment concentration and grain size in the Colorado River in Grand Canyon using a laser acoustic system. In Proceedings of the 9th International Symposium on River Sedimentation, Yichang, China, 2507–14.

Topping, D. J., Wright, S. A., Melis, T. S. & Rubin, D. M. (2006) High-resolution monitoring of suspended-sediment concentration and grain size in the Colorado River using laser-diffraction instruments and a three-frequency acoustic system. In Proceedings, 8th Federal Interagency Sedimentation Conference, April 2–6 2006, Reno, Nevada, CD-ROM, ISBN 0-9779007-1-1.

Topping, D. J., Wright, S. A., Melis, T. S. & Rubin, D. M. (2007) High-resolution measurement of suspended-sediment concentrations and grain size in the Colorado River in Grand Canyon using a multi-frequency acoustic system. In Proceedings of the 10th International Symposium on River Sedimentation, Moscow, Russia, vol. 3, 330–39.

Traykovski, P., Latter, R. & Irish, J. D. (1999) A laboratory evaluation of the LISST instrument using natural sediments: *Marine Geology* **159**, 355–67.

Uhrich, M. A. (2002) Determination of total and clay suspended-sediment loads from instream turbidity data in the North Santiam River Basin, Oregon 1998–2002. In Proceedings of the Federal Interagency Workshop on Turbidity and Other Sediment Surrogates, April 30-May 2 2002, Reno, Nevada, J. R. Gray & G. Douglas Glysson (eds). US Geological Survey Circular 1250, 3pp., http://water.usgs.gov/pubs/circ/2003/circ1250/.

Uhrich, M. A. & Bragg, H. M. (2003) Monitoring instream turbidity to estimate continuous suspended-sediment loads and yields and clay-water volumes in the Upper North Santiam River Basin, Oregon 1998–2000. US Geological Survey Water-Resources Investigations Report 03-4098, 44pp., http://pubs.usgs.gov/wri/WRI03-4098/.

Urick, R. J. (1975) *Principles of Underwater Sound*, 2nd edn. New York: McGraw Hill, 384pp.

US Department of Agriculture. (1965) Proceedings of the Federal Inter-Agency Sedimentation Conference: Miscellaneous Publication No. 970, 933pp.

US Environmental Protection Agency. (2008) National section 303-d list fact sheet, causes of impairments: total maximum daily loads. http://iaspub.epa.gov/waters10/attains_nation_cy.control?p_report_type=T#causes_303d

US Geological Survey. (2004) Revision of NFM Chapter 6, Section 6.7 – Turbidity: Office of Water Quality Technical Memorandum 2004.03, http://water.usgs.gov/admin/memo/QW/qw04.03.html.

US Geological Survey. (2005) Real-time water-quality concentrations and loads estimated using regression analysis, http://nrtwq.usgs.gov/explore/plot?site_no=07144100&pcode=80154&period=ytd_all&timestep=uv.

US Geological Survey. (2008a) National Streamflow Information Program: Home Page, http://water.usgs.gov/nsip/.

US Geological Survey. (2008b) USGS water data for the nation: National Water Information System Home Page, http://waterdata.usgs.gov/nwis.

US Geological Survey. (2008c) USGS water data for the nation: National Stream Quality Accounting Network Home Page, http://water.usgs.gov/nasqan/.

US Geological Survey. (2008d) US Geological Survey implements new turbidity data-reporting procedures, in, US Geological Survey National Field Manual for the collection of water-quality data: US Geological Survey Techniques of Water-Resources Investigations, Book 9, Ch. 6.7, http://water.usgs.gov/owq/turbidity/TurbidityInfoSheet.pdf.

Wall, G. R., Nystrom, E. A. & Litten, S. (2006) Use of an ADCP to computer suspended-sediment discharge in the tidal Hudson River, New York, US Geological Survey, Scientific Investigations Report 2006-5055, 16pp.

Walling, D. E. (1977) Assessing the accuracy of suspended sediment rating curves for a small basin, *Water Resources Research*, **13**, 531–38.

Ziegler, A. C. (2003) Breakout session 1 – Definition of optical methods for turbidity and data reporting. In Proceedings of the Federal Interagency Workshop on Turbidity and Other Sediment Surrogates, April 30–May 2 2002, Reno, Nevada, J. R. Gray & G. Douglas Glysson (eds), 9–13. US Geological Survey Circular 1250, http://water.usgs.gov/pubs/circ/2003/circ1250/.

2 Surrogate technologies for monitoring bed-load transport in rivers

John R. Gray[1] & Jeffrey W. Gartner[1] (editors)
Jonathan S. Barton[2], Janet Gaskin[3], Smokey A. Pittman[4] & Colin D. Rennie[3]
[1]*United States Geological Survey, USA*
[2]*National Aeronautics and Space Administration, USA*
[3]*University of Ottawa, Canada*
[4]*Graham Matthews and Associates, USA*

Surrogate technologies for bed-load transport monitoring are being evaluated toward eventually supplanting traditional data-collection methods that require routine collection of physical samples and subsequent field or laboratory analyses. Commercially available and prototype technologies based on active- and passive-hydroacoustic principles are the foci of much of the current research on bed-load surrogate techniques, and are the subjects of this chapter. Field and laboratory tests of bed-load surrogate-monitoring techniques using active hydroacoustics (acoustic Doppler current profilers (ADCPs)) in sand- and gravel-bed rivers or passive hydroacoustics (various sensors) in gravel-bed rivers have been shown to provide useful data in a limited number of flume and field tests, and some are the subject of continuing research. Research on other technologies including tracer-tracking (visual, radioactive, magnetic, and radio); sonar, load-cell, videography, particle-tracking, ground-penetrating radar, and magnetic techniques is ongoing in several countries.

Similar to choices for monitoring suspended-sediment transport, selection of an appropriate technology for bed-load transport monitoring usually entails an analysis of the advantages and limitations associated with each technique, the monitoring objective, and the physical and dynamic sedimentary characteristics at each deployment site. Some factors that may limit or enhance the efficacy of a surrogate technology used to monitor bed-load transport include cost (purchase, installation, operation, calibration, and data analysis), reliability, robustness, accuracy, size and location of the instantaneous and time-integrated measurement realm, and range in size of bed-load particles. Most if not all surrogate technologies for monitoring bed load, including passive and active hydroacoustics, require periodic site-specific calibrations to infer transport rates occurring over the entire channel cross section.

Should bed-load surrogate technologies prove successful in a wide range of applications, the monitoring capability could be unprecedented, providing the prospect of obtaining continuous records of bed-load discharge potentially qualified by estimates of uncertainty. As with suspended-sediment surrogate technologies, the potential benefits could be enormous, providing for more frequent and consistent, less expensive, and arguably more accurate bed-load data obtained with reduced personal risk for use in managing the world's sedimentary resources.

2.1 Introduction

Bed load is the part of total-sediment load that is transported by rolling, skipping, or sliding on the riverbed (ASTM International 1998) (Fig. 2.1). Historically, bed-load data for US rivers have been produced by gradation and gravimetric analyses performed on samples obtained with manually deployed samplers (Edwards & Glysson 1999; Kuhnle 2008). As with suspended sediment, traditional bed-load data-collection methods tend to be expensive, labor intensive, time-consuming, difficult, and under some conditions, hazardous. Specialized instruments and considerable training in their proper deployment are prerequisites for obtaining reliable bed-load samples.

Sedimentology of Aqueous Systems, 1st edition.
Edited by Cristiano Poleto and Susanne Charlesworth.

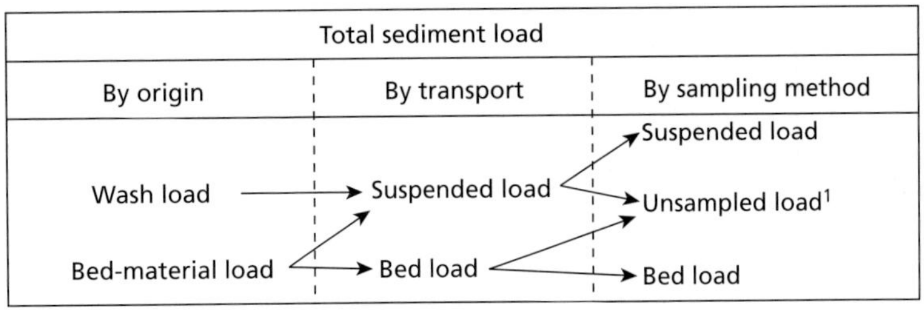

[1]That part of the sediment load that is not collected by the depth-integrating suspended-sediment and pressure-difference bed-load samplers used, depending on the type and size of the sampler(s). Unsampled-zone sediment can occur in one or more of the following categories: (a) sediment that passes under the nozzle of the suspended-sediment sampler when the sampler is touching the streambed and no bed-load sampler is used; (b) sediment small enough to pass through the bed-load sample's mesh bag; (c) sediment in transport above the bed-load sampler that is too large to be sampled reliably by the suspended-sediment sampler; and (d) material too large to enter the bed-load-sampler nozzle.

Fig. 2.1 Components of total-sediment load considered by origin, by transport, and by sampling method.
From Diplas *et al.* (2008).

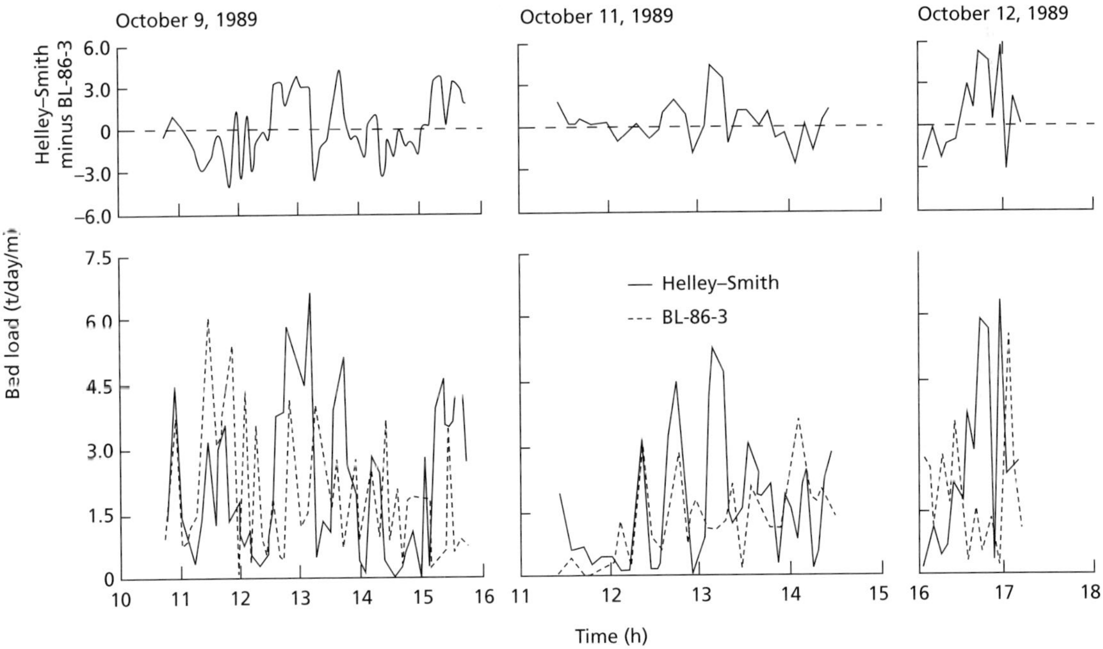

Fig. 2.2 Variability in sand bed-load transport rates measured 2 meters apart by a Helley–Smith bed-load sampler and a BL-86-3 bed-load sampler (the latter identical to the US BL-84 bed-load sampler), at the U.S. Geological Survey (USGS) streamgage on the Colorado River above National Canyon near Supai, Arizona, USA, October 1989.
From Gray *et al.* (1991).

The spatiotemporal distribution of bed material transport is a complicated, non-linear function of sediment supply, bed state, and fluid forcing (Gomez 1991). Figure 2.2 shows variations in bed-load transport rates measured by two types of pressure-difference sampler deployed at fixed locations 2 meters apart during steady flows near the middle of the sand-bedded Colorado River above National Canyon near Supai, Arizona, USA (Gray *et al.* 1991). Such variability is more or less typical for at-a-point bed-load measurements. However, after collection of 390 discrete bed-load transport samples using two types

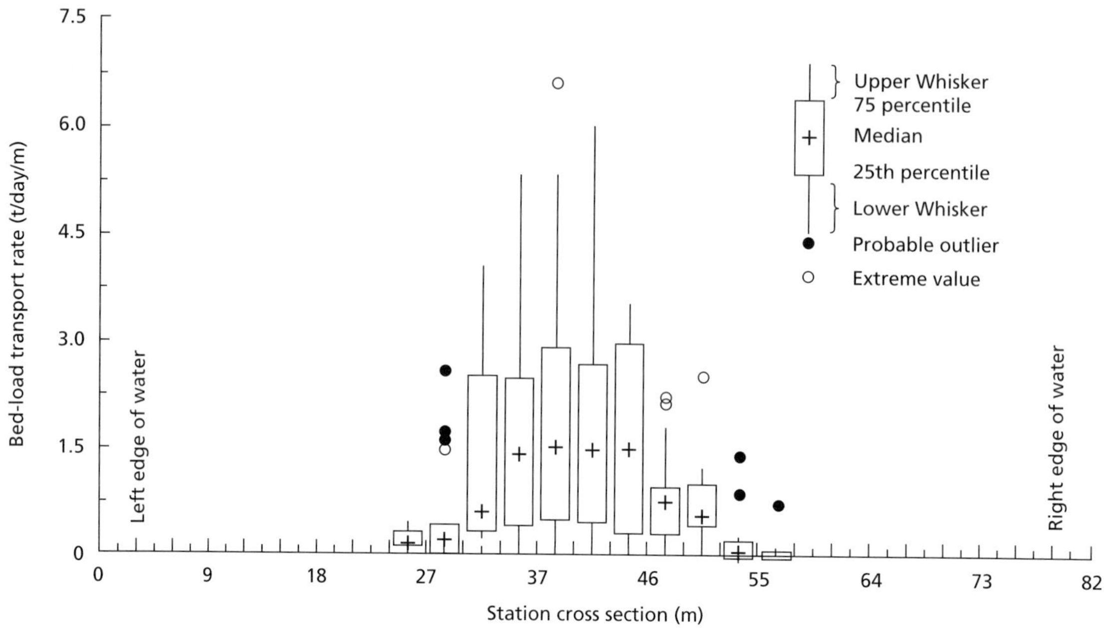

Fig. 2.3 Spatially averaged transport rates computed from 390 bed-load samples collected by a Helley–Smith bedload sampler and a BL-86-3 bedload sampler (the latter identical to the US BL-84 bed-load sampler), at the USGS streamgage on the Colorado River above National Canyon near Supai, Arizona, USA, October 1989.
From Gray *et al.* (1991).

of pressure-difference sampler from points across the channel, a pattern in bed-load transport became evident with most bed load occurring in the center third of the river (Fig. 2.3). These data are illustrative of the fact that bed-load data collected by traditional manual techniques as part of periodic or runoff-initiated site visits are rarely sufficient to reliably characterize the spatiotemporal variability in bed-load transport rates over periods exceeding a fraction of a day.

Lacking a reliable means for developing a bed-load transport time series, practitioners often revert to estimations based on stochastic techniques, such as a bed-load transport equation or an empirically derived bed-load transport curve with instantaneous water discharge as the independent variable (Glysson 1987; Gray and Simões 2008). However, the uncertainty associated with bed-load-discharge estimates is rarely quantified or quantifiable, and is more often the subject of speculation rather than reliable calculation. Thus, considerable interest and effort has been directed toward surrogate measurements that may potentially provide a bed-load time series that is representative of the cross section or reach of interest.

Sediment-surrogate technologies are defined as instruments coupled with operational and analytical methodologies that enable acquisition of temporally and (or) spatially dense fluvial-sediment data sets without the need for routine collection and analysis of physical samples other than for periodic calibration purposes. Bed-load surrogate technologies have been addressed as part of at least three workshops held since 2002, namely:

- Erosion and Sediment Transport Measurements in Rivers: Technological and Methodological Advances, June 19–21 2002, Oslo, Norway, convened by the International Commission of Continental Erosion of the International Association for Hydrological Sciences, and sponsored by the Norwegian Water Resources and Energy Directorate (Bogen *et al.* 2003).
- Federal Interagency Sediment Monitoring Instrument and Analysis Research Workshop, September 9-11 2003, Flagstaff, Arizona, USA, sponsored by the Advisory Committee on Water Information's Subcommittee on Sedimentation (Gray 2005).
- International Bedload Surrogate Monitoring Workshop, April 11-14 2007, Minneapolis, Minnesota, USA, sponsored by the Advisory Committee on Water Information's Subcommittee

on Sedimentation (Gray *et al.* 2007; Laronne *et al.* 2007).

The 2002 workshop in Oslo, Norway, included 13 papers under the category, "bed-load monitoring and transport processes." The workshop paper by Ergenzinger and DeJong (2003) listed and briefly described each of, "... the well known measuring techniques of sediment trapping and sampling, tracing, and surveying using both conventional techniques and remotely sensed images." Those techniques that qualify as "bed-load surrogate technologies" include passive hydroacoustics; visual, radioactive, magnetic, and radiotracers; magnetic detectors; underwater video cameras; load-cell traps; and analyses of scanned or photographic images.

Breakout session II from the 2003 workshop in Flagstaff, Arizona, USA, was entitled, "Bedload-Transport Measurements: Data Needs, Uncertainty, and New Technologies" (Ryan *et al.* 2005). Among other information, the table in that report section (reproduced herein as Table 2.1 without annotation) lists eight bed-load surrogate technologies: active and passive hydroacoustic sensors; gravel impact sensors; magnetic tracers, and sensors; topographic differencing with sonar; sonar-measured debris basin; and underwater video cameras. The breakout group identified characteristics associated with the ideal bed-load sampling device or technology, as paraphrased below.

Surrogate technologies should:

- provide accurate measurements and precise data on the amounts and sizes of bed-load material over a wide range of flow conditions;
- be reliable, safe to operate, and used without wading in streams at high flow;
- be foolproof, easy to calibrate, and not disrupt the local transport field to the extent that it affects measurements,
- be rugged, durable, and able to withstand occasional collisions with large grains;
- have minimal and tractable power requirements for use in remote environments;
- automatically provide continuous record;
- be scalable; and
- be affordable.

The 2003 workshop summary (Gray, 2005) included a matrix that compared and contrasted selected characteristics of bed-load surrogate technologies to other types of sediment-surrogate technologies, and to related data-management and flux-computation issues. This matrix is reproduced herein as Table 2.2. About 50 participants from nine countries attended the 2007 workshop in Minneapolis, Minnesota, USA; others participated by video link. The 25 papers submitted to the workshop identified passive- and active-hydroacoustic, magnetic-tracer and magnetic-sensor, load-cell trap, topographic differencing with sonar, particle-tracking, gravel-impact sensors, and ground-penetrating radar technologies to infer bed-load transport.

This chapter presents descriptions, progress in, and examples of applications of active and passive hydroacoustics considered by the editors to be among the most promising of the aforementioned bed-load surrogate technologies. This observation is in part based on the fact that no fewer than a combined 14 papers presented at the three workshops listed above described passive- and active-hydroacoustics research results. In comparison, the next most prevalent topic among these workshops was magnetic- and radiotracer studies, described in four of the papers. It was also noted that in many cases hydroacoustic technologies are affordable, portable, and relatively robust. Additionally, results from some techniques that are not based on, or calibrated with integrated cross-section bed-load measurements, such as some of the tracer technologies and some impact sensors, can be relatively difficult to interpret quantitatively. However, it is important to note that selected technologies other than the hydroacoustics techniques presented below have a potential monitoring niche, and should not be ignored. Those interested in non-hydroacoustic bed-load surrogate technologies are encouraged to peruse the relevant papers from these workshops and from other publications on this subject.

The *in situ* technologies presented in this chapter require periodic site-specific calibrations to infer the bed-load transport characteristics representative of the entire channel cross section or reach segment. This requirement is expected to be substantial for new river-monitoring applications, but may diminish as comparative data accumulate.

None of the technologies represents a panacea for bed-load monitoring in all rivers under all flow and sediment-transport conditions. To make the transition from research to operational monitoring applications, these new technologies must be rigorously tested with respect to accuracy and reliability in different physiographic and (or) laboratory

Table 2.1 Comparison of characteristics of different bed-load sampling technologies (Ryan et al. 2005). See the original table for all annotations.

Bed-load sampling technology	Stream type	Requires wading or retrieval during high flows	Physical sample obtained for sieving	High percentage of channel width sampled	Large opening relative to grain size	Relatively long sampling duration	Stream excavation required	Relative ease of use	Disruptive to flow fields	Status of development (2003)	Potential use as calibration standard
1. Instream Installations											
Birkbeck sampler (weighable pit trap)	Narrow gravel bed channel	No	No, automatically weighs mass in stream	Typically not; depends on slot width	Depends on slot width	Continuous	Yes	Easy	May change with fill level	Additional testing and modifications	High
Vortex sampler	Gravel bed channel	No	Yes	Yes	Yes	Continuous	Yes	Depends on flow conditions	Depends on experimental setup	Additional testing and modifications	High
Pit traps, unweighable	Gravel bed channel	Yes	Yes	Typically not	Possibly	Possibly	Yes, small scale	Depends on flow conditions	Slightly	Additional testing	Probably not
Net-frame sampler	Gravel bed channel	Possibly	Yes	Yes	Yes	Yes	Depends on experimental setup	Can be difficult	Depends on experimental setup	Completed	Possible
Sediment detention basins/weir ponds	Sand-gravel bed channels	No	Periodically	Yes	Yes	Yes	Yes	Relatively easy	No	Completed	High
2. Portable/physical devices											
Pressure-difference samplers (small openings)	Sand-gravel bed channel	Yes	Yes	No	No	No	No	Depends on flow conditions	Slightly	Additional verification	Additional verification needed

Pressure-difference samplers (large openings)	Gravel bed channel	Yes	Yes	No	Yes	No	No	Depends on flow conditions	Highly	Additional verification	Additional verification needed
Baskets (suspended or instream)	Gravel bed channel	Yes	Yes	Depends on design	Depends on design	Yes	No	Depends on flow conditions	Depends on experimental setup	Completed	Moderate
Bedload traps	Gravel bed channel	Yes	Yes	Depends on number of traps deployed	Yes	Yes	Minor	Depends on flow conditions	Slightly	Completed; testing of modifications	Moderate with additional verification
Tracer particles (painted, magnetic, signal emitting rocks)	Gravel bed channel	Possibly	No	Depends on tracer placement	N/A	Yes	No	Easy	No	Additional verification	Low
Scour chains; scour monitor; scour core	Sand-gravel bed channel	Possibly	No	No	N/A	Yes	Yes	Easy	No	Completed	Low
Bedload collector (Streamside Systems)	Sand-gravel bed channel	No	Yes	Depends on number and size of devices deployed	Depends on design of device	Yes	Yes	Operation is easy once installed	Unknown	Needs verification	Needs to be tested
3. Surrogate technologies											
ADCP – acoustic Doppler current profiler	Sand bed rivers, experimental in larger gravel bed channels	No	No	Yes	N/A	Continuous	No	Logistics and data reduction are complex	No	Moderate (sand systems) early (gravel systems)	Additional verification for gravel bed systems

Table 2.1 *Continued*

Bed-load sampling technology	Stream type	Requires wading or retrieval during high flows	Physical sample obtained for sieving	High percentage of channel width sampled	Large opening relative to grain size	Relatively long sampling duration	Stream excavation required	Relative ease of use	Disruptive to flow fields	Status of development (2003)	Potential use as calibration standard
Hydrophones (active and passive acoustic sensor)	Gravel bed channel	No	No	Depends on deployment	N/A	Continuous	Possibly	Easy	No	Early	Additional development needed
Gravel impact sensor	Gravel bed channel	Yes, for hand-held model	No	Not as currently designed	N/A	Continuous	Yes for instream model	Easy under many conditions	In fast flow	Early	Additional development needed
Magnetic tracers	Gravel bed with naturally magnetic particles	No	No	Yes	N/A	Continuous	Yes	Relatively easy	Depends on experimental setup	Additional testing	Possible at appropriate locations
Magnetic sensors	Gravel bed channel	No	No	Yes	N/A	Continuous	Yes	Easy under many conditions	Minor; flush with stream bottom	Early	Additional verification needed
Topographic differencing	Sand-gravel bed channel	No	No	Yes	N/A	Episodically or continuous	No	Easy	No	Early?	Additional verification for gravel bed systems
Sonar-measured debris basin	Gravel bed channel	No	No	Yes	N/A	Continuous	With debris basin installation	Easy under many conditions	N/A	Early	High
Underwater video cameras	Relatively clear flow	Used from bridges or boats	No	No	N/A	Continuous	No	Easy under right lighting conditions	Slightly	Early	Additional verification needed

N/a, Not applicable.

Table 2.2 Matrix of selected information gleaned from the four breakout sessions as compiled in the second plenary session of the Federal Interagency Sediment Monitoring Instrument and Analysis Research Workshop, September 9–11, 2003 Flagstaff, Arizona, USA. Empty boxes indicate that the topic was not addressed in the breakout or second plenary sessions, or was not applicable to the category.

	Breakout session I: suspended sediment	Breakout session II: bedload	Breakout session III		Breakout session IV	
			Bed material	Bed topography	Data management	Flux computations
Data						
Continuous time-series/ greater data amount, density	Needed	Needed	Needed	Needed	Need to store original data	Critically needed
Ancillary information	Needed	Needed	Needed	Needed	Considerable need	Considerable need
Physical calibration samples	Needed	Needed	Needed	Needed	Needed	Critically needed
Accuracy criteria	Have some	Needed	Needed	Needed	Needed (to accept/reject data)	Needed
Uncertainty estimates	Needed; available in some cases	Needed		Needed	Needed; also need storage capabilities	Needed in some cases
Protocols for data collection, computation & storage	Available for traditional technologies	Available for traditional technologies	Available for traditional technologies	Needed	Databases generally insufficient	Needed for computations
Clearinghouse, data standards		Establish clearinghouse, data standards			Establish clearinghouse, data standards	
Scale limitations	Yes	Yes	Yes	Yes		Yes
Traditional techniques						
Extant	Yes	Not for all conditions	Not for unwade-able gravel bed	For most conditions	Yes	Yes
Accuracy	Relatively accurate	Accuracy uncertain	Mostly acceptable	Mostly acceptable	None available	(Standards for computations)

Table 2.2 *Continued*

	Breakout session I: suspended sediment	Breakout session II: bedload	Breakout session III		Breakout session IV	
			Bed material	Bed topography	Data management	Flux computations
Surrogate techniques						
Availability of instruments	Many, commercially available	Few, mostly research, in early development	Some, but not for unwadeable gravel bed	Several, Government or commercially available	Create data gaps/ problems; need qualifying data	Need protocols for computations
Quantify accuracy	Some need	Major need	Needed	Some need	Major need	Major need
Applicable environments	Fluvial, coastal zone, estuaries	Fluvial, marine and coastal zones	Freshwater, marine and coastal zones	Freshwater, marine and coastal zones		
Models						
Uses and needs		Accurate data needed for better models		Uses and needs		Uses and needs
Research and oversight						
Basic research sought	Yes	Yes, considerable	Yes	Yes	Yes	Yes
White paper sought		Past, present, future technologies				
Extant focus of current research venues or entities	Many field sites	Need national calibration standard sites	Online interest groups		Need sediment database management task group	
SMIAR Program needed	Yes	Yes	Yes	Yes	Yes	Yes
Organizational oversight of a SMIAR Program	FISP, or FISP-type organization	FISP, or FISP-type organization	FISP, or FISP-type organization	FISP, or FISP-type organization	FISP, or FISP-type organization	FISP, or FISP-type organization

SMIAR, Sediment Instrument and Analysis Research; FISP, the Federal Interagency Sedimentation Project.

settings as appropriate. Their performances must be compared with laboratory-control data and (or) field measurements by traditional techniques. In most cases, performance comparisons should include collection of concurrent data by traditional and new techniques for a sufficient period – probably years – to identify potential bias and minimize differences in precision between the old and new technologies. However, with careful matching of surrogate-monitoring technologies to selected river reaches and objectives, it may be possible in the future to remotely, continuously, and accurately monitor bed-load discharges, possibly by particle-size class. Qualifying the derived transport data with reliable uncertainty assessments may also be possible.

These are revolutionary concepts in sedimentology when considered from an operational perspective. The benefits of such applied capability could be enormous, providing for safer, more frequent and consistent, arguably more accurate, and ultimately less expensive fluvial-data collection for use in managing the world's sedimentary resources.

This chapter begins with an overview of traditional instruments and techniques used for measuring bed load, against which the surrogate technologies using hydroacoustics are evaluated. Descriptions of the theory, applications, some advantages, limitations, and costs of each surrogate technology are presented and compared. A subjective evaluation of the efficacy of each technology concludes this chapter. Use of firm, brand, or trade names are for identification purposes only and do not constitute endorsement by the US Government.

2.1.1 Background: traditional bed-load sediment-sampling techniques

Published records of bed-load sampler use dates back to at least the late 1800s, and published attempts at bed-load sampler calibration date to at least the early 1930s (Carey 2005). As with the development of isokinetic suspended-sediment samplers, the Federal Interagency Sedimentation Project (FISP) endeavored to address problems and needs related to bed-load data collection starting in the later 1930s (Federal Interagency Sedimentation Project 1940). However, development and calibration of reliable portable bed-load samplers capable of sampling a wide range of particle sizes and transport rates remains a work in progress (Marr *et al.* in press). No single apparatus or procedure has been universally accepted as completely adequate for the determination of bed-load discharges over the wide range of sediment and hydraulic conditions found in nature (ISO 1992).

Bed-load samplers fall under one or a combination of the following four categories: Box or basket samplers; pan, tray, or slot samplers; pressure-difference samplers; and trough or pit samplers (Hubbell 1964). Box or basket samplers retain sediment deposited in the sampler owing to a reduction in the flow velocity and (or) capture by the sampler screen (Hubbell 1964). Pan, tray, or slot samplers retain the sediment that drops into one or more slots after the material has rolled, slid, or skipped up an entrance ramp (Hubbell 1964). Pressure-difference samplers are designed so that the sampler's entrance velocity is about equal to or somewhat larger than the ambient stream velocity. They collect material that is small enough to enter the nozzle but too large to pass through the mesh collection bag. Figure 2.4 shows selected pressure-difference bed-load samplers. Trough or pit samplers are rectangular holes constructed in the streambed, into which bed-load particles drop. Troughs are usually continuous across the channel, whereas pits cover only a part of the streambed (Hubbell 1964). Troughs and pits tend to provide the most reliable bed-load data (Federal Interagency Sedimentation Project 1940; Hubbell 1964; Emmett 1980; Carey 2005).

There can be substantial differences in calibration and deployment between the trough and other types of sampler. The trough-type samplers are the most difficult to construct and operate but the least challenging to calibrate. In contrast, no universally agreed-upon method has been developed for calibrating portable bed-load samplers, but they are the easiest to deploy (Carey 2005).

The efficiency of a bed-load sampler is the ratio of the sampled bed-load mass divided by the mass that would have been transported in the same section and time in the absence of the bed-load sampler. Unlike FISP isokinetic suspended-sediment samplers which are designed for isokinetic efficiencies within about 10% of unity (Federal Interagency Sedimentation Project 1940, 2008; Gray *et al.* 2008), known or potential bias in efficiencies of bed-load samplers can cast doubt upon the reliability of their derivative

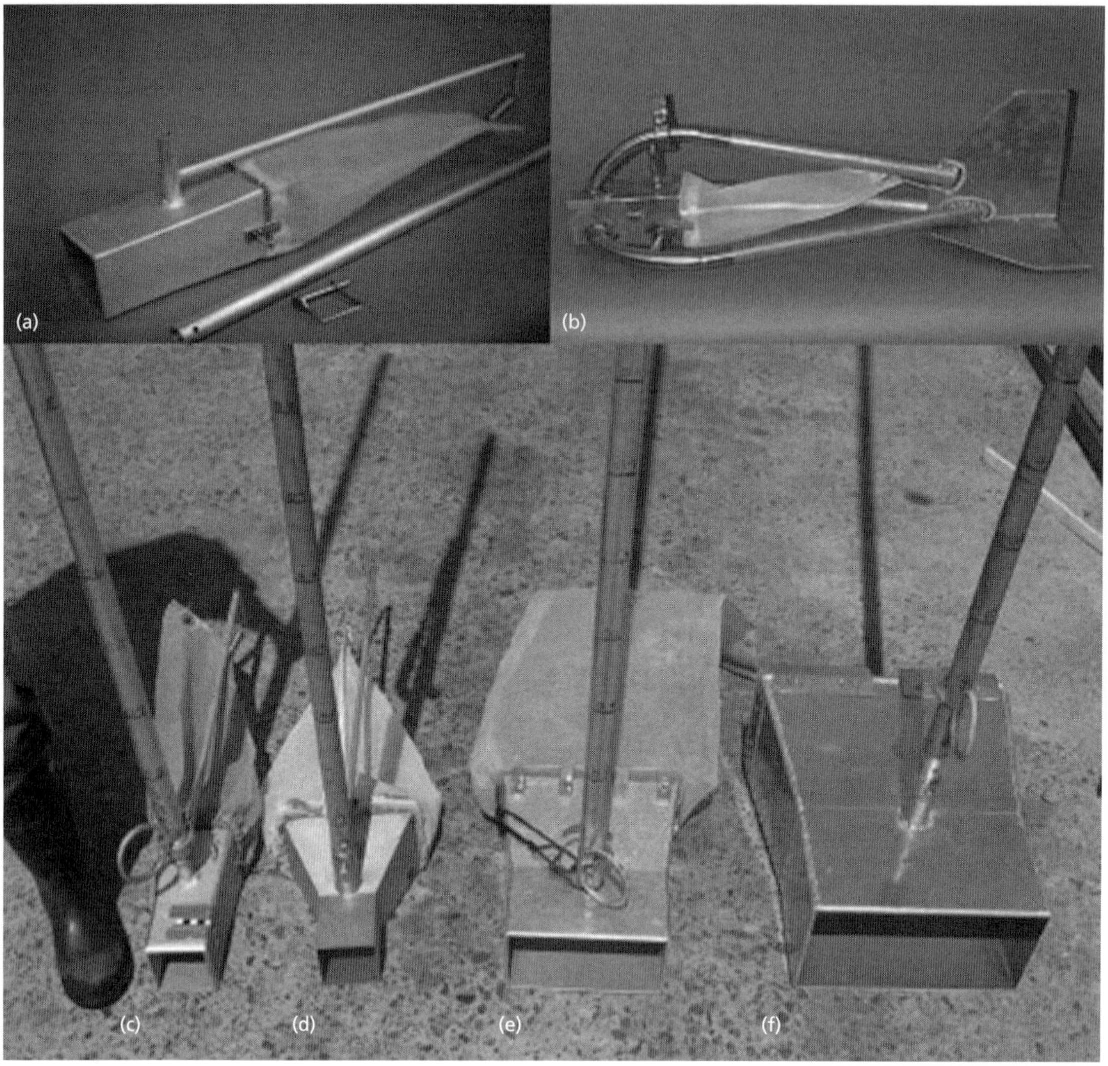

Fig. 2.4 Pressure-difference bed-load samplers. (a) and (c) Hand-held US BLH-84; (b) Cable-suspended US BL-84; (d) hand-deployed Helley–Smith; (e) hand-deployed Elwha; (f) hand-deployed Toutle River-2 (TR-2) without bag (although only one cable-suspended sampler is shown, all of these bed-load samplers are also available in cable-suspension configurations). Lower photograph courtesy of Kristin Bunte, Colorado State University, USA.

data. Bed-load sampler calibrations are complicated by a fundamental dichotomy, to wit: an innate inability to quantify the bed-load transport rate that would have occurred in a stream section in the absence of a deployed bed-load sampler, unless the bed-load sampler's efficiency is known *a priori*.

Most calibration studies have been performed in laboratory flumes where bulk bed-load transport rates can be controlled. Although flume bed-load transport-rate measurements – often referred to as "ground truth" measurements – can be quite accurate, they do not represent natural river conditions well. Leopold & Emmett (1997) observed that a river's ability to adjust its cross section to a variety of flows is a characteristic not shared by a fixed-wall flume. Riverine sediment transport is determined by the geological and physical setting of the river and river basin; thus, sediment is not a controllable variable. The variety of conditions controlled in a laboratory experiment cannot be established in a natural river.

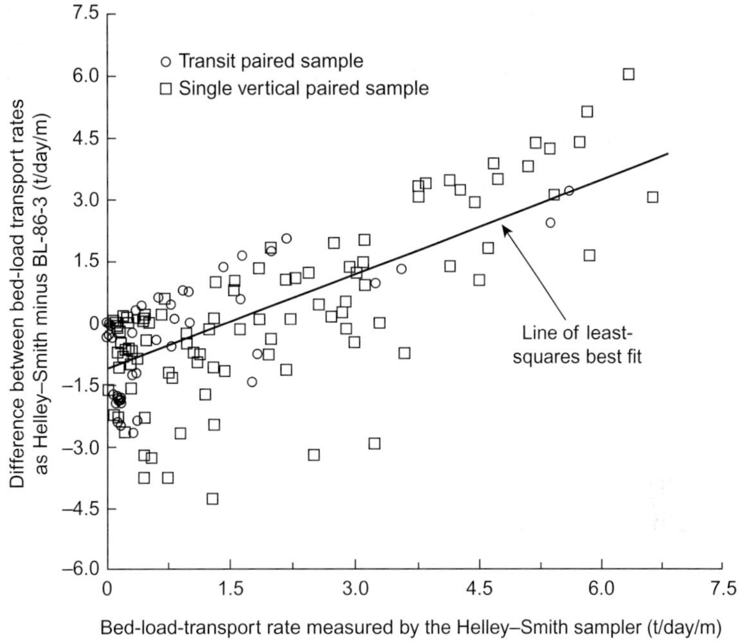

Fig. 2.5 Relation between sand bed-load transport rates measured 2 meters apart by a Helley–Smith bed-load sampler and a BL-86-3 bed-load sampler (the latter identical to the US-BL-84 bedload sampler), at the USGS streamgage on the Colorado River above National Canyon near Supai, Arizona, USA, October 1989. From Gray *et al.* (1991).

Flume bed-load sampler calibrations are subject to at least two serious problems: First, even with a stable mean bed-load transport rate, the instantaneous rate normally varies widely about the mean value (Hamamori 1962; Carey 2005; Gray & Simões 2008). Second, transport conditions in the section of the flume in which the bed-load sampler is deployed may differ from those at the flume ground-truth measuring point, such as a slot sampler.

Emmett's (1980) solution to these problems was to construct a conveyor-belt bed-load trap in a concrete trough across the bed of the East Fork River, Wyoming, USA. The trap caught all of the bed load that dropped into the trough, conveyed it to the stream bank for weighing and sampling, and returned it to the river downstream from the trough. This apparatus was used to collect bed-load data for seven years and to field-calibrate the Helley–Smith bed-load sampler (Helley & Smith 1971), the precursor to the US BLH-84 and US BL-84 bed-load samplers. This work is as notable for its considerable success in quantifying the bed-load characteristics of the East Fork River and calibrating the Helley–Smith bed-load sampler as it is in highlighting difficulties and the considerable expense of obtaining reliable bed-load data.

Field-based comparisons between sequentially or side-by-side deployed bed-load samplers cannot be used to identify the absolute sampling efficiency of any bed-load sampler without ground-truth data. However, such comparisons are useful to infer the relative efficiency of two or more bed-load samplers. Childers (1999) compared the relative sampling characteristics of six pressure difference bed-load samplers in high-energy flows of the Toutle River at Coal Bank Bridge near Silver Lake, Washington, USA. The sampling ratio of each pair of samplers tested was computed by dividing the mean bed-load transport rate determined for one sampler by the mean rate for a second sampler. Ratios of bed-load transport rates between measured bed-load sample pairs ranged from 0.40 to 5.73, or more than an order of magnitude over the relative range of bed-load sampling efficiencies. Gray *et al.* (1991) demonstrated that two pressure-difference bed-load samplers exhibited divergent sampling efficiencies when deployed simultaneously 2 meters apart in the thalweg of the 76-m-wide sand-bedded Colorado River above National Canyon, near Supai, in Grand Canyon, Arizona, USA, under steady low-flow conditions (Fig. 2.5).

The accuracy quantified for any bed-load surrogate technology can only be as reliable as the accuracy of its calibration data. Because bed-load surrogate technologies require empirical calibrations with data collected by physical bed-load samplers, it should come as no surprise that careful calibration with the most appropriate bed-load sampler is a prerequisite for reliable bed-load transport-surrogate monitoring in rivers.

2.1.2 Information germane to surrogate technology costs

After surrogate-technology efficacy is resolved, cost considerations are often of penultimate interest. The cost of producing reliable, quality-assured bed-load data can be separated into four categories:

- the purchase price of the instrument;
- other capital costs associated with installation, and initial operation of the instrument;
- operational costs to maintain and calibrate the instrument;
- analytical costs to evaluate, reduce, compute, review, store, and publish the derivative data.

Of these four categories, only the current purchase price is relatively straightforward to quantify. The others are dependent on several factors, including site location and physical characteristics, hydrological and sedimentological regime, availability of electrical power, limitations associated with accessibility, safety considerations, and the time and complexity associated with data analysis. Additionally, any such information inevitably becomes obsolete due, in part, to technological advances, marketing competition, and changes in currency valuation. Costs referred to in the ensuing sections might be placed in perspective considering that the cost to compute, store, and provide daily suspended-sediment-discharge data at a United States Geological Survey (USGS) streamgaging station in 2001 (adjusted for inflation in 2008 dollars) ranged from US$24,000 to US$78,000 (Gray 2003). No comparable cost statistics were available for acquisition of time-series bed-load data.

2.2 Technological advances in bed-load surrogate monitoring

Unlike daily suspended-sediment records, which have been collected and computed for the better part of a century in the USA, bed-load transport is rarely measured on a continuous basis. Hence, any technology capable of providing a time-series of bed-load transport, even with a relatively large coefficient of variation, would represent a major technological advance. The following sections describe theoretical principles, selected examples of field or laboratory applications, and advantages and limitations of two bed-load surrogate technologies considered to be the most promising by the USGS.

2.2.1 Active hydroacoustics with a acoustic doppler current profiler

Janet Gaskin & Colin D. Rennie

2.2.1.1 Background and theory

Active hydroacoustics refers herein to the use of an acoustic emission and reception system to infer and quantify the mobility of the riverbed. In this case, an ADCP is used to perform a fast, non-intrusive measurement of an apparent bed velocity, which yields a spatial distribution of relative bed-load transport when the ADCP is deployed from a boat. Apparent bed velocity is defined as the difference between the boat velocity measured by the bottom track pulse, biased by near-bed sediment movement, and the absolute boat velocity measured by a global positioning system (GPS). The bottom track boat velocity is determined from the Doppler shift of the returning acoustic echoes of the bottom track pulse. The measurement realm comprises the locations of the conical beams' "footprints" on the riverbed (Rennie *et al.* 2002).

The technology generally requires manual deployment. The cost of a commercially available, manually deployable ADCP is about US$20,000 in 2008. Because quantification of bed-load transport is typically difficult and problematic even in sand-bed rivers, any surrogate means for providing quantifiably reliable sand bed-load data is desirable. Because the technology is heretofore manually deployed, there is no routine field-maintenance cost.

An ADCP transmits sound pulses into the water from either three or four transducers and measures the Doppler shift of the echoes that reflect off particles in the flow. The particles that scatter the acoustic signal are assumed to be traveling at the speed of the filament of flow in which they are suspended. The Doppler shift is thereby related to the velocity of the water relative to the instrument. The Doppler shift is defined as:

$$F_{\mathrm{d}} = 2F_{\mathrm{s}}\left(\frac{V}{c}\right) \quad (1)$$

where: F_d is the Doppler shift frequency; F_s is the frequency of the ADCP; c is the speed of sound (~1500 m/s); and V is the relative velocity of the scatterers.

Velocities measured along each slanted beam are coordinate-transformed to estimate a three-dimensional velocity for separate segments of the water column, namely bins in the vertical profile. The algorithm used to determine the velocity components assumes homogeneous conditions over the area encircling those ensonified by the transducer beams. This assumption becomes more tenuous as the distance from the ADCP increases.

Bottom track is a Doppler sonar measurement designed to measure the relative velocity between the instrument, or the boat to which it is attached, and an immobile bed. In the case of a mobile bed, the bottom-track velocity is biased by the movement of the sediment along the bed; a differential global positioning system (DGPS) system is required to measure the velocity of the boat relative to the Earth. The difference between the biased bottom track velocity and the DGPS velocity is known as the apparent bed velocity. The apparent bed velocity is considered a measure of the bed-load transport rate.

$$v_b = v_{DGPS} - v_{bt} \quad (2)$$

where: v_b is the apparent bed velocity; v_{DGPS} is the velocity of the ADCP relative to the Earth; and v_{bt} is the bottom track velocity of the ADCP relative to the bed.

It is essential that the ADCP internal compass is properly calibrated, such that both v_{DGPS} and v_{bt} are measured in the same coordinate system. The beam homogeneity assumption is especially significant for the apparent bed velocity because flow depths can be large, bed topography can be irregular, and bed-load particle transport can be locally variable.

The bottom-track pulse measures the echoes from a volume, not an area. The echoes from the bed consist of echoes from particles moving in the bed layer as well as echoes from immobile sections of the bed. Backscatter, from particles moving just above the bed, contributes positively to the signal and is known as water bias. The distance above the bed to which particle movement influences the signal depends on the pulse length selected (Rennie & Millar 2004).

The average surface velocity (v_{pa}) of the bed-load layer depends on the various sizes and velocities (v_p) of bed-load particles. Apparent bed velocity (v_b) should be representative of the average surface velocity within the ensonified volume, except that v_b is weighted by the relative backscatter strength of all individual mobile and immobile particles in the sample volume. The relative backscatter strength of mobile particles depends in part on the frequency of the instrument and the characteristic size of the particles. Acoustic backscatter strength, relative to particle size, is greater for particles with a diameter equal to or greater than $2/\pi$ times the wavelength of the instrument's sound wave (Thorne *et al.* 1995). Thus, for a 1200-kHz ADCP, backscatter from particles with diameters equal to or greater than 0.8 mm is emphasized, and the weighting of these particles in the apparent bed velocity should be greater. The relative contribution of mobile particles versus the stationary bed is discussed further below.

For a sand bed where the depth and porosity of the active layer can be assumed constant, the bed-load transport rate can be calculated as (Rennie *et al.* 2002):

$$g_b = \overline{v_p} d_a (1 - \lambda_a) \rho_s \quad (3)$$

where: g_b is the bed-load transport rate; $\overline{v_p}$ is the average particle velocity; d_a is the depth of active bed layer; λ_a is the porosity of active bed layer; and ρ_s is the density of sediment.

2.2.1.2 Example field applications

The active-hydroacoustic technology has been applied to both stationary and moving-boat studies. Stationary measurement of apparent bed velocity has been conducted in sand- and gravel-bed reaches of Canada's Fraser River, and in a sand-bed reach in the lower Missouri River, USA. Apparent bed velocity was correlated to bed-load transport measured by physical bed-load samplers in the Fraser River. A kinematically calculated bed-load transport rate has also been correlated to that measured with physical samplers. Apparent bed velocity was also correlated to bed-load transport measured by dune tracking in the lower Missouri River, USA. Coherent patterns existed between spatial distributions of apparent bed velocity and the flow's near-bed velocity, depth-averaged velocity, and shear velocity in two reaches of the Fraser River, Canada. Use of a statistical deconvolution technique has allowed successful modeling of the distribution of actual bed velocity and of instrument noise for measured data from two gravel bed sites. The use of ADCP-measured apparent bed velocity as a surrogate for bed-load transport is a technique that shows considerable potential for characterizing bed-load transport, although calibration is required for each site, and instrument noise is substantial.

2.2.1.2.1 Stationary boat studies. Initial studies of apparent bed velocity correlated the bed velocity with bed-load transport rates measured by a physical sampler and by dune tracking. The first study was conducted in 2000 (Rennie *et al.* 2002). Apparent bed velocities were correlated with bed-load transport rates, measured by concurrent physical bed-load sampling, in the Agassiz gravel bed reach in the Fraser River, British Columbia, Canada. This was the first indication that apparent bed velocity could serve as a useful measure of bed-load transport.

Apparent bed velocity (v_b) and concurrent bed-load transport rate (g_b) measured by physical samplers were compared for five data sets from three reaches in Canada's Fraser River (Rennie & Villard 2004). Sea Reach and Canoe Pass were sand-bed reaches near the river mouth. The third reach was the gravel bed Agassiz site. A Helley–Smith bed-load sampler (Helley & Smith 1971) was used for sand and a VUV pressure-difference-type sampler (Novak 1957; Hubbell 1964; Cashman 1988) was used for gravel. In the sand-bed reaches, measurements were performed on the stoss sides of dunes to reduce spatial heterogeneity. In the gravel-bed reach, several 5-minute VUV bed-load transport samples were collected and averaged during a single ADCP measurement (see Rennie *et al.* 2002). The ADCP samples lasted between 2 and 112 minutes, (two 2-minute samples were taken when the boat could not be tethered to maintain position). The "long average" samples refer to these measurements (Table 2.3). Furthermore, individual 5-minute ADCP measurements contemporaneous with single VUV samples are referred to as "5-minute averages".

The apparent bed velocity was strongly correlated with measured bed-load transport rate for the long average Agassiz data and the Sea Reach data, and less well for 5-minute averaged Agassiz data and both Canoe Pass data sets (Fig. 2.6; Table 2.2). Larger values of bed-load transport existed for the Agassiz data than for the Sea Reach data for similar values of apparent bed velocity; for particles travelling at the same average velocity, the larger the particle the higher the mass-transport rate. In Canoe Pass, similar bed velocities were measured in 2000 and 2001, despite lower bed-load transport rates measured in 2001. Equivalent apparent bed velocity despite lower bed-load transport in 2001 may have resulted from use of a longer ADCP bottom-track pulse length for ADCP bottom track measurement that increased the influence of suspended scatterers on apparent bed velocity. The variations in the regression equations between sites suggested that the relation between apparent bed velocity and bed-load transport is site-specific, thus apparent bed velocity must be calibrated for each site. Similar to the relations shown in Table 2.2, correlations of measured bed-load transport and that calculated kinematically with measured v_b varied for these data sets. Variations resulted from differences in particle-size distributions, suspended-sediment concentrations, and ADCP operating parameters.

All available data were plotted together using non-dimensionalized bed-load transport rate, g_b^*, correlated with non-dimensionalized apparent bed velocity, v_b/u^*, where u^* is shear velocity calculated

Table 2.3 Linear regression and functional relations for measured g_b versus measured v_b, Fraser River, Canada.

Location	N	r^2	Regression	Functional relation	95% CL[a]
Agassiz long avg.	9	0.89	$g_b = 1.2v-0.037$	$g_b = 1.2v-0.041$	0.91–1.7
Agassiz 5 min.	13	0.52	$g_b = 2.0v-0.059$	$g_b = 2.6v-0.088$	0.60–7.8
Sea Reach	68	0.76	$g_b = 0.057v-0.0007$	$g_b = 0.062v+0.0005$	0.062–0.062
Canoe Pass 2000	49	0.38	$g_b = 0.23v+0.001$	$g_b = 0.36v-0.00008$	0.34–0.38
Canoe Pass 2001	15	0.42	$g_b = 0.090v-0.0003$	$g_b = 0.14v-0.0004$	0.0043–0.18
Non-dimensional	127	0.42	$g_b^* = 0.043(v/u)^{0.85}$	$g_b^* = 0.045(v/u)^{0.90}$	0.74–2.6

[a] 95% confidence limits for functional relation slope.
From Rennie & Villard (2004).

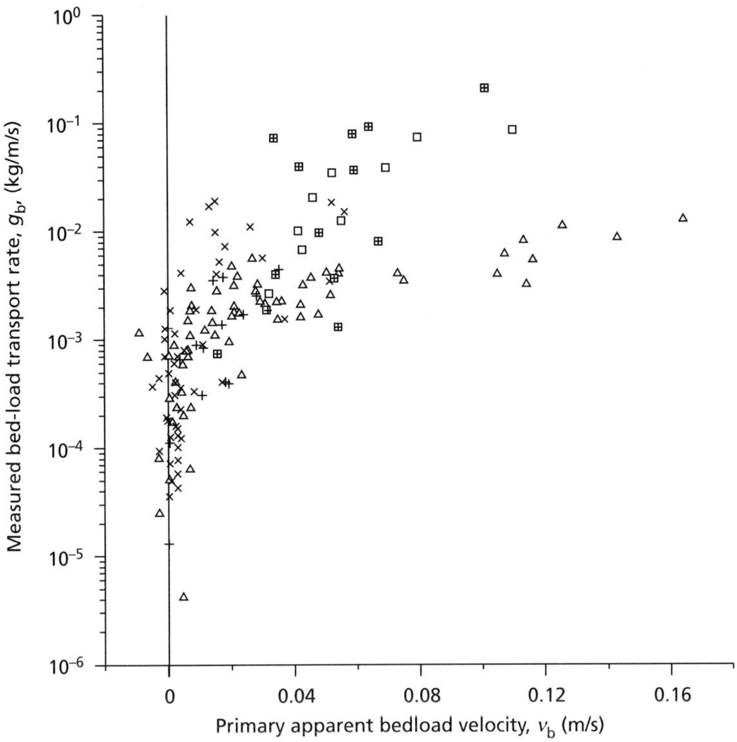

Fig. 2.6 Site-specific measured bed-load transport rate versus measured bed-load velocity. Symbols: □ Agassiz (gravel bed) long averages; ⊞ Agassiz (gravel bed) 5-minute samples; × Canoe Pass 2000 (sand bed); * Canoe Pass 2001 (sand bed); △ Sea Reach (sand bed).
From Rennie & Villard (2004).

from the log-law Keulegan equation (see below). Bed-load transport rate was non-dimensionalized using Einstein's formula (Einstein 1950):

$$g_b^* = \frac{g_b}{\rho_s\sqrt{(S_s - 1)gd_{50}^3}} \tag{4}$$

where: S_s is the sediment specific gravity; g is the gravitational acceleration; and d_{50} is the bed-load median grain size. It was found that 42% of the variance in g_b^* was explained by variance in v_b/u^*.

Apparent bed velocity was correlated to bed-load transport rate from physical sampling and dune tracking in the lower Missouri River (Gaeuman & Jacobson 2007). Measurements were taken in the thalweg, which consisted of a sand bed with dunes. Physical bed-load sampling used a Helley–Smith sampler in 2004 and a US BL-84 sampler (Kuhnle 2008) in 2005. Apparent bed velocity was correlated with g_b measured from dune tracking for values lower than 0.9 kg/(m-s), whereas large variability above that value resulted from localized values of g_b being measured over large dunes. No correlation existed between v_b and g_b measured from physical sampling. It was suggested that physical sampling was an unsatisfactory method for characterizing g_b at the higher transport rates found in the lower Missouri River, USA.

Gaeuman & Jacobson (2006) also modeled the relation between the average particle velocity, v_p, and the apparent bed velocity measured by the ADCP. The average particle velocity was calculated using the van Rijn (1984) formula, a shear stress approach. The spatially averaged surface particle velocity (v_{pa}) can be assumed to vary from a value much lower than the calculated v_p near entrainment (because much of the bed surface is immobile) to a value approaching the calculated v_p at higher transporting conditions (Gaeuman & Jacobson 2006).

$$v_b = v_p w_b w_f \tag{5}$$

where: v_p is the particle velocity calculated from van Rijn (1984); w_b is the weighting factor for percentage of bed mobile; w_f = weighting factor for position over bedform.

The weighting function, w_b, evaluates the proportion of the bed particles that are moving and accounts for the relative strength of the backscatter from the immobile bed particles versus mobile particles. Gaeuman & Jacobson (2006) considered particles moving in different layers of the active bed, with the immobile bed consisting of those bed particles that are not acoustically blocked by moving particles in any layer above them.

$$w_b = \left(\frac{b_p}{b_p + b_b F} \right) \quad (6)$$

where: b_p is the fraction of bed area with moving bed particles; b_b is the fraction of immobile bed "visible" to transducer beam; F is the relative strength of echoes reflected from immobile bed.

The bed fractions depend on the particle concentration in the bed-load layer and the height of the top of the bed load layer, both calculated according to van Rijn (1984). The value of F was assumed to be roughly 10. An additional scaling factor, w_f, was proposed, but not defined, to account for spatial differences due to the influence of bedform morphology. As expected, the ratio of v_b/v_p increased with the transport stage, T^*, (the ratio of non-dimensional shear stress to critical non-dimensional shear stress) and the modeled v_b was found to be close to the measured v_b.

Ramooz & Rennie (in press) performed calibration tests on bed velocity versus bed-load transport rates at St. Anthony's Falls Laboratory at the University of Minnesota, USA, in 2006. Apparent bed velocity was reasonably correlated with bed-load transport rate from physical sampling using a continuous-weighing slot sampler and from dune tracking for the sand bed runs. This was the only study to evaluate the sensitivity of v_b correlation with g_b to the ADCP transmit frequency (600 kHz versus 1200 kHz) and bottom track pulse length. Of the operating parameters tested, the most reliable results were obtained with the 1200 kHz ADCP with bottom track pulse length set to the default value of 20% of range to the bottom. This configuration yielded the highest correlation with measured transport rates in the sand-bed runs, and was least sensitive to positive bias at low transport rates in the gravel-bed runs. The results confirmed that longer pulse lengths are more subject to water bias.

Instrument error constitutes most of the measurement error for apparent bed velocity (Rennie *et al.* 2002). The probability density function (PDF) of particle velocities measured in the ensonified beam areas of gravel beds at Agassiz and Norrish Creek was modeled by deconvolving the PDF of the instrument error from that of the measured data (Rennie & Millar 2007). In gravel-bed reaches, bed-load transport occurs as discrete events. A large percentage of the bed is immobile at any given time, with the bed velocity assumed to be an average of moving and stationary particles. Two velocity distributions were used to model the actual bed velocities, a compound Poisson-gamma distribution and an empirically fit gamma distribution. There was good fit between the modeled and measured distributions. However, each of many possible particle velocity distributions yielded a reasonable fit, owing to the strong influence of instrument noise on the measured signal. The compound Poisson-gamma distribution was found to fit better with optimized parameters. The particle- and bed-velocity distributions were positively skewed, which would result from a few high values among mostly low values, as expected for partial transport of gravel. The instrument noise was found to be 0.21 m/s for the Agassiz (adjusted to single ping) and 0.31 m/s for the single ping Norrish Creek data. This error was similar to that for water velocity measurement, estimated to be 0.23 m/s for a 1-second average (nine pings) with 0.20 m pulse length (bin size) for the narrowband ADCP utilized.

2.2.1.2.2 Studies from moving boats. Three studies of the spatial distribution of apparent bed velocity in a reach have been conducted: Rennie & Millar (2004), Gaeuman and Jacobson (2006), and Rennie & Church (2007). In the studies led by Rennie, kriging was used to smooth the raw data to produce coherent distributions from moving-boat apparent bed-velocity measurements. Assessment of these distributions was achieved by comparison to those of shear velocity, depth, near-bed water velocity, and depth-averaged water velocity.

The near-bed velocity was measured in the bin located between 25–50 cm above the bed. The bed shear velocity was calculated by Rennie *et al.* (2002), Rennie & Millar (2004), and Rennie & Church (2007) by fitting the vertical profile of local streamwise water velocity measured with the ADCP to the log law:

$$u = \frac{u_*}{\kappa}\ln\left(\frac{30h}{k_s}\right) \tag{7}$$

where u is the velocity at h; h is the elevation above the bed; $u_* = \sqrt{\tau/\rho}$ is the shear velocity; τ is the bed shear stress; ρ is the fluid density; κ is the von Karman constant (0.41); and k_s is the bed roughness.

Significant variations existed in the shear velocity distributions mapped in Sea Reach, a sand-bed estuarine distributary of the Fraser River, Canada. (Rennie & Millar 2004). Both the near-bed water velocities and the depth-averaged water velocities were correlated with the apparent bed velocities for spatial lags up to about 10 m. Similarly, areas with high shear velocity matched those with high apparent bed velocities. High shear velocities were found to stretch from the upper left side to the lower right side of the reach.

Velocity distributions were produced for a 5.5-km-long gravel-bed reach of the Fraser River, Canada, about 150 km upstream from the river mouth (Rennie & Church 2007). Vertical velocity profiles, averaged over a width of 7.7 m, were fitted to the log law to calculate the shear velocity. Apparent bed velocities were interpolated by kriging onto a 25-m grid to yield the spatial distribution. The distributions of flow depth, depth-averaged water velocity, and shear velocity were generated likewise. The distributions for depth, depth-averaged water velocity (Fig. 2.7a), shear velocity, and apparent bed velocity (Fig. 2.7b) were very coherent. Maximum values of shear stress were found in the deep bend pools of the thalweg just downstream from areas of flow convergence. Areas of flow separation and over shallow point bars had lower shear stress. Apparent bed velocity matched bed shear except in a deep pool adjacent to a rapidly eroding bank, where highly turbulent flow existed. This pool was located downstream from the river's confluence with a major side channel. The highest apparent bed velocities were measured here with the erosion due to high 3-dimensional turbulence in a region of flow separation. The shear velocity, which is calculated from mean velocity profiles, was not estimated to be high at this location.

2.2.1.3 Summary: active hydroacoustics as bed-load surrogate technology

Stationary measurements of apparent bed velocity in sand and gravel reaches have been correlated to bed-load transport rates measured concurrently from physical sampling, dune tracking (for sand-bed rivers), and bed shear. Apparent bed velocity distributions measured from a moving boat have been correlated to concurrent distributions of near-bed water velocity, depth averaged water velocity, shear velocity, and channel depth.

Error is a significant limitation of computation of apparent bed velocity. Instrument error constitutes the majority of the error (Rennie *et al.* 2002). Raw bed velocities are computationally very noisy, and must be averaged. The error of the bottom track velocity for a mobile bed is the same order of magnitude as that for water velocity (Rennie & Millar 2007). Measurements taken from moving boats use the inherent averaging of kriging to reduce error (Rennie & Millar 2004; Rennie & Church 2007). Another limitation of apparent bed velocity computation is that the technique needs calibration for each site. The calibration is a function of the bed-load sediment size and the operating parameters of the ADCP, and can be influenced by near-bed suspended transport (water bias). The ADCP requires manual deployment, and can be purchased for about four-fold the price of a turbidimeter.

Bottom track velocity is calculated using proprietary firmware. Improvements to the firmware used to determine apparent bed velocity would be helpful. The spectrum of returned echoes could be used to determine the range of velocities contributing to the signal instead of estimating a spectral peak from the autocovariance function to represent an apparent average velocity.

Apparent bed velocity measurement using an ADCP is a fast and non-intrusive surrogate technique for computing bed-load transport. One major advantage of using an ADCP to characterize bed-load transport rates is the ability to measure the spatial distribution of relative bed-load transport. From a more general perspective, because quantification of bed-load transport is typically difficult and problematic even in sand-bed rivers, any surrogate means for providing quantifiably reliable sand bed-load data is desirable.

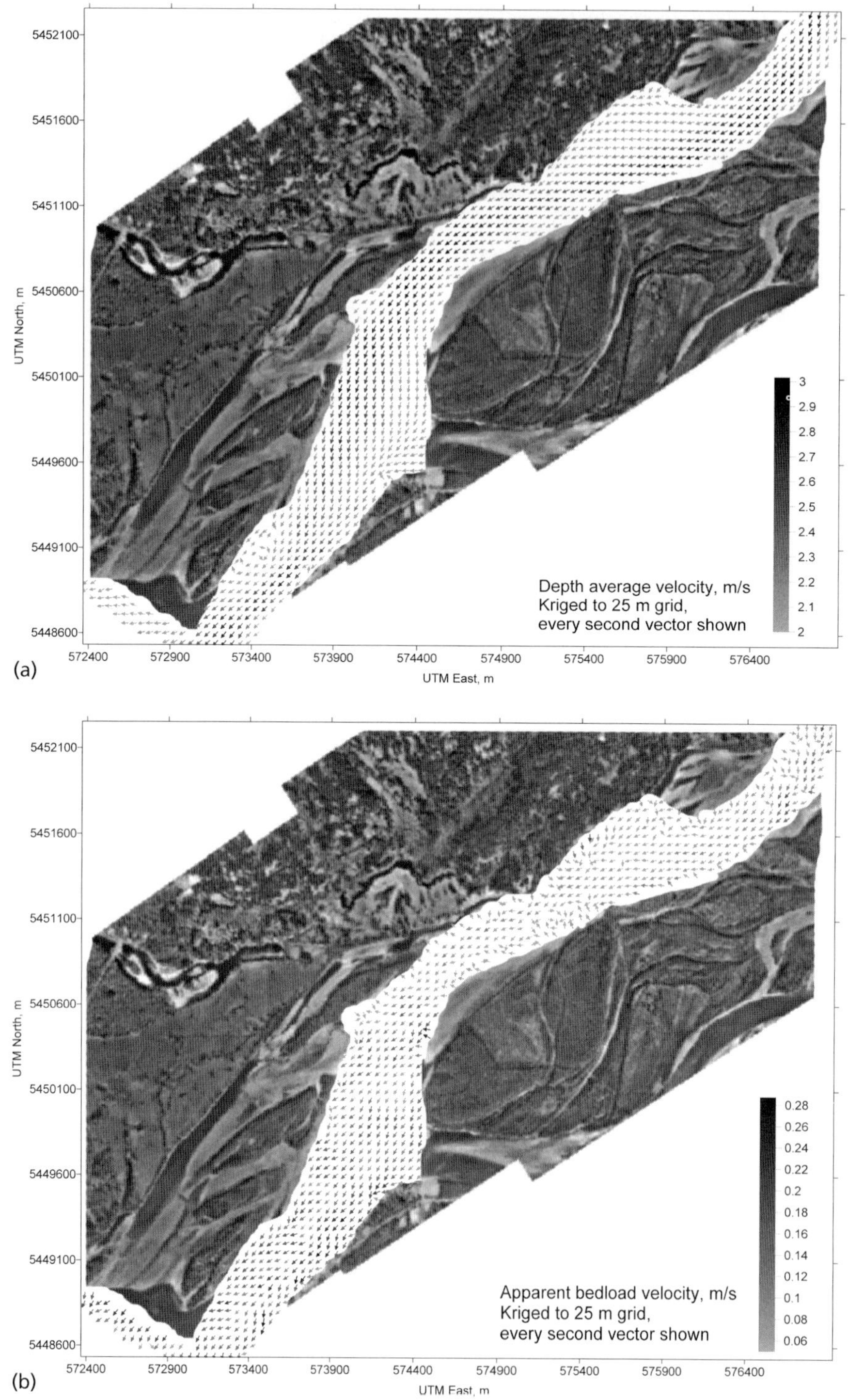

Fig. 2.7 Velocity distributions measured in m/s on the Fraser River, Canada. (a), depth-averaged water velocity, and (b) apparent bed-load velocity.
Modified from Rennie & Church (2007).

2.2.2 Passive-transducer Hydroacoustics

Jonathan S. Barton & Smokey A. Pittman

2.2.2.1 Background and theory

Investigations into the quantification of bed-load transport using acoustic signals have steadily increased in number and in complexity as researchers seek a tractable surrogate for measuring and predicting bed-load discharge. Use of passive hydroacoustic signals is attractive compared with many traditional sampling methods because of:

- relative ease of deployment;
- lower data-collection cost;
- lower hydraulic impact, and perhaps most importantly;
- continuous measurement capability, a characteristic that enables quantification of the considerable variability inherent in the bed-load transport process.

Some technologies also offer the potential for characterizing the bed-load particle size distribution. Passive hydroacoustic technologies can be grouped by the transducer type used in the measurement device. Five acoustic transducer deployments are in current use for the study of bed-load transport: hydrophones (measuring acoustic pressure fluctuations in water), microphones (measuring acoustic pressure fluctuations in air), accelerometers (measuring acceleration of a mass), velocity transducers (measuring velocity of a mass), and pressure plates (measuring impact pressure). The hydrophone is usually deployed in a protective enclosure in quiet water away from the main flow. Microphones are generally deployed within pipes installed on or in the streambed. Accelerometers are usually deployed on the underside of metal plates installed on the bed of the stream. Velocity transducers can be deployed in one of two ways: In the same fashion as accelerometers, or in geophone arrays, as in seismic surveys, along the edge of a river. Pressure plates are typically deployed perpendicular to the streambed (angled to the flow vector), as either an installed system or as a portable device.

Minimum costs associated with passive surrogate technologies for monitoring bed load are about US$5000. These technologies are relatively robust and, in theory, installations will require minimal field maintenance. The performance of the instruments have been calibrated to bed-load samples manually collected in the cross section or in flume studies (e.g. Barton et al., in press, and Møen et al., in press).

The method of using acoustic energy to derive bed-load transport rates is predicated on theories of impact based on that of Hertz (Goldsmith 2001). Depending on the specific application, the appropriate theory may involve: the collision of two irregular solids (hydrophone, velocity transducer as seismic array); the collision of an irregular solid with a cylinder (microphone); or the collision of an irregular solid with a plate (accelerometer, plate-mounted velocity transducer, pressure plate). In all cases, empirical calibration is necessary to convert to an estimate of bed-load transport rate; in most cases, this calibration must be done *in situ*, though the accelerometer has been calibrated in a flume.

Acoustic measurement of bed-load transport is not a new idea. The earliest measurements were made by Mühlhofer (1933), on Austria's Inn River using a watertight steel box containing a microphone. Bed-load collisions with the box were counted manually through the use of headphones. The Grenoble Laboratory (Labaye 1948) placed a triangular steel plate on the streambed, with a microphone in a steel box above it, and the noise of sediment striking the plate was transmitted to the microphone through a steel bar connecting the plate to the microphone membrane (no results were reported). This system was modified by Braudeau (1951), who used a brass plate and deployed the microphone in direct contact with the plate. The resulting sound was amplified and transmitted to headphones. Braudeau (1951) was able to determine the critical discharge for incipient motion to within 1 m^3/s, but did not attempt to quantify the transport rate. Bedeus & Ivicsics (1964) used a directional microphone in a boat-mounted steel housing to remotely record sediment-generated noise on the Danube River, Hungary. They compared estimates of lateral variability in transport, and results were compared with sampler data from the same cross sections. Johnson & Muir (1969) reported on flume experiments with a piezoelectric microphone, in which they calibrated an empirical relation between bed-load transport and microphone output based on the Meyer-Peter & Müller (1948) gravel-transport relation, the Hertz law of contact, and a saltation-length formula from Einstein (1950), which they also showed to improve insignificantly on a power-law fit to the data.

Froehlich (2003; in press) installed a set of microphones encased in steel pipes, and recorded the signals generated by gravel collisions with the pipes. He was able to quantify the relation between the number of cumulative gravel-pipe interactions and cumulative bed-load discharge captured in sediment basins. Mizuyama *et al.* (2003; two papers in press) and others installed a similar system, consisting of a single pipe containing a microphone deployed on a Sabo-type dam, designed to retard the propagation of debris flows. Mizuyama *et al.* (2003) found good correlation between counted impacts and bed-load transport rate at intermediate- to high-transport rates, with lower correlations at very low transport rates and at extremely high transport rates.

Hinrich (1970) modified the Grenoble sensor to use a hydrophone instead of the microphone, and a brass plate instead of a steel plate. Hinrich (1970) also installed a hydrophone on an Arnhem sampler (Hubbell 1964) and used it to verify the sampler data. Although Hinrich (1970) could recognize incipient motion, he was unable to calculate transport rates. Anderson (1976) based his microphone system on that of Johnson & Muir (1969), and suggested that moving sand generates noise dominated by frequencies above 38 kHz, based on directionality arguments relating to the microphone that he used. Anderson also observed 15- and 6-minute periodicity in the acoustic record. Richards & Milne (1979) modified Anderson's (1976) system to allow frequency analysis and in two field sites, observed that the Froude number of the flow may impact the sensor volume, and that the scatter in the acoustic amplitude was much higher in sand-bed streams than in gravel-bed streams.

In the marine literature, Thorne and colleagues (see, for example, Thorne *et al.* 1984, 1989; Thorne 1986a,b, 1987, 1993; Thorne & Foden 1988; Voulgaris *et al.* 1995) began with a hydrophone recording the noise generated by glass spheres in a rotating drum, then created a theoretical relation based on the Hertz law of contact, and ultimately created a field platform where the agreement of acoustic signals with video recordings and comparisons with Doppler velocity transducer current measurements led the authors to conclude that second-scale temporal variability of gravel transport is dominated by turbulent bursting events.

Barton (2006) and Barton *et al.* (2005, 2006, in press) have expanded upon this work, examining the effectiveness of a hydrophone for fluvial bed-load monitoring. Their hydrophone was mounted in near-bank slack waters of the Trinity River, California, USA, providing protection from impacts with sediment and debris, and separation from turbulent noise. Continuous data were collected concomitant with manual bed-load measurements using pressure-difference samplers (Fig. 2.4). Barton *et al.* (2006) found a significant relation between bed-load transport and the noise generated by the process; the acoustic signals were exploited to predict the bed-load discharge between pressure-difference sampling measurements. Smith (Graham Matthews and Associates 2006, 2007, 2008) has continued this work, collecting data at the same location on the Trinity River.

Rickenmann (1997), Rickenmann *et al.* (1997), Rickenmann & Fritschi (in press), and Hegg & Rickenmann (1998, 2000), building on earlier work by Bänziger & Burch (1990), have shown the effectiveness of accelerometer and geophone (velocity transducer) installations (mounted beneath a metal plate installed on the bed) for long-term bed-load monitoring in the Swiss Alps. Bogen & Møen (2003) and Møen *et al.* (in press), using a system similar to that of Rickenmann (1997), but with different frequency sensitivity, have shown that an accelerometer with a narrow frequency band is heavily influenced by sediment grain size, and that with appropriate calibration, a wideband accelerometer may be able to account for changes in the grain size. Richardson *et al.* (2003) also mounted an accelerometer beneath a steel plate, and found that although the relation between sediment impact rate and transport rate was nonlinear (particularly at high transport rates), the relation was consistent with theory based on shear stress.

Govi *et al.* (1993) counted impacts recorded by geophones (velocity transducers) buried in the streambed immediately upstream from a weir, and were able to establish streamflow discharges corresponding to initiation and cessation of bed-load motion, but did not calculate transport rates. Burtin *et al.* (2008) used a high-density seismic array in the Himalayas to monitor the bed-load flux qualitatively in the narrow and deeply incised Trisuli River, Nepal. Although they were unable to separate contributions to the seismic signal completely owing to turbulence in the flow, they were able to record a hysteresis loop in the seismic rating curve, indicating a significant contribution to the seismic signal from sources other than streamflow.

Downing & Ryan (2001), Downing *et al.* (2003) and Downing (in press) describe a manually deployed pressure-plate device that, when impacted by a moving sediment grain, produces a charge that is proportional to the force applied, which through integration yields the momentum flux. They derived a pulse-count record of the bed-load interaction with the plate above a minimum threshold impact value. Application of the device requires *a priori* knowledge of the size distribution in motion. Unlike the other devices discussed here, this device interacts with the flow, and so a calibration involving the hydraulic efficiency is required. Downing (in press) showed, for two floods on the same river, that assuming a constant calibration coefficient would result in an error in the calculated transport rate of only ±20%.

2.2.2.2 Example field application

A single hydrophone (Geospace Technologies MP-18) system was installed 250 m downstream from the USGS streamgage on the Trinity River at Douglas City, California, USA (Barton *et al.* in press). Acoustic data were collected from May 6 to May 19 2005; total acoustic power ranging from 0.01 to 14.8 kHz over 1-minute intervals was calculated from the data. Sample data collected using a Toutle River-2 (TR-2) bed-load sampler, a modified version of the BL-84-type bed-load sampler capable of collecting medium-size gravel (Childers 1999; Pittman 2005) (Fig. 2.4) deployed from a tethered raft system, were compared with the temporal average of acoustic data collected during a sampling interval (Fig. 2.8). The resulting regression was applied to the 1-minute data (Fig. 2.9). Barton *et al.* (in press) indicate that the range of the acoustic data is consistent with the range of most Toutle River-2 bed-load sampler data. Spectral analysis of the 1-minute data shows discrete frequency peaks, the lowest of which falls within the frequency range reported for bed-load sheet movement.

2.2.2.3 Summary: passive hydroacoustics as bed-load surrogate technology

This technology is applicable for continuous bed-load monitoring in gravel-bed systems where the acoustic energy emitted by contacts of bed-load particles larger than a minimum grain-size threshold can be measured. In all cases, this minimum size is not clearly

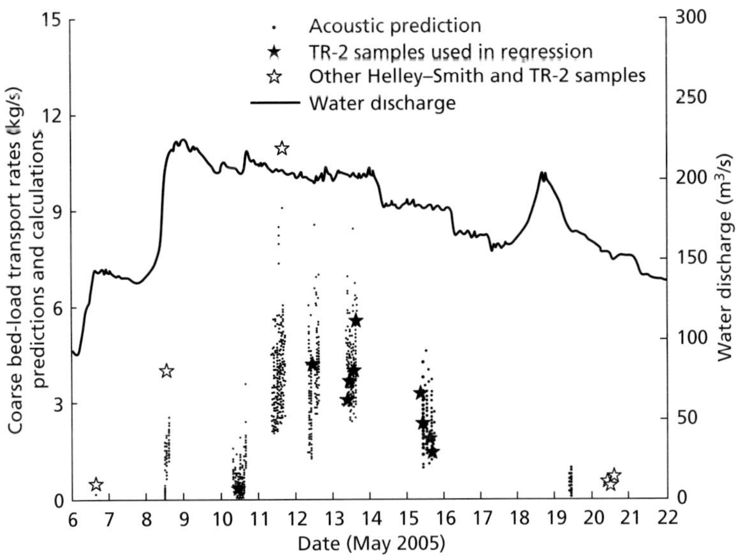

Fig. 2.8 Predictions of coarse (>8 mm) bed-load transport rates from one-minute-averaged acoustic power (small dots) over the study interval plotted with the water discharge (solid line), Trinity River at Douglas City, California, USA, and bed-load transport rates from data collected by Helley–Smith and TR-2 bed-load samplers (solid and hollow stars). The solid stars represent data used in the least-squares regression shown in Fig. 2.9.
From Barton *et al.* (in press).

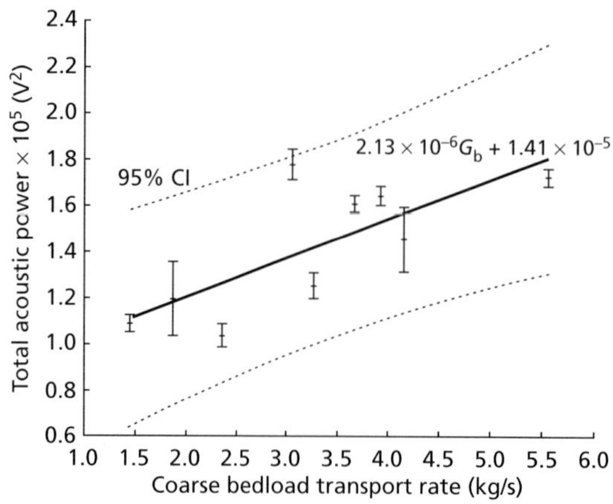

Fig. 2.9 Correlation plot between temporally averaged total acoustic power (totaled over the frequency range of 0.01–14.8 kHz) and bed-load transport rate from the Toutle River 2 sampler. Error bars show ±2 standard errors of the temporal mean. The Pearson's correlation coefficient R is 0.758, with a p-value of 0.0180. Confidence interval for the regression parameters assumes Gaussian error. From Barton *et al.* (in press).

defined; in many cases, size thresholds may depend on the specifics of the surrogate technology installation.

The technique relies entirely on calibrations to cross-section bed-load samples. This technology can be used to infer the incipient motion, and with calibration by reliable bed-load samplers, to infer mass transport. Most parts are available off the shelf at a cost similar to that for a fully equipped *in situ* turbidimeter. Specific advantages and limitations of each type of sensor follow.

2.2.2.3.1 Advantages of passive hydroacoustic technologies

Hydrophone:
- By integrating over a large area of the streambed, the hydrophone allows estimation of average transport rate, compensating for spatial variability in the transport rate.
- Taking advantage of the high acoustic conductivity of water, the hydrophone can be placed in slack water adjacent to the main flow.
- The hydrophone can be installed at minimal cost, requiring no excavation of the bed and can be installed during high flow.

Microphone:
- Isolation of electronics from the water leads to improved reliability and maintainability.
- Sensors can operate unattended for long intervals with minimal maintenance.
- Method is robust for monitoring fine gravel to small boulder transport.

Plate-mounted accelerometer or velocity transducer:
- Sensors can operate unattended for long intervals with minimal maintenance.
- Technique has a 15-year operational history;
- Technique has ability to differentiate grain sizes with sufficiently high-frequency data acquisition and advanced processing techniques.
- Flume calibration may be sufficient.

Velocity transducer as seismic array:
- Sensors are deployed outside the river channel; Burtin *et al.* (2008) showed that sensors as much as 2 km away from the river channel still showed significant sensitivity to river hydraulics.
- Integrated bed-load transport measurements are on the reach-to-basin scale.
- Two-dimensional array deployment may allow watershed-scale transport analysis of regions of high transport using seismic tomography techniques.

Pressure plate:
- Technique can be used as either permanent (installed) system or portable (wading-stick mounted) system.
- Calibration has been shown to be fairly stable (±20% variation) for two floods on the same stream.
- System is effective for grain sizes as small as 4 mm in diameter (the largest size that will not damage the instrument has not been reported).

2.2.2.3.2 Limitations of passive hydroacoustic technologies. All passive hydroacoustic technologies for bed load require site-specific calibrations. Other limitations include the following.

Hydrophone:
- Only single-instrument systems have been tested, and evidence suggests that this arrangement may be sensitive to changes in spatial distribution of bed-load transport. Array deployment may help to reduce this sensitivity.
- Technique is only appropriate for medium-gravel to large-boulder applications. Fine gravel and sand

produce high frequency noise, which may be problematic to separate from flow noise.

Microphone:

- Technique has limited applicability for extremely low or extremely high sediment discharges. Long-term averaging at low discharges can improve signal-to-noise ratio.
- High-flow performance depends upon half-burying the pipe in the bed.

Plate-mounted accelerometer or velocity transducer:

- Selection of placement site is strongly influenced by river geometry, as some sites may be susceptible to deposition at certain flows, which could cover the instrument.
- Installation may be expensive, and possibly require excavation.

Velocity transducer as seismic array:

- An array such as that used by Burtin *et al.* (2008) is expensive to purchase and deploy. Effectiveness of the technology is uncertain if scaled down.
- Studies thus far have focused only on qualitative evaluation of transport. No quantitative information is available yet.
- The minimum particle size to which the system is sensitive has not been determined.

Pressure plate:

- Instrument projects into flow, which changes the local hydraulics, and subsequently the local bed-load transport, leading to scour or deposition.
- Technique requires *a priori* knowledge of size distribution in transport.

2.3 Summary and conclusions

One active (ADCP) and several passive (hydrophone or geophone) acoustic surrogate technologies for monitoring bed-load transport that have been described in this chapter are being tested and evaluated for use in large-scale operational sediment-transport monitoring programs. Active and passive hydroacoustics are but two of more than a dozen bed-load surrogate technologies described in the literature. However, hydroacoustics technologies are considered by the editors to be among the most promising of the bed-load surrogate technologies with which they are familiar.

With the potential exception of some passive bed-load hydroacoustic technologies in selected streams, the *in situ* technologies do not directly measure the constituent of interest over the entire cross section. Hence, the technologies require cross-section calibration with reliable bed-load samplers.

The technique of monitoring bed load using active acoustics has been tested in sand- and gravel-bed systems. Like the passive acoustic technology, site-specific, empirically derived relations using data from an ADCP and a bed-load sampler are required. For active acoustics, the calibration is a function of the sediment size and the operating parameters of the ADCP.

Stationary measurements of apparent bed velocity utilizing manually deployed ADCPs have been correlated with concurrent measurements of bed-load transport and bed shear stress in sand and gravel reaches, and to dune tracking in sand-bed rivers. Distributions of apparent bed velocity measured by ADCP from a moving boat have been correlated to concurrent distributions of near-bed water velocity, depth-averaged water velocity, shear velocity, and channel depth. Instrument measurement variance constitutes the majority of error in the technique. The variance of the bottom track velocity for a mobile bed is the same order of magnitude as that for water velocity.

Apparent bed-velocity measurements made by using active acoustics is a fast and non-intrusive technique for computing bed-load transport. One advantage of using an ADCP to characterize bed-load transport is the ability to measure the spatial distribution of apparent bed velocity. The method also benefits from substantial averaging of measurements. However, lack of spatial homogeneity of apparent bed velocity in the region sampled by the acoustic beams may cause increased variance in bed-load computations. The cost of the technology (ADCP) is about US$20,000, in addition to the costs of a GPS, boat, and other equipment necessary for deployment.

Passive acoustic techniques are limited to applications in gravel-bed systems where bed-load particles are sufficiently large for the acoustic energy emitted by contacts to be measured. In all cases, this particle size is not clearly defined; in many cases, size thresholds may depend on the specifics of the installation.

Many of these techniques, designed to function remotely, can be used to infer incipient bed motion, and with calibration by samples collected manually with reliable bed-load samplers, to infer mass transport. As with the active-acoustic technology, empirical site-specific relations between acoustic signal strength (or other acoustic parameters) and bed-load sampler data must be developed and used with the continuous acoustic signal to compute continuous bed-load transport. The minimum cost of a passive-acoustic instrument is about US$5000.

Five types of passive-acoustic system have been tested: hydrophones, microphones, plate-mounted accelerometers or velocity transducers, pressure plates, and velocity transducers as seismic arrays. Hydrophones, submerged in a relatively quiescent location, integrate the acoustic energy over a large area of the streambed, in effect inferring an average bed-load transport rate. Only single-instrument systems have been tested, and they may respond differentially to changes in the spatial distribution of bed-load transport. The technology is only appropriate for applications where bed-load particle sizes range from medium gravel to large boulders. Fine gravel and sand produce high-frequency noise, which is computationally difficult to separate from ambient noise. When deployed in slack water areas adjacent to the main flow, the system is relatively robust.

Microphones, which measure acoustic pressure fluctuations in air, isolate the instrument's electronics from the water resulting in improved long-term reliability and maintainability. These systems are considered robust for monitoring fine gravel to small boulder transport, but their performance is inferior to other passive acoustic systems at extremely low or extremely high bed-load discharges.

Plate-mounted accelerometers or velocity transducers have proven, over a one- to two-decade operational history, to operate unattended for long intervals with minimal maintenance. The technology can differentiate among grain sizes given sufficiently high-frequency data acquisition and advanced processing techniques. Flume calibration may be sufficient. Instrument placement is strongly influenced by river geometry, as some sites may be susceptible to deposition that could cover the instrument. It is one of the more expensive of the passive-acoustic technologies because installation may require excavation.

Velocity transducers as seismic arrays integrate bed-load transport on the reach-to-basin scale. Sensors are deployed outside the river channel, with sensors installed as much as 2 km from the river channel showing sensitivity to river hydraulics. Two-dimensional array deployment may allow watershed-scale transport analysis of regions of high bed-load transport using seismic tomography techniques. The system can be expensive to purchase and deploy, and the effectiveness of its scaled-down performance is unknown. Only qualitative information is available, and the minimum particle size to which the system is sensitive has not been determined.

Pressure plates can be used as either an installed system or as a manually deployed wading-stick mounted portable device. System calibration has been shown to be somewhat stable (within a range of ±20%) for two floods on the same stream. It is effective for grain sizes as small as 4-mm diameter but the upper size limit is unknown. *A priori* knowledge of size distribution in transport is required. The instrument projects into flow, which changes the local hydraulics, and subsequently the local bed-load transport rate, potentially leading to local scour or deposition.

2.4 Prospects for operational surrogate monitoring of bed-load transport in rivers

This chapter has described an active hydroacoustic and several passive hydroacoustic technologies for monitoring characteristics important to understanding properties of bed-load transport in rivers. Some characteristics common to these technologies include the following:

- All address measurement of bed-load characteristics that are difficult, expensive, and (or) dangerous to directly measure with sufficient frequency to adequately define their spatial and temporal variability.
- At least some are relatively affordable, costing between US$5000 and US$20,000. Some, such as cross-channel impact-plates installations, may cost substantially more.
- Most if not all require site-specific calibrations equating values recorded by the surrogate instrument to the mean cross-section constituent value.
- All require additional testing and evaluation before deployment in operation sediment-transport programs.

None of the technologies is suitable for monitoring bed-load transport under all flow and sediment-transport conditions. Nevertheless, if care is exercised in matching surrogate technologies to appropriate river and sedimentological conditions, it may be eventually possible to remotely and continuously monitor bed-load transport in a variety of rivers over a range of flow and sedimentary conditions within acceptable accuracy limits. This is a revolutionary concept in fluvial sedimentology; benefits of such applied capability could be enormous, providing for safer, more frequent and possibly more accurate, and ultimately less expensive data for use in managing the world's sedimentary resources.

Acknowledgments

This chapter benefited from the contributions and efforts of several individuals other than the authors. The manuscript was improved by the reviews provided by Michael Singer, University of St Andrews, UK, and James D. Fallon and Broderick E. Davis, USGS, Minneapolis, Minnesota, USA, and Vicksburg, Mississippi, USA, respectively. Annette L. Ledford, USGS, Reston, Virginia, USA, devoted considerable effort in the development of the chapter's figures and tables.

References and further reading

Anderson, M. G. (1976) An inexpensive circuit design for the acoustic detection of oscillations in bedload transport in natural streams. *Earth Surface Processes*, **1**, 213–17.

ASTM International. (1998) Terminology for fluvial sediment. Designation D 4410-98, 6p.

Atkinson, E. (1994) Vortex-tube sediment extractors. I: Trapping efficiency. *Journal of Hydraulic Engineering*, **120**(10), 1110–25.

Bänziger, R. & Burch, H. (1990) Acoustic sensors (hydrophones) as indicators for bed load transport in a mountain torrent. In *Hydrology in Mountainous Regions I – Hydrological Measurements; The Water Cycle*, H. Lang & A. Musy (eds), 207–14. Lausanne: International Association of Hydrological Sciences, Lausanne.

Bänzinger, R. & Burch, H. (1991) Geschiebetransport in Wildbächen: Messung mittels eines neuartigen Sensors. *Schweizer Ingenieur und Architekt*, **24**, 576–79.

Barton, J. S. (2006) *Passive acoustic monitoring of bedload in mountain streams*. Ph.D. thesis, The Pennsylvania State University, University Park, PA, 107pp.

Barton, J. S., Slingerland, R. L., Gabrielson, T. B. & Johnson, P. A. (2005). Listening to bedload—a flume study relating acoustic response to bedload motion. In *Federal Interagency Sediment Monitoring Instrument and Analysis Research Workshop*, J. R. Gray (ed.) 2005, Proceedings of the Federal Interagency Sediment Monitoring Instrument and Analysis Workshop, September 9-11 2003, Flagstaff, Arizona: US Geological Survey Circular 1276, http://water.usgs.gov/osw/techniques/sediment/sedsurrogate2003workshop/barton.pdf.

Barton, J. S., Slingerland, P., S. & Gabrielson, T. B. (2006) Passive acoustic monitoring of coarse bedload transport on the Trinity River. In *Eighth Federal Interagency Sedimentation Conference, Subcommittee on Sedimentation of the Advisory Committee on Water Information, Reno, NV, USA*.

Barton, J. S., Slingerland, R. L., Pittman, S. & Gabrielson, T. B. (in press) Monitoring coarse bedload transport with passive acoustic instrumentation: A field study. In *Proceedings of the International Bedload-Surrogate Monitoring Workshop, April 11–14, 2007, Minneapolis, Minnesota*, J. R. Gray, J. B. Laronne & J. D. G. Marr (eds).

Bedeus, K. & Ivicsics, L. (1964) Observation of the noise of bed load. In *General Assembly, Commission on Hydrometry, International Association of Hydrological Sciences, Berkeley, CA, USA*, 384–90.

Beschta, R. L. (1981) Increased bag size improves Helley–Smith bed load sampler for use in streams with high sand and organic matter transport. In *Erosion and Sediment Transport Measurement*, 17–25. Oslo, Norway: International Association of Hydrological Sciences, Publication 133.

Bogen, J., Fergus, T. & Walling, D. E. (2003) *Erosion and Sediment Transport in Rivers, Technological and Methodological Advances*. Oslo, Norway: International Association of Hydrological Sciences, Publication 283, 238pp.

Bogen, J. & Møen, K. (2003) Bed load measurements with a new passive ultrasonic sensor. In *Erosion and Sediment Transport Measurement in Rivers—Technological and Methodological Advances*, J. Bogen, T. Fergus & D. Walling (eds), 181–92. Oslo, Norway: International Association of Hydrological Sciences, Publication 283.

Braudeau, G. (1951) Quelques techniques pour l'étude et la mesure du débit solid. *La Houille Blanche*, **6**, 243–52.

Bransington, J., Rumsby, B. T. & McVey, R. A. (2000) Monitoring and modelling morphological change in a braided gravel-bed river using high resolution GPS-based survey. *Earth Surface Processes and Landforms*, **25**, 973–90.

Bunte, K. (1992) Particle number grain-size composition of bedload in a mountain stream. In *Dynamics of Gravel Bed Rivers*, P. Billi, R. D. Hey, C.R. Thorne & P. Tacconi (eds), 55–72. Chichester, UK: John Wiley.

Bunte, K. (1996) Analyses of the temporal variation of coarse bedload transport and its grain size distribution (Squaw Creek, Montana): English translation of Ph.D. dissertation submitted to the Freie Universität Berlin, Germany. USDA Forest Service, Rocky Mountain Forest and Range Experiment Station, General Technical Report RM-GTR-288, 123pp.

Bunte, K. (1997) Development and field testing of a bedload trap for sand and fine gravels in mountain gravel-bed streams (South Fork Cache la Poudre Creek, CO). Report prepared for the Stream Systems Technology Center, USDA Forest Service, Rocky Mountain Forest and Range Experiment Station, Fort Collins, Colo., 53pp.

Bunte, K. (1998) Development and field testing of a stationary net-frame bedload sampler for measuring entrainment of pebble and cobble particles. Report prepared for the Stream Systems Technology Center, USDA Forest Service, Rocky Mountain Forest and Range Experiment Station, Fort Collins, Colo., 74pp.

Bunte, K. (1999) Field testing of bedload traps for measuring entrainment of pebbles and cobbles at Little Granite Creek, WY. Report prepared for the Stream Systems Technology Center, USDA Forest Service, Rocky Mountain Forest and Range Experiment Station, Fort Collins, Colo., 78pp.

Bunte, K. (2001) Field testing the sampling efficiency of bedload traps at East St. Louis Creek, CO. Report submitted to the Stream Systems Technology Center, USDA Forest Service, Rocky Mountain Research Station, Fort Collins, Colo., 84pp.

Bunte, K. (2002) Testing the sampling efficiency of the bedload traps by comparison with sediment collected in the debris basin at East St. Louis Creek. Report submitted to the Stream Systems Technology Center, USDA Forest Service, Rocky Mountain Research Station, Fort Collins, Colo., 44pp.

Bunte, K. & Abt, S. R. (2003) Sampler size and sampling time affect measured bedload transport rates and particle sizes measured with bedload traps in gravel-bed streams. In *Erosion and Sediment Transport Measurement in Rivers—Technological and Methodological Advances*, J. Bogen, T. Fergus and D. Walling (eds), 126–33. Oslo, Norway: International Association of Hydrological Sciences, Publication 283.

Bunte, K. & Swingle, K. (2002) Results from testing the bedload traps at Little Granite Creek 2002: effect of sampling duration and sampler type on bedload transport rates and systematic variability of rating curves with basin area and stream bed parameters. Report submitted to the Stream Systems Technology Center, USDA Forest Service, Rocky Mountain Research Station, Fort Collins, Colo., 110pp.

Bunte, K. & Swingle, K. (2003) Field testing the sampling efficiency of bedload traps at East St. Louis Creek, May – July (2003: Comparison of annual gravel load between bedload traps and debris basin. Report submitted to the Stream Systems Technology Center, USDA Forest Service, Rocky Mountain Research Station, Fort Collins, Colo., 86pp.

Bunte, K. & Swingle, K. (2004) Flume and field measurements of flow velocity profiles near the bedload trap entrance. Report submitted to the Stream Systems Technology Center, USDA Forest Service, Rocky Mountain Research Station, Fort Collins, Colo., 40pp.

Bunte, K., Abt, S. R. & Potyondy, J. P. (2001) Portable bedload traps with high sampling intensity for representative sampling of gravel transport in wadable mountain streams. In *Proceedings of the 7th Federal Interagency Sedimentation Conference, Reno, Nevada.*, vol. 2, pp. III 24–III 31, http://water.usgs.gov/pubs/misc_reports/FISC_1947-2001/, accessed August 31, 2004.

Bunte, K., Abt, S. R., Potyondy, J. P. & Ryan, S. E. (2004) Measurement of coarse gravel and cobble transport using a portable bedload trap: *Journal of Hydraulic Engineering*, **130**(9), 879–93.

Burtin, A., Bollinger, L., Vergne, J., Cattin, R. & Náb lke, J. L. (2008) Spectral analysis of seismic noise induced by rivers: a new tool to monitor spatiotemporal changes in stream hydrodynamics. *Journal of Geophysical Research*, **113**, B05301, doi:10.1029/2007JB005034.

Busskamp, R. (1994a) Erosion, einzellaufwege und ruhephasenanalyse und modellierung der stochastischen parameter des grobgeschiebetransportes. [Erosion, individual travel distances, and rest phases: analysis and modeling of the stochastic parameters of coarse bedload transport.]. Ph.D. dissertation, Department of Earth Sciences, Freie Universität Berlin, Germany.

Busskamp, R. (1994b) The influence of channel steps on coarse bed load transport in mountain torrents—case study using the radio tracer technique 'PETSY'. In *Dynamics and Geomorphology of Mountain Rivers*, P. Ergenzinger & K.-H. Schmidt (eds), 129–39. Lecture Notes in Earth Sciences. Berlin: Springer Verlag, Berlin.

Busskamp, R. & Ergenzinger, P. (1991) Neue analysen zum transport von grobgeschiebe—messung lagrangescher parameter mit der radiotracertechnik (PETSY). [New analyses of coarse bedload transportmeasurements of Lagrangian parameters with the radio tracer technique (PETSY)]. *Deutsche Gewässerkundliche Mitteilungen*, **35**(2), 57–63.

Busskamp, R., Gintz, D. (1994, Geschiebefrachterfassung mit hilfe von tracern in einem wildbach (Lainbach/Oberbayern). [Bedload transport determination using tracers in a mountain torrent (Lainbach/Upper Bavaria)]. In *Messungen in fluvialen Systemen. Feld- und Laboruntersuchungen zur Erfassung des Wasser- und Stoffhaushaltes*, D. Barsch, R. Mäusbacher, K.-H. Pörtge and K.-H. Schmidt, (eds), 179–93 Heidelberg: Springer Verlag.

Butler, P. R. (1977) Movement of cobbles in a gravel-bed stream during a flood season. *Geological Society of America Bulletin*, **88**, 1072–74.

Carey, W. P. (2005) A brief history of the development and calibration of bedload samplers. National Center for Earth-surface Dynamics – US Geological Survey 2005 Bedload Monitoring Technologies Workshop, December

1–2, 2005, Minneapolis, Minnesota, papers and manuscripts, 18pp.

Cashman, M. A. (1988) Sediment survey equipment catalogue. Environment Canada, Inland Waters Directorate, Water Resources Branch, Water Survey of Canada Division, Sediment Survey Section, IWD-HQ-WRB-SS-88-9-90 p.

Chacho, E. F., Burrows, R. L. & Emmett, W. W. (1989) Detection of coarse sediment movement using radio transmitters. In *International Association for Hydraulic Research, Proceedings of the 23rd Congress, Technical Session B*, pp. B367–73.

Chacho, E. F., Emmett, W. W. & Burrows, R. L. (1994) Monitoring gravel movement using radio transmitters. In *Hydraulic Engineering '94, Proceedings of the 1994 Conference, ASCE, New York*, vol. 2, G. Cotroneo & R.R. Rumer (eds), 785–89.

Chacho, E. F., Emmett, W. W. & Burrows, R. L. (1996) Motion characteristics of coarse sediment in a gravel-bed river. In *Proceedings of the 6th Federal Interagency Sedimentation Conference, Las Vegas, Nevada*, V 1–V 8, http://water.usgs.gov/pubs/misc_reports/FISC_1947-2001/, accessed August 31 2004.

Childers, D. (1991) Sampling differences between the Helley–Smith and BL-84 bedload samplers. In *Proceedings of the 5th Federal Interagency Sedimentation Conference, Las Vegas, Nevada*, VI 31 – VI 38, http://water.usgs.gov/pubs/misc_reports/FISC_1947-2001/, accessed August 31, 2004.

Childers, D. (1999) Field comparison of six-pressure-difference bedload samplers in high energy flow: US Geological Survey Water-Resource Investigations Report 92-4068, 59 pp., http://pubs.er.usgs.gov/usgspubs/wri/wri924068.

Childers, D., Kresch, D. L., Gustafson, S. A., Randle, T. J., Melena, J. T. & Cluer, B. (2000) Hydrologic data collected during the 1994 Lake Mills drawdown experiment, Elwha River, Washington: US Geological Survey Water-Resources Investigation Report 99-4215, 115pp.

Church, M., Wolcott, J. F., Fletcher, W. K. (1991) A test of equal mobility in fluvial sediment transport—behavior of the sand fraction. *Water Resources Research*, **27**(11), 2941–51.

D'Agostino, Lenzi, M. A. & Marchi, L. (1994) Sediment transport and water discharge during high flows in an instrumented watershed. In *Dynamics and Geomorphology of Mountain Rivers*, P. Ergenzinger and K.-H. Schmidt (eds), Lecture Notes in Earth Sciences, vol. **52**, 67–81. Berlin: Springer Verlag.

DeVries, P., Burges, S. J., Daigneau, J. & Stearns, D. (2001) Measurement of the temporal progression of scour in a pool-riffle sequence in a gravel-bed stream using an electronic scour monitor: *Water Resources Research*, **37**(11), 2805–16.

Dinehart, R. L. (1992) Evolution of coarse gravel bed forms: field measurement at flood stage: *Water Resources Research*, **28**(10), 2667–89.

Dinehart, R. L. (2001) Bedform mapping in the Sacramento River. In *Proceedings of the 7th Federal Interagency Sedimentation Conference, Reno, Nevada*, vol. 1, p. III 55 – III 62, http://water.usgs.gov/pubs/misc_reports/FISC_1947-2001/, accessed August 31, 2004.

Diplas, P., Kuhnle, R. A., Gray, J. R., Glysson, G. D. & Edwards, T. E. (2008) Sediment transport measurements. In *Sedimentation Engineering*, M. H. García (ed.), American Society of Civil Engineering Manuals and Reports on Engineering Practice, no. **110**, 307–53.

Dixon, M. & Ryan, S. E. (2001) Using an underwater video camera for observing bedload transport in mountain stream. In *Proceedings of the 7th Federal Interagency Sedimentation Conference, Reno, Nevada*, poster-70-73, accessed, at http://water.usgs.gov/pubs/misc_reports/FISC_1947-2001/, accessed August 31, 2004.

Downing, J. (in press) Acoustic gravel-momentum sensor. In Gray J. R., Laronne J. B., and Marr, J. D. G., in press, Bedload-surrogate monitoring technologies: U.S. Geological Survey Scientific Investigations Report, in press.

Downing, J., Farley, P. J., Bunte, K., Swingle, K., Ryan, S. E. & Dixon, M. (2003) Acoustic gravel-transport sensor. In *Erosion and Sediment Transport Measurement in Rivers—Technological and Methodological Advances*, J. Bogen, T. Fergus and D. Walling (eds), 193–200, Oslo, Norway: International Association of Hydrological Sciences, Publication 283.

Downing, J. & Ryan, S. E. (2001) Field comparisons of the Elwha bedload sampler and an acoustic gravel-transport sensor. Middle Fork of the Piedra River, Colorado, USA, EOS Trans. AGU, Fall Meet. Suppl., H21B-0299.

Druffel, L., Emmett, W. W., Schneider, R. & Skinner, J. V. (1976) Laboratory hydraulic calibration of the Helley–Smith bedload sediment sampler. US Geological Survey Open-File Report 76-752.

Duizendstra, H. D. (2001a). Measuring, observing and pattern recognition of sediment transport in an armoured river using a sampler with attached underwater camera. *Physics and Chemistry of the Earth B*, no. 26, vol. **1**, 59–64.

Duizendstra, H. D. (2001b) Determination of the sediment transport in an armoured gravel-bed river. *Earth Surface Processes and Landforms*, **26**, 1381–93.

Edwards, T. E. & Glysson, G. D. (1999) Field methods for collection of fluvial sediment. US Geological Survey Techniques of Water-Resources Investigations Book 3, Chapter C2, 89pp. http://water.usgs.gov/osw/techniques/Edwards-TWRI.pdf.

Einstein, H. A. (1950) Bed-load functions for sediment transport in open channel flows. Technical Bulletin 1026. Washington, DC: US Department of Agriculture.

Emmett, W. W. (1980) A field calibration of the sediment-trapping characteristics of the Helley–Smith bedload sampler. US Geological Survey Professional Paper 1139, 44pp, http://pubs.er.usgs.gov/usgspubs/pp/pp1139.

Emmett, W. W. (1981) Measurement of bed load in rivers. In *Erosion and Sediment Transport Measurements*, 3–15, Oslo, Norway: International Association of Hydrological Sciences, Publication 133.

Emmett, W. W., Burrows, R. L. & Chacho, E. F. (1996) Coarse-particle transport in a gravel-bed river. *International Journal of Sediment Research*, **11**(2), 8–21.

Engel, P. & Lau, Y. L. (1981) The efficiency of basket type bed load samplers. In *Erosion and Sediment Transport Measurement*, 27–34, Oslo, Norway: International Association of Hydrological Sciences, Publication 133.

Ergenzinger, P. & De Jong, C; (2003) Perspectives on bed load measurement. In *Erosion and Sediment Transport in Rivers: Technological and Methodological Advances*, J. Bogen, T. Fergus & D. E. Walling (eds), 113–25. Oslo, Norway: International Association of Hydrological Sciences, Publication 283.

Ergenzinger, P., De Jong, C. & Christaller, G. (1994a) Interrelationships between bedload transfer and river bed adjustment in mountain rivers—an example from Squaw Creek, Montana. In *Process Models and Theoretical Geomorphology*, M. J. Kirkby (ed.), 141–58. New York: John Wiley and Sons.

Ergenzinger, P., De Jong, C., Laronne, J. B. & Reid, I. (1994b) Short term temporal variations in bedload transport rates: Squaw Creek, Montana, USA, and Nahal Yatir and Nahal Estemoa, Israel. In *Dynamics and Geomorphology of Mountain Rivers*, P. Ergenzinger and K.-H. Schmidt (eds), 251–64, Lecture Notes in Earth Sciences, Berlin: Springer Verlag.

Federal Interagency Sedimentation Project. (1940). Field practice and equipment used in sampling suspended sediment. Iowa City, Iowa University Hydraulics Laboratory, Interagency Report no. 1, 175pp., http://fisp.wes.army.mil/Report%201.pdf.

Federal Interagency Sedimentation Project. (2008) http://fisp.wes.army.mil/.

Ferguson, R. I., Hoey, T. B., Wathen, S. J., Werritty, A., Hardwick R. I. & Smith, G. H. S. (1998) Downstream fining of river gravels: integrated field, laboratory and modeling study. In *Gravel-Bed Rivers in the Environment*, P.C. Klingeman; R. L. Beschta; P. D. Komar; J.B. Bradley (eds), 85–114. Highlands Ranch, CO: Water Resources Publications LLC.

Ferguson, R. I., Wathen, S. J. (1998) Tracer-pebble movement along a concave river profile—virtual velocity in relation to grain size and shear stress: *Water Resources Research*, **34**(8), 2031–38.

Froehlich, W. (2003) Monitoring bedload transport using acoustic and magnetic devices. In *Erosion and Sediment Transport Measurement: Technological and Methodological Advances*, J. Bogen, T. Fergus & D. Walling (eds), 201–10. Oslo, Norway: International Association of Hydrological Sciences Publication 283.

Froehlich, W. (in press) Monitoring of bed load transport within a small drainage basin in the Polish Flysch Carpathians. In Gray J. R., Laronne J. B., and Marr, J. D. G., in press, Bedload-surrogate monitoring technologies: U.S. Geological Survey Scientific Investigations Report, in press.

Gaeuman, D. & Jacobson, R. B. (2006) Acoustic bed velocity and bedload dynamics in a large sand-bed river. *Journal of Geophysical Research*, **111**, F02005.

Gaeuman, D. & Jacobson, R. B. (2007) Field assessment of alternative bed-load transport estimators. *Journal of Hydraulic Engineering*, **133**(12), 1319–28.

Garcia, C., Laronne, J. B., Sala, M. (2000) Continuous monitoring of bedload flux in a mountain gravel-bed river. *Geomorphology*, **34**, 23–31.

Gao, H. (1991) The comparison tests of gravel bed load samplers. In *Proceedings of the 5th Federal Interagency Sedimentation Conference, Las Vegas, Nevada*, VI 55 – VI 62, http://water.usgs.gov/pubs/misc_reports/FISC_1947-2001/, accessed at August 31, 2004.

Gaudet, J. M., Roy, J. B. & Best, J. B. (1994) Effects of orientation and size of Helley–Smith sampler on its efficiency: *Journal of Hydraulic Engineering*, **120**(6), 758–66.

Gintz, D., Hassan, M. A. & Schmidt, K. H. (1996) Frequency and magnitude of bedload transport in a mountain river: *Earth Surface Processes and Landforms*, **21**, 433–45.

Glysson, G. D. (1987) Sediment-transport curves. US Geological Survey Open-File Report 87-218, 47pp., http://pubs.er.usgs.gov/usgspubs/ofr/ofr87218.

Goldsmith, W. (2001) *Impact: The Theory and Physical Behaviour of Colliding Solids*. Mineola, NY, USA: Dover Publications, Inc., **379** pp.

Gomez, B. (1991) Bedload transport. *Earth-Science Reviews*, **31**, 89–132.

Gottesfeld, A. S. & Tunnicliffe, J. (2003) Bed load measurements with a passive magnetic induction device. In *Erosion and Sediment Transport Measurement: Technological and Methodological Advances*, J. Bogen, T. Fergus & D. Walling (eds), 211–21, Oslo, Norway: International Association of Hydrological Sciences, Publication 283.

Govi, M., Maraga, F. & Moia, F. (1993) Seismic detectors for continuous bed load monitoring in a gravel stream. *Hydrological Sciences Journal*, **38**, 123–32.

Graham Matthews and Associates. (2006) Trinity River Restoration Program – Annual Sediment Transport Monitoring Report, US Bureau of Reclamation, Northern California Area Office.

Graham Matthews and Associates. (2007) Trinity River Restoration Program – Annual Sediment Transport Monitoring Report, US Bureau of Reclamation, Northern California Area Office.

Graham Matthews and Associates. (2008) Trinity River Restoration Program – Annual Sediment Transport

Monitoring Report, US Bureau of Reclamation, Northern California Area Office.

Gray, J. R. (2003) US Geological Survey suspended-sediment surrogate research, Part I: Call for a sediment monitoring instrument and analysis research program. In *Proceedings of the Virginia Water Research Symposium* (2003, Water Resource Management for the Commonwealth, Virginia Polytechnic Institute and State University, Blacksburg, October 8–10, 2003, pp.53–57.

Gray, J. R. (ed.). (2005) *Proceedings of the Federal Interagency Sediment Monitoring Instrument and Analysis*. Workshop, September 9-11 (2003, Flagstaff, Arizona: US Geological Survey Circular 1276, 46p., http://water.usgs.gov/pubs/circ/2005/circ1276/. Appendix 4 is available only online at: http://water.usgs.gov/osw/techniques/sediment/sedsurrogate2003workshop/listofpapers.html.

Gray, J. R., Glysson, G. D. & Edwards, T. E. (2008) Suspended-sediment samplers and sampling methods. In Sediment transport measurements. In Garcia, M. (ed.), *Sedimentation Engineering – Processes, Measurements, Modeling, and Practice*, American Society of Civil Engineers, Manual 110, Chapter 5.3, 320–39.

Gray, J.R., Laronne, J.B. & Marr, J.D.G. (2007) Outcomes from the International Bedload-Surrogate Monitoring Workshop, Minneapolis, Minnesota, USA, April 11–14, 2007. *American Geophysical Union, EOS*, 88(45), November 6, 2007, 471.

Gray, J. R. & Simões, F. J. M. (2008) Estimating sediment discharge. In *Sedimentation Engineering – Processes, Measurements, Modeling, and Practice*, M. Garcia (ed.), 1067–88. American Society of Civil Engineers, Manual 110, Appendix D.

Gray, J. R., Webb, R. H. & Hyndman, D. W. (1991) Low-flow sediment transport in the Colorado River. In *Proceedings of the Fifth Federal Interagency Sedimentation Conference: Interagency Advisory Committee on Water Data, Subcommittee on Sedimentation, Las Vegas*, vol. 1, pp.4–63 to 4–71, http://water.usgs.gov/pubs/misc_reports/FISC_1947-2001/.

Habersack, H. M. (2001) Radio-tracking gravel particles in a large braided river in New Zealand: a field test of the stochastic theory of bedload transport proposed by Einstein. *Hydrological Processes*, **15**, 377–91.

Habersack, H. M. (2003) Use of radio-tracking techniques in bedload transport investigations. In *Erosion and Sediment Transport Measurement: Technological and Methodological Advances*, J. Bogen, T. Fergus & D. Walling (eds), 172–80. Oslo, Norway: International Association of Hydrological Sciences, Publication 283.

Habersack, H. M. & Laronne, J. B. (2001) Bed load texture in an alpine gravel bed river: *Water Resources Research*, 37(12), 3359–70.

Habersack, H. M. & Laronne, J. B. (2002) Evaluation and improvement of bedload discharge formulas based on Helley–Smith sampling in an Alpine gravel-bed river: *Journal of Hydraulic Engineering*, **128**(5), 484–99.

Habersack, H. M., Nachtnebel, P. N. & Laronne, J. B. (2001) The continuous measurement of bedload discharge in a large alpine gravel bed river with a slot sampler. *Journal of Hydraulic Research*, **39**, 125–33.

Hamamori, A. (2003). A theoretical investigation on the fluctuations of bedload transport. Report R4, Delft Hydraulics Laboratory, The Netherlands, 21pp.

Hardardottir, J. & Snorrason, A. (2003) Sediment monitoring of glacial rivers in Iceland: new data on bed load transport. In *Erosion and Sediment Transport Measurement: Technological and Methodological Advances*, J. Bogen, T. Fergus & D. Walling (eds), 154–63. Oslo, Norway: International Association of Hydrological Sciences, Publication 283.

Harris, T. & Richards, K.S. (1995) Design and calibration of a recording bedload trap: *Earth Surface Processes and Landforms*, **20**, 711–20.

Haschenburger, J. K. & Church, M. (1998) Bed material transport estimated from the virtual velocity of sediment: *Earth Surface Processes and Landforms*, **23**, 791–808.

Hassan, M. A. (1990) Scour, fill, and burial depth of coarse material in gravel bed streams: *Earth Surface Processes and Landforms*, **15**, 341–56.

Hassan, M. A. & Church, M. (1992) The movement of individual grains on the streambed. In *Dynamics of Gravel Bed Rivers*, P. Billi, R.D. Hey, C. R. Thorne & P. Tacconi, (eds), 159–75. Chichester, UK: John Wiley.

Hassan, M. A. & Church, M. (2001) Sensitivity of bedload transport in Harris Creek – seasonal and spatial variation over a cobble-gravel bar: *Water Resources Research*, 37(3), 813–25.

Hassan, M. A. & Ergenzinger, P. (2003) Use of tracers in fluvial geomorphology. In *Tools in Fluvial Geomorphology*, G. M. Kondolf & H. Piégay, (eds), 397–423. Chichester, UK: John Wiley.

Hassan, M. A., Church, M. & Ashworth, P. J. (1992) Virtual rate and mean distance of travel of individual clasts in gravel-bed channels: *Earth Surface Processes and Landforms*, **17**, 617–27.

Hassan, M. A., Church, M. & Schick, A. P. (1991) Distance of movement of coarse particles in gravel bed streams: *Water Resources Research*, 27(4), 503–11.

Hassan, M. A., Schick, A. P. & Shaw, P. (1999) The transport of gravel in an ephemeral sandbed river: *Earth Surface Processes and Landforms*, **25**, 623–40.

Hayes, S. K., Montgomery, D. R., Newhall, C. G. (2002) Fluvial sediment transport and deposition following the 1991 eruption of Mount Pinatubo: *Geomorphology*, **45**, 211–24.

Hayward, J. A. (1980) Hydrology and stream sediments in a mountain catchment. In *Tussock Grasslands and*

Mountain Lands Institute: Special Publication 17, 236pp.

Hayward, J. A. & Sutherland, A. J. (1974) The Torlesse stream vertex-tube sediment trap. *Journal of Hydrology*, **13**(1), 41–53.

Hegg, C. & Rickenmann, D. (1998) *Short-time relations between runoff and bedload transport in a steep mountain torrent.* Oslo, Norway: International Association of Hydrological Sciences, Publication 249, pp.317–24.

Hegg, C. & Rickenmann, D. (2000) Geschiebetransport in Wildbächen – Vergleich zwischen Feldmessungen und einer Laborformel, *Interpraevent* **2000**, pp.117–29.

Helley, E. J. & Smith, W. (1971) Development and calibration of a pressure-difference bedload sampler. US Geological Survey Open-File Report, 38pp., http://pubs.er.usgs.gov/usgspubs/ofr/ofr73108.

Hinrich, H. (1970) Bedload measurement by means of bottom plates and bedload samplers with hydrophone attachments. In *Hydrometry Symposium*, International Association of Hydrological Sciences, Koblenz, pp.538–43.

Hubbell, D. W. (1964) Apparatus and techniques for measuring bedload: US Geological Survey Water Supply Paper 1748, http://onlinepubs.er.usgs.gov/lizardtech/iserv/browse?cat=WSP&item=/wsp_1748.djvu&style=simple/view-dhtml.xsl&wid=750&hei=600&props=img(Name,Description)&page=1, accessed January 26, 2005.

Hubbell, D. W., Stevens, H. H. & Beverage, J. P. (1985) New approach to calibrating bed load samplers. *Journal of Hydraulic Engineering*, **111**(4), 677–94.

Hubbell, D. W., Stevens, H. H., Skinner, J. & Beverage, J. P. (1987) Laboratory Data on Coarse-Sediment Transport for Bedload-Sampler Calibrations: US Geological Survey Water-Supply Paper 2299, http://onlinepubs.er.usgs.gov/lizardtech/iserv/browse?cat=WSP&item=/wsp2299.djvu&style=simple/view-dhtml.xsl&wid=750&hei=600&props=img(Name,Description)&page=7, accessed January 26, 2005.

ISO. (1992) Measurement of liquid flow in open channels – Method for measurement of bedload discharge. Technical report 92-9212, 20p.

Johnson, P. & Muir, T. C. (1969) Acoustic detection of sediment movement. *Journal of Hydraulic Research*, 7, 519–40.

Johnson, C. W., Engleman, R. L., Smith, J. P. & Hansen, C. L. (1977) Helley–Smith bed load samplers. *Journal of the Hydraulics Division, ASCE*, **103** (HY10), 1217–21.

Kondolf, G. M. & Matthews, W. V. G. (1986) Transport of tracer gravels on a coastal California River. *Journal of Hydrology*, **85**, 265–80.

Kuhnle, R. (2008) Bed load samplers. In *Sedimentation Engineering – Processes, Measurements, Modeling, and Practice*, M. Garcia (ed.), American Society of Civil Engineers Manual 110, Appendix D, 339–46.

Labaye, G. (1948) Note sur le débit solide des cours d'eau. *La Houille Blanche*, 3.

Laronne, J. B. & Carson, M. A. (1976) Interrelationships between bed morphology and bed material transport for a small, gravel-bed channel. *Sedimentology*, **23**, 76–85.

Laronne, J. B., Alexandrov, Y., Bergman, N., *et al.* (2003) The continuous monitoring of bedload flux in various fluvial systems. In *Erosion and Sediment Transport Measurement in Rivers—Technological and Methodological Advances*, J. Bogen, T. Fergus & D. Walling (eds), 134–45. Oslo, Norway: International Association of Hydrological Sciences, Publication 283.

Laronne, J. B., Marr, J. D. G. & Gray, J. R. (2007) Synopsis of outcomes from the International Bedload-Surrogate Monitoring Workshop, Minneapolis, Minnesota, USA, April 11–14 2007. US Forest Service Stream Notes, October 2007, pp. 5–7 (http://www.stream.fs.fed.us/news/streamnt/pdf/SN_10_07.pdf).

Laronne, J. B., Outhet, D. N., Duckham, J. L. & McCabe, T. J. (1992) Determining event bedload volumes for evaluation of potential degradation sites due to gravel extraction, N.S.W., Australia. In *Erosion and Sediment Transport Monitoring in River Basins*, 87–94. Oslo, Norway: International Association of Hydrological Sciences, Publication 210.

Lenzi, M. A., Marchi, L. & Scussel, G. R. (1990) Measurement of coarse sediment transport in a small Alpine stream. In *Hydrology in Mountainous Regions*, **1**, 223–30. Oslo, Norway: International Association of Hydrological Sciences, Publication 193.

Lenzi, M. A., D'Agostino, & Billi, P. (1999) Bedload transport in the instrumented catchment of the Rio Cordon. Part I: Analysis of bedload records, conditions and threshold of bedload entrainment. *Catena*, **36**, 171–90.

Lewis, J. (1991) An improved bedload sampler. In *Proceedings of the 5th Interagency Sedimentation Conference, Las Vegas, Nevada*, VI 1 – VI 8, http://water.usgs.gov/pubs/misc_reports/FISC_1947-2001/, accessed August 31, 2004.

Leopold, L. B. & Emmett, W. W. (1997) Bedload and river hydraulics – Inferences from the East Fork River, Wyoming. US Geological Survey Professional Paper 1583, 52pp., http://pubs.er.usgs.gov/usgspubs/pp/pp1583.

Marr, J. D. G., Gray, J. R., Davis, B. E., Ellis, C. & Johnson, S. (in press) Large-scale laboratory testing of bedload-monitoring technologies – overview of the Streamlab06 experiments. In Gray J. R., Laronne J. B., and Marr, J. D. G., in press, Bedload-surrogate monitoring technologies: U.S. Geological Survey Scientific Investigations Report, in press.

McBain & Trush. (2004) Coarse sediment management plan for the lower Tuolumne River, revised final report. Prepared for the Tuolumne River Advisory Committee, McBain and Trush, Arcata, CA, 347pp., http://www.

delta.dfg.ca.gov/afrp/Project.asp?code=2000-41, accessed January 19, 2005.

McNamara, J. P., Bordon, J. C. & Fitzgerald, F. (2001) Radio-tracking of cobbles in a mountain stream in southwest Idaho. In *Proceedings of the 7th Federal Interagency Sedimentation Conference, Reno, Nevada*, 1: p. III 63 – III 69, http://water.usgs.gov/pubs/misc_reports/FISC_1947-2001/, accessed August 31 2004.

Milhous, R. T. (1973) *Sediment transport in a gravel-bottomed stream*. Ph.D. dissertation, Oregon State University, Corvallis.

Meyer-Peter, E. & Müller, R. (1948) Formula for bedload transport. In *Proceedings of the 2nd Meeting of the IAHR, Stockholm*, pp.39–64.

Mizuyama, T., Fujita, M. & Nonaka, M. (2003) Measurement of bedload with use of hydrophone in mountain torrents. In *Erosion and Sediment Transport Measurement: Technological and Methodological Advances*, J. Bogen, T. Fergus & D. Walling (eds), 222–27. Oslo, Norway: International Association of Hydrological Sciences, Publication 283.

Mizuyama, T., Laronne, J. B., Nonaka, M., *et al.* (in press). Calibration of a passive acoustic bedload monitoring system in Japanese mountain rivers. In Gray J. R., Laronne J. B., and Marr, J. D. G., in press, Bedload-surrogate monitoring technologies: U.S. Geological Survey Scientific Investigations Report, in press.

Mizuyama, T., Oda, A., Laronne, J. B., Nonaka, M. & Matsuoka, M. (in press). Laboratory tests of a Japanese pipe hydrophone for continuous acoustic monitoring of coarse bedload. In Gray J. R., Laronne J. B., and Marr, J. D. G., in press, Bedload-surrogate monitoring technologies: U.S. Geological Survey Scientific Investigations Report, in press.

Møen, K., Bogen, J., Zuta, J. F., Ade, P. K. & Esbensen K. (in press). Bed load measurements in rivers using passive acoustic sensors. In Gray J. R., Laronne J. B., and Marr, J. D. G., in press, Bedload-surrogate monitoring technologies: U.S. Geological Survey Scientific Investigations Report, in press.

Mühlhofer, L. (1933) Untersuchungen uber der Schwebstoff und Geschiebefurung des Inn nachst Kirchbichl, Tirol. *Wasserkraft und Wasserwirtschaft*, 1–6.

Nankervis, J. (1994) Summary of fluvial study site data collection. Water Division 1 Water Rights Adjudication, Greeley, CO. Report prepared for the Stream Systems Technology Center, USDA Forest Service, Rocky Mountain Forest and Range Experiment Station, Fort Collins, Colo., 24pp.

Nanson, G. C. (1974) Bedload and suspended-load transport in a small, steep, mountain stream. *American Journal of Science*, **274**, 471–86.

Novak, P. (1957) Bed load meters – development of a new type and determination of their efficiency with the aid of scale models. In *Proc., Trans. Int. Assoc. Hyd. Res. 7th General Meeting, Lisbon*, A9-1–A9-11.

O'Leary, S. J. & Beschta, R. L. (1981) Bed load transport in an Oregon Coast Range stream. *Water Resources Bulletin*, **17**(5), 886–94.

Pitlick, J. (1988) Variability of bed load movement: *Water Resources Research*, **24**(1), 173–77.

Pittman, S. (2005) Toutle River-2 Bedload Sampler, WY2005 Technical Memo no. 1, US Bureau of Reclamation, Northern California Area Office.

Powell, D. M. & Ashworth, P. J. (1995) Spatial pattern of flow competence and bedload transport in a divided gravel bed river: *Water Resources Research*, **31**(3), 741–52.

Powell, D. M., Reid, I., Laronne, J. B. & Frostick, L. E. (1998) Cross stream variability of bedload flux in narrow and wider ephemeral channels during desert flash floods. In *Gravel-Bed Rivers in the Environment*, P.C. Klingeman, R.L. Beschta, P.D. Komar & J.B. Bradley, (eds), 177–96. Littleton, CO, USA: Water Resources Publications.

Ramooz, R. & Rennie, C. D. (in press). Laboratory measurement of bedload with an ADCP, in Bedload-Surrogate Monitoring. In Gray J. R., Laronne J. B., and Marr, J. D. G., in press, Bedload-surrogate monitoring technologies: U.S. Geological Survey Scientific Investigations Report, in press.

Reid, I. & Frostick, L. E. (1986) Dynamics of bedload transport in Turkey Brook, a coarse grained alluvial channel: *Earth Surface Processes and Landforms*, **11**, 143–55.

Reid, I. & Laronne, J. B. (1995) Bed load sediment transport in an ephemeral stream and a comparison with seasonal and perennial counterparts: *Water Resources Research*, **31**(3), 773–81.

Reid, I., Frostick, L. E. & Layman, J. T. (1985) The incidence and nature of bedload transport during flood flows in coarse-grained alluvial channels. *Earth Surface Processes and Landforms*, **10**, 33–44.

Reid, I., Layman, J. T. & Frostick, L. E. (1980) The continuous measurements of bedload discharge. *Journal of Hydraulic Research*, **18**(3), 243–49.

Rennie, C. D. & Church, M. (2007) ADCP shear stress and bedload transport in a large wandering gravel-bed river. 32nd IAHR Congress 2007, Venice, 10pp.

Rennie, C. D. & Millar, R. G. (2004) Measurement of the spatial distribution of fluvial bedload transport velocity in both sand and gravel. *Earth Surface Processes and Landforms*, **29**(10), 1173–93.

Rennie, C. D. & Millar, R. G. (2007) Deconvolution technique to separate signal from noise in gravel bedload velocity data. *Journal of Hydraulic Engineering*, **133**(8), 845–56.

Rennie, C. D. & Villard, P. (2004) Site specificity of bed load measurement using an acoustic Doppler current profiler. *Journal of Geophysical Research, (Earth Surface)*, **109**, F03003.

Rennie, C. D., Millar, R. G. & Church, M. A. (2002) Measurement of bed load velocity using an acoustic

Doppler current profiler. *Journal of Hydraulic Engineering*, **128**(5), 473–83.

Richards, K. S. & Milne, L. M. (1979) Problems in the calibration of an acoustic device for the observation of bedload transport. *Earth Surface Processes*, **4**, 335–46.

Richardson, K., Benson, I. & Carling, P.A. (2003) An instrument to record sediment movement in bedrock channels. In *Erosion and Sediment Transport Measurement: Technological and Methodological Advances*, J. Bogen, T. Fergus & D. Walling (eds), 228–35. Oslo, Norway: International Association of Hydrological Sciences, Publication 283.

Rickenmann, D. (1994) Bedload transport and discharge in the Erlenbach stream. In *Dynamics and Geomorphology of Mountain Rivers*, P. Ergenzinger & K.-H. Schmidt (eds), 53–66, *Lecture Notes in Earth Sciences*, **52** Berlin: Springer Verlag.

Rickenmann, D. (1997) Sediment transport in Swiss torrents. *Earth Surface Processes and Landforms*, **22**, 937–51.

Rickenmann, D. & Dupasquier, P. (1994) Messung des feststofftransportes im Erlenbach. [Sediment transport measurement in the Erlenbach stream]. *Beiträge zur Geologie der Schweiz – Hydrologie*, **35**, 134–44.

Rickenmann, D. & Fritschi, B. (in press). Bedload transport measurements using piezoelectric impact sensors and geophones. In Gray J. R., Laronne J. B., and Marr, J. D. G., in press, Bedload-surrogate monitoring technologies: U.S. Geological Survey Scientific Investigations Report, in press.

Rickenmann, D., Hofer, B. & Fritschi, B. (1997) Geschiebemessungen mittels Hydrophon. *Österrechische Wasser- und Abfallwirtschaft*, **49**.

Rosenfeld, C. L., Castro, J. M. & Childers, E. S. (1996) Individual gravel tracking using a passive radio transponder system. In *Proceedings of the 6th Federal Interagency Sedimentation Conference, Las Vegas, Nevada*, V47 – V51, http://water.usgs.gov/pubs/misc_reports/FISC_1947-2001/, accessed August 31, 2004.

Rouse, H. L. (1994) Measurement of bedload gravel transport: the calibration of a self-generated noise system. *Earth Surface Processes and Landforms*, **19**, 789–800.

Rubin, D. M., Tate, G. B., Topping, D. J. & Anima, R. A. (2001) Use of rotating side-scan sonar to measure bedload. In *Proceedings of the 7th Federal Interagency Sedimentation Conference, Reno, Nevada*, vol. 1, III 139 – III 143, http://water.usgs.gov/pubs/misc_reports/FISC_1947-2001/, accessed August 31, 2004.

Ryan, S. E. (1998) Sampling bedload transport in coarse-grained mountain channels using portable samplers. In *Proceedings of the Federal Interagency Workshop "Sediment Technology for the 21st Century", St. Petersburg, FL*, http://water.usgs.gov/osw/techniques/sedtech21/ryan.html, accessed September 2, 2004.

Ryan, S. E. (2001) The influence of sediment supply on rates of bedload transport: a case study of three streams on the San Juan National Forest. In *Proceedings of the 7th Federal Interagency Sedimentation Conference, Reno, Nevada*, vol. 1, III 48 – III 54, http://water.usgs.gov/pubs/misc_reports/FISC_1947-2001/, accessed August 31, 2004.

Ryan, S. E. & Dixon, M. (2002) Bedload movement in a mountain gravel-bed stream. CD-ROM Video, Technology Transfer, Stream Systems Technology Center, Fort Collins, CO.

Ryan, S. E. & Emmett, W. W. (2002) The nature of flow and sediment movement in Little Granite Creek near Bondurant, WY. USDA Forest Service, Rocky Mountain Research Station, General Technical Report RMRS-GTR-90.

Ryan, S. E. & Porth, L. S. (1999) A field comparison of three pressure-difference bedload samplers. *Geomorphology*, **30**, 307–22.

Ryan, S. E. & Troendle, C. A. (1997) Measuring bedload in coarse-grained mountain channels: procedures, problems, and recommendations. In *Water Resources Education, Training, and Practice: Opportunities for the Next Century, American Water Resources Association Conference, Keystone, CO*, 949–58.

Ryan, S. E., Bunte, K. & Potyondy, J. P. (2005) Breakout session II, bedload-transport measurements: data needs, uncertainty, and new technologies. In *Proceedings of the Federal Interagency Sediment Monitoring Instrument and Analysis, Workshop, September 9-11 2003, Flagstaff, Arizona*, J. R. Gray (ed.), US Geological Survey Circular 1276, 16–28. http://water.usgs.gov/pubs/circ/2005/circ1276.

Schmidt, K. H. & Ergenzinger, P. (1992) Bedload entrainment, travel lengths, step lengths, rest periods, studied with passive (iron, magnetic) and active (radio) tracer techniques. *Earth Surface Processes and Landforms*, **17**, 147–65.

Schmidt, K. H. & Gintz, D. (1995) Results of bedload tracer experiments in a mountain river. In *River Geomorphology*, E. J. Hickins (ed.), 37–54. Chichester, UK: John Wiley.

Sear, D. A. (1996) Sediment transport processes in pool-riffle sequences. *Earth Surface Processes and Landforms*, **21**, 241–62.

Sear, D. A. (2003) Event bed load yield measurement with load cell bed load traps and prediction of sediment yield from hydrograph shape. In *Erosion and Sediment Transport Measurement: Technological and Methodological Advances*, J. Bogen, T. Fergus & D. Walling (eds), 146–53. Oslo, Norway: International Association of Hydrological Sciences, Publication 283.

Sear, D. A., Damon, W., Booker, D. J. & Anderson, D. G. (2000) A load cell based continuous recording bedload trap. *Earth Surface Processes and Landforms*, **25**, 689–72.

Sear, D. A., Lee, M. W. E., Carling, P. A., Oakley, R. J. & Collins, M. B. (2003) An assessment of the accuracy of the Spatial Integration Method (S.I.M.) for estimating coarse bedload transport in gravel-bedded streams using passive tracers. In *Erosion and Sediment Transport Measurement: Technological and Methodological Advances*, J. Bogen, T. Fergus & D. Walling (eds), 164–71. Oslo, Norway: International Association of Hydrological Sciences, Publication 283.

Sterling, S. M. & Church, M. (2002) Sediment trapping characteristics of a pit trap and the Helley–Smith sampler in a cobble gravel-bed river. *Water Resources Research*, 38(6), doi:10.1029/2000WR000052 2002.

Tacconi, P. & Billi, P. (1987) Bed load transport measurement by a vortex-tube trap on Virginio Creek, Italy. In *Sediment Transport in Gravel-Bed Rivers*, C. R. Thorne, J. C. Bathurst & R. D. Hey (eds), 583–615. Chichester, UK: John Wiley.

Taniguchi, S., Itakura, Y., Miyamoto, K. & Kurihara, J. (1992) A new acoustic sensor for sediment discharge measurement. In *Erosion and Sediment Transport Monitoring in River Basins*, 135–42. Oslo, Norway: International Association of Hydrological Sciences, Publication 210.

Thompson, D. M., Wohl, E. E. & Jarrett, R. D. (1996) A revised velocity-reversal and sediment sorting model for a high-gradient, pool-riffle stream. *Physical Geography*, **17**(2), 142–56.

Thorne, P. D., Heathershaw, A. D. & Triano, L. (1984) Acoustic detection of seabed gravel movement in turbulent tidal currents. *Marine Geology*, **54**, M43–48.

Thorne, P. D. (1986a) An intercomparison between visual and acoustic detection of gravel movement. *Marine Geology*, **72**, 11–31.

Thorne, P. D. (1986b) Laboratory and marine measurements on the acoustic detection of sediment transport. *Journal of the Acoustical Society of America*, **80**, 899–910.

Thorne, P. D. (1987) The acoustic measurement of gravel transport. Paper presented at Fifth International Conference on Electronics for Ocean Technology, IERE, Edinburgh, UK, 24–26 March 1987.

Thorne, P. D. (1993) Seabed saltation noise. In *Natural Physical Sources of Underwater Sound*, B. R. Kerman (ed.), 721–44. Dordrecht, The Netherlands: Kluwer Academic.

Thorne, P. D. & Foden, D. J. (1988) Generation of underwater sound by colliding spheres. *Journal of the Acoustical Society of America*, **84**, 2144–52.

Thorne, P. D., Waters, K. R. & Brudner, T. J. (1995) Acoustic measurements of scattering by objects of irregular shape. *Journal of the Acoustical Society of America* **97**(1), 242–51.

Thorne, P. D., Williams, J. J. & Heathershaw, A. D. (1989) In situ acoustic measurements of marine gravel threshold and transport. *Sedimentology*, **36**, 61–74.

Troendle, C. A., Nankervis, J. M. & Ryan, S. E. (1996) Sediment transport from small, steep-gradient watersheds in Colorado and Wyoming. In *Proceedings of the 6th Federal Interagency Sedimentation Conference, Las Vegas, Nevada*, vol. 2, p. IX 39 – IX 45, http://water.usgs.gov/pubs/misc_reports/FISC_1947-2001/, accessed August 31 2004.

Tunnicliffe, J., Gottesfeld, A. S. & Mohamed, M. (2000) High resolution measurement of bedload transport. *Hydrological Processes*, **14**, 2631–43.

van Rijn, L. C. (1984) Sediment transport. Part I: Bed load transport. *J Journal of Hydraulic Engineering*, **110**(10), 1431–56.

Voulgaris, G., Wilkin, M. P. & Collins, M. B. (1995) The in situ passive acoustic measurement of shingle movement under waves and currents – instrument (TOSCA) development and preliminary results. *Continental Shelf Research*, **15**, 1195–11.

Wathen, S. J., Hoey, T. B. & Werritty, A. (1995) Unequal mobility of gravel and sand in weakly bimodal river sediments: *Water Resources Research*, **31**(8), 2087–96.

Whitaker, A. C. (1997) The initiation of coarse bed load transport in gravel bed streams. Ph.D. dissertation, University of Montana, Missoula.

Whitaker, A. C. & Potts, D. F. (1996) Validation of two threshold models for bedload initiation in an upland gravel-bed stream. In *Watershed Restoration Management – Physical, Chemical, and Biological Considerations: Proceedings of the Annual Symposium 1996*, American Water Resources Association, 85–94.

Wilcock, P. R. (2001) Toward a practical method for estimating sediment transport rates in gravel-bed rivers: *Earth Surface Processes Landforms*, **26**, 1395–408.

Xiang, Z. & Zhou, G. (1992) Measuring techniques of bed load in the Yangtze River. In *Erosion and Sediment Transport Monitoring in River Basins*, 175–80. Oslo, Norway: International Association of Hydrological Sciences, Publication 210.

Yang, X. & Gao, H. (1998) Development of AYT gravel bed-load sampler and method for bed-load measurement. In *Modelling Soil Erosion, Sediment Transport and Closely Related Hydrological Processes*, 345–52 Oslo, Norway: International Association of Hydrological Sciences, Publication 249.

3 Sediment characterization

Edson Campanhola Bortoluzzi[1], Maria Alice Santanna dos Santos[2] & Marcos Antonio Villetti[2]

[1]*Passo Fundo University, Brazil*
[2]*Federal University of Santa Maria, Brazil*

3.1 Introduction

The landscape today, with its topography of valleys and mountains, was mostly modeled by erosional process. Water, as a vector of this process, is capable of carrying materials in either suspended or dissolved forms. Over geological time, sediments impact land-surface evolution as products of the erosion of rocks and soil, as well in the formation of new materials.

On the other hand, it is human activity that promotes the production of sediments in urban areas, in mining regions, or in areas under agricultural production (Minella *et al.* 2007; Poleto 2007). Anthropogenic activities also facilitate the transport of a significant amount of various types of organic and inorganic pollutants from terrestrial to aquatic systems, namely pesticides, nutrients, heavy metals, and microorganisms (Accioly & Siqueira 2000). Thus, human activity over the land negatively affects the quality of the soil, water, and, therefore, the function of natural ecosystems (Gonçalves *et al.* 2005). The main consequences associated with excess sediments are eutrophication, siltation of lakes and rivers, high costs associated with treating potable water, and public health problems due to the presence of pathogens and pollutants.

Sediment is composed of particles that are heterogeneous in their form, size, and nature, sourced from sites with a variety of geological and pedological contexts and different soil management (Stumm 1993; Minella *et al.* 2007; Bortoluzzi & Petry 2008). According to FAO–WRB (2006), there are 31 groups of soils in the world, each with the potential for sediment production. Despite the fact that a sediment contains a range of particle sizes, its mineralogy is mainly based on silicates (Si_4^+ in a lattice). The layered silicates, called phyllosilicates, are the most common and important minerals in soils and thereby in sediments. Some structure-related properties of phyllosilicates, such as the specific surface area and the ion-exchange capacity, give rise to the different affinities between sediments and pollutants, which are responsible for the sediment sorption capacity (Schulze 1989). Thus, the characterization of sediments and an understanding of their properties aid researchers in predicting the behavior of sediments (Horowitz 1991; Lin *et al.* 2002; Citeau *et al.* 2006). Sediments' properties can be determined with relatively simple analytical techniques using well-known methodologies. However, sediment characterization should be undertaken by understanding the particles they comprise, such as their particle size distributions and mineralogy.

Careful construction of a sampling strategy and an understanding of the temporal and spatial variations in sediment concentration and makeup, as well as preparation of samples for analysis, are essential for rigorous characterization of the sediments (Bortoluzzi & Poleto 2006). Minerals and organic particles are capable of complex associations, such as aggregates of oxide-clay minerals or of microorganisms and minerals, which demonstrates this complexity and the necessity for an interdisciplinary approach (Chenu 2001; Chenu & Plante 2006).

Relating particle size information and the mineralogical nature of fine particles composing the sediment is a basic strategy to understanding their behavior and properties, and ultimately their origin (Hsieh 1984), the forms of pollutants associated with particles, and possibly prediction of their mobility

Sedimentology of Aqueous Systems, 1st edition.
Edited by Cristiano Poleto and Susanne Charlesworth.

and bioavailability (Entwistle *et al.* 2003; Citeau *et al.* 2006; Buffle 2006).

In this context, this chapter aims to (i) comment on the most important properties of fine particles in order to contextualize the methodologies used in studies of sediment mineralogy, and (ii) present some methodologies used the characterization of mineral particles making up sediments.

3.2 Behavior of particles in water

3.2.1 Colloids

Aquatic suspended particles comprise a continuous particle-size distribution, including several at submicrometer levels, such as clays, iron (Fe) and aluminum (Al) (hydro)oxides and humic substances. Particles measuring between 1 and 1000 nm in at least one direction are called colloids (Stumm 1993). Although colloidal particles are made up of many atoms or molecules, they are still too small to see using optical microscopy. Colloids are widespread in fresh surface waters, groundwaters, and interstitial soil and sediment waters. They pass through most paper filters, but can be observed by light scattering and sedimentation.

The word colloid (meaning glue-like) was coined by Thomas Graham (1805–69), a Scottish chemist, who studied diffusion through membranes separating pure water from aqueous solutions of several substances. Graham observed that most salts in solution diffused freely, but some substances, such as gelatins and Arabic gum, had low mobility in water, as well as a tendency to stick to the membrane. Graham referred to these species as colloids and he became convinced that these particles were aggregates of small molecules (Graham 1861; van Olphen 1977).

Colloidal solutions prepared from organic macromolecular substances such as natural and synthetic gums were initially classified as hydrophilic colloids, owing to the great affinity observed between the particles and water. This affinity is due to the chemical similarity between the colloid particle and the solvent, for example a hydroxyl group that can bind with water using hydrogen bonds.

With the growing knowledge of colloidal systems, it has been recognized that this kind of colloid should be better considered as a true solution of macromolecules or macro-ions. Therefore, hydrophilic colloids became known as macromolecular colloids or polyelectrolyte solutions. Their colloidal properties are a consequence of the large size of dispersed molecules compared with the size of molecules in the liquid medium (van Olphen 1977).

In contrast with the spontaneous creation of macromolecular colloids (hydrophilic), colloidal dispersions of a material such as gold in water are difficult to obtain. Hence, they are called hydrophobic colloids, meaning that the colloidal particles repel water. However, this term is misleading to some extent, because actually at least one or two monomolecular layers of water are held tightly by the particle surface, owing to adsorption forces.

Clay solutions are hydrophobic colloids that are widespread in natural waters; they are actually homogeneous dispersions of very small particles. Depending on the dimensions of the particles, they may not settle within a reasonable time; in this case they are called a colloidal solution or sol. If the particles are large enough, settling is faster and the dispersion is called a suspension. The distinction between a sol and a suspension is based on the different settling rates of the particles. Usually, particles with an equivalent spherical radius (Stokes' radius) smaller than 1 micrometer are placed in the colloidal size range. The equivalent spherical radius of a particle of any shape is obtained using Stokes' law for spherical particles to compute its settling velocity.

A remarkable difference between hydrophobic and macromolecular (hydrophilic) colloidal solutions is the way in which they are affected by the addition of salt. In the presence of small amounts of salt, hydrophobic sols flocculate, whereas macromolecular sols are rather insensitive. Several macromolecular compounds remain dissolved even in highly concentrated salt solutions.

Colloidal particles in hydrophobic sols are small enough to undergo Brownian motion, which results in collisions between particles. These collisions can cause particle aggregation. An explanation was therefore required to address the fact that some hydrophobic sols are stable for relatively long periods. The explanation came in the theory of the stability of hydrophobic sols, developed independently by four individual researchers: Derjaguin, Landau, Verwey, and Overbeek, in the middle of the 20th century. In honor of these authors, it is called the DLVO theory, according to which, interparticle forces are regarded as the sum of electrostatic repulsion and van der Waals attraction.

The electrostatic repulsion force assumes that the colloidal particles are charged, which has been established by electrophoresis, when colloidal particles move under the influence of an electric field. In clay sols, the particles move toward the positive electrode, indicating that the clay particle is negatively charged. On the other hand, ferric hydroxide sols are positively charged, because the particles move toward the negative electrode.

The charge on the colloidal particle is compensated for in solution, because the whole hydrophobic sol must be electrically neutral, like an ionic solution. The idea of the electric double layer was used to explain the internal balance of charges in a soil.

3.2.2 The electric double layer

The electric double layer can be described as a charge located on the surface of the particle and a counter-ion charge in the surrounding liquid phase. The counter-ions undergo two opposite tendencies: they are attracted by the oppositely charged surface; and they have a tendency to move away from the surface toward the bulk solution, where their concentration is lower. The two opposite tendencies (electrostatic attraction and diffusion in the other direction) result in an equilibrium distribution of the counter-ions near the surface of the particle. This theory was first introduced by Gouy in 1910 and then Chapman in 1913, in a model that considers the diffuse character of the counter-ion atmosphere and is referred to as the diffuse layer or Gouy–Chapman layer. In the diffuse layer, the non-uniform distribution of ions with the same charge as the particle is also considered, because these ions are depleted from the region near the surface owing to electrostatic repulsion.

Mathematically, the Gouy–Chapman layer uses both electrostatic repulsion and diffusion (the Poisson–Boltzmann equation) to obtain the exact distribution of positive and negative ions as a function of distance from the surface. The average electric potential is also computed: starting with a maximum value at the surface and decreasing roughly exponentially with distance from it.

The origin of the double-layer is therefore the surface charge on the particle. The question then arises as to why the particle is charged; there are two possibilities. Firstly, when the charge on the particle originates from interior crystal imperfections such as isomorphic substitution in clay minerals, the charge per unit surface area (the charge density) is a fixed quantity, or a permanent charge. The second possibility, whereby a particle does not have any interior crystal imperfections, is that the surface charge can be created by adsorption or chemical reaction of species from the solution on the reactive sites of the particle. This is called a variable charge.

The addition of an electrolyte to a stable hydrophobic sol changes the electric double layer configuration, leading to compression of the diffuse counter-ion atmosphere at the surface. The degree of compression of the double layer is proportional to the increase in electrolyte concentration. This effect is determined mostly by the concentration and valency of the counter ions, whereas the influence of co-ions is comparatively small. This phenomenon is based on the empiric Schulze–Hardy rule, which established that counter ions with higher valency are more efficient flocculating agents for hydrophobic colloids (Atkins 1994).

3.2.3 Double-layer repulsions

The thermal kinetic energy of colloidal particles in a hydrophobic sol gives rise to Brownian motion, which can bring two particles so near each other that their diffuse counter-ion atmospheres begin to overlap, causing electrostatic repulsion. The repulsive potential energy is the amount of work required to bring the particles from infinite separation to a given distance between them. From the DLVO theory, it is possible to plot the repulsive potential energy (V_R) as a function of distance, which gives a roughly exponentially decreasing value of V_R with increasing particle separation. Such plots are called potential energy curves.

The range of repulsive influence is considerably diminished by the increase in electrolyte concentration, owing to compression of the double layer.

3.2.4 van der Waals attractions

For flocculation to occur, attractive forces must overcome the double-layer repulsion. Attractive interactions are attributed to van der Waals forces, which have comparable range and magnitude to the electrostatic double-layer repulsion. Van der Waals forces can even occur between non-charged particles, owing to the attraction between mutually induced dipoles generated by charge fluctuations in the interacting atoms. Because van der Waals forces are additive, the total attraction between two particles with many atoms is the sum of all the attractive forces between every atom in one particle and every atom in the other particle. Therefore, van der Waals strength increases with the number of atoms in the interacting particles.

The attractive interaction between two particles can be described by a potential energy curve, whereby the attractive potential energy (V_A) decays with increasing particle distance, following a hyperbolic function.

In contrast with the behavior of the double-layer repulsive forces, van der Waals attraction between particles are not affected by changes in electrolyte concentration.

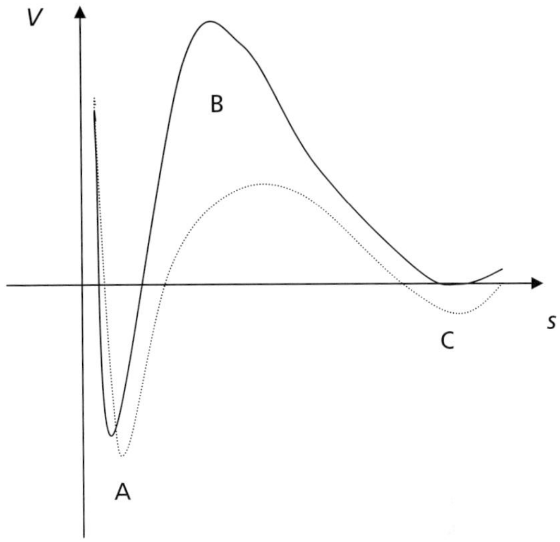

Fig. 3.1 Schematic potential energy curve for the interaction between two spherical particles separated by a distance "*s*" in a hydrophobic colloidal solution before (solid curve) and after (dotted curve) the addition of an indifferent electrolyte. Point A, coagulation; point B, energy barrier; point C, flocculation.

3.2.5 Net potential energy curve

Interparticle forces are therefore the sum of attractive and repulsive interactions. The net potential energy curve of particle interaction is obtained by adding the attractive and repulsive potential energy at each particle distance. By convention, attractive potential energies are negative and repulsive ones are positive.

However, the net potential energy curve must consider an additional repulsion force, acting at a very short range. Two kinds of contribution can be responsible for this short-range repulsion. First, there is the so-called Born repulsion, related to the resistance between the particles crystal lattices. A second short-range repulsion is the result of specific adsorption forces between the crystal surface and water molecules. Work is required to remove this water when the two interacting particles approach each other less than the thickness of the adsorbed water layers in both particles (<1 nm or 10 Å). Sharply increasing potential energy curves at very small particle separation is the consequence of this short-range repulsion.

The net potential energy curve is a result of both attractive and repulsive forces: the attraction is dominant at small and large distances, and an energy barrier may occur in between (Fig. 3.1). The colloidal solution is stable when the barrier is high enough that it cannot be overcome by the kinetic energy of the particles, namely at the minimum of the curve (point A in Fig. 3.1). This occurs at small separation distances where irreversible aggregation of the particles or coagulation occurs. On the other hand, for concentrated solutions (high ionic strength) the potential energy curve may present a secondary minimum (point C in Fig. 3.1) at large separation distances. Aggregation of particles promoted by the stabilizing effect of this minimum is called flocculation. Simple agitation can disperse the flocculated particles, because the potential energy well (point C in Fig. 3.1) is to shallow to support an aggregated state and can be easily overcome by the kinetic energy of the particles (Atkins 1994). If this happens, the system will present energy potential equal zero, which means non-interacting particles.

An indifferent electrolyte consists of ions that do not specifically adsorb to the particle surface, although they do contribute to the ionic strength of the solution. The addition of an indifferent electrolyte makes the diffuse double-layer smaller, reducing the height of the energy barrier (point B in Fig. 3.1).

As a result, the coagulation rate rises owing to an increase in collisions, resulting in aggregation of particles.

Raindrop impact on the soil surface can promote detachment of colloidal particles from soil aggregates. Most of this fine particulate matter is dispersible in water, giving rise to hydrophobic colloidal solutions that can remain stable for some time (from minutes to hundreds of years) (Seta & Karathanasis 1996). At high flow rates, particle release can be due to hydraulic shear stress on larger particles.

At low flow rates (in soil or natural subsurface water), the causes of particle release are more likely due to changes in soil solution or groundwater chemistry. For instance, particle detachment can be the result of changes in pH or ionic strength, which modify the balance of forces at the particle–grain interface (Seta & Karathanasis 1996).

In conclusion, water dispersible colloids in soil play an important role in soil erosion. Moreover, dispersed soil colloids that remain stable in subsurface moisture are potential carriers of contaminants to groundwater because they are abundant in the subsurface. The surface sites of these particles can bind contaminants with low solubility, such as radionuclides and hydrophobic organic compounds by aqueous-phase transport models (Czigány *et al.* 2005).

Factors such as pH and salinity are important in the dispersibility of natural particles. A critical salt concentration (CSC) can be experimentally determined, meaning the salt concentration below which fine particles are released from the matrix surface (Nowicki & Nowicka 1991; Blume *et al.* 2005). According to the DLVO theory, this situation can be achieved at low indifferent electrolyte concentration, when repulsive forces between colloidal particles and the matrix surface surpass binding forces, stabilizing the colloidal dispersion.

Blume *et al.* (2005) show the importance of determining the CSC in understanding the behavior of a highly radioactive waste deposit, which had been leaking into the vadose zone of the Handford Formation (USA) for several years. The hypersaline waste solution (>5M Na^+) included radionuclides and other toxic metals. According to the authors, migration of this solution from the leaking tanks through the soil and into the vadose zone had been accompanied by substantial dilution. The decrease in salt concentration is important because it gave rise to detachment of the fine particles from the sediments, enhancing radionuclide migration or reducing the permeability of the formation (owing to the settling of these colloidal particles within fine layers).

In addition, pH plays a role in the dispersibility of natural particles, because part of the variable charge on fine particles is due to surface reactive sites that are dependent on pH, such as carboxylic or phenolic groups in humic acids, hydroxilic groups in clays or Fe and Al oxyhydroxides. The zero point of charge of the particle (pH_{zpc}) represents the pH where it has no net charge; at pHs lower than the pH_{zpc} the particle would be positively charged, owing to the addition of protons to some of the reactive surface groups. Above the pH_{zpc}, the loss of protons at these sites makes the particle negatively charged. Therefore, when the soil pH is different from the pH_{zpc}, the net surface charge of the particle will be either positive or negative. Similar particles have the same charge, giving rise to double-layer repulsive forces between them, enhancing their dispersibility. At pHs near the zero point of charge, van der Waals attractive forces prevail and the colloidal solution is no longer stable; that is, water-dispersible particles are at a minimum.

3.2.6 Organic and inorganic carbon

The carbon content in soils, sediments, and natural particulates can be present as inorganic or organic forms; the former is largely found in carbonate minerals, whereas the latter is present in organic matter. In most cases, inorganic carbon in river sediments is found as calcite ($CaCO_3$) and dolomite ($CaCO_3 \cdot MgCO_3$), derived from sedimentary rocks. Sometimes other forms of carbonate, such as siderite ($FeCO_3$), are also present (Galy *et al.* 2007). In soils derived from calcareous parent material under arid conditions, the inorganic C concentration can be higher than organic carbon (Nelson & Sommers 1996). In marine sediments, significant amounts of carbonates can be present, occurring mostly as calcite and aragonite (anhydrous $CaCO_3$) from organisms such as molluscs, and dolomite, incorporated into the sediments from weathered soil parent materials and transported to the sea by river flow (Schubert & Nielsen 2000).

The organic carbon content of sediments and soils is from animal and plant residues at different stages of decomposition, including stable humic substances, and elemental forms of carbon in highly carbonized compounds such as charcoal, coal and graphite (Nelson & Sommers 1996).

The organic matter in marine sediments is a significant source of nutrients for benthic organisms, and the study of organic enrichment (natural or anthropogenic) is important to evaluate disturbance in the benthos (Luczak *et al.* 1997). Furthermore, in estimating the global carbon budget it is fundamental to measure organic carbon correctly in sediments (Byers *et al.* 1978). Another important goal of marine sedimentological study is to distinguish between marine and terrestrial organic matter contributions, which can be achieved using carbon isotope analysis. The source of carbon for photosynthesis in marine phytoplankton is bicarbonate dissolved in seawater, whereas land plants take carbon from atmospheric carbon dioxide (Schubert & Nielsen 2000). The stable carbon isotope ratio $^{13}C/^{12}C$ is represented by δ (Coplen 1996):

$$\delta^{13}C(‰) = [(R_{sample}/R_{standard}) - 1] \times 1000 \qquad (1)$$

Bicarbonate in seawater has a $\delta^{13}C$ value of 0‰. VPDB (Vienna PeeDee Belemnite), the reference standard for $\delta^{13}C$, and atmospheric carbon dioxide have $\delta^{13}C$ values of −7‰ (Schubert & Nielsen 2000; Dickens *et al.* 2006).

Particulate inorganic carbon (C_{inorg}) and particulate organic carbon (C_{org}) have distinct isotopic signatures (Lorrain *et al.* 2003). Therefore it is necessary to remove carbonates from organic carbon samples to avoid contamination of the isotopic signal. The procedure used to remove this carbonate should seek to preserve not only the organic matter content but also its composition. The elucidation of the molecular structure of organic matter in sediments allows some assessment of its age or origin, e.g. lignin is an indicator of terrestrial origin.

Differentiation between C_{org} and C_{inorg} is usually based on preliminary decarbonation performed by leaching the sample with acetic or hydrochloric acid (Galy *et al.* 2007). Acid dissolution of carbonates can be summarized in the reactions below, for calcite and dolomite minerals (Loeppert & Suarez 1996):

$$CaCO_3 + 2H^+ \rightarrow Ca^{2+} + CO_2 + H_2O$$

$$CaMg(CO_3)_2 + 4H^+ \rightarrow Ca^{2+} + Mg^{2+} + 2CO_2 + 2H_2O$$

In soils, quantification of carbonates based on these reactions can be obtained either by analysis of the amount of reactant H^+ consumed or the products Ca^{2+} and Mg^{2+} or CO_2 released.

In studies of the organic carbon in sediments and aquatic particulates, determination of total organic carbon content must be preceded by the removal of carbonates. There are several methods involving dissolution of carbonates by acid-treatment, which causes the release of carbonate carbon as carbon dioxide (Schubert & Nielsen 2000). Ideally, the removal of carbonate should be complete and the amount and composition of organic carbon preserved. However, it may be impossible to achieve both conditions simultaneously. Liquid acidification with HCl is commonly used, although it may be accompanied by solubilization of a significant portion of the organic carbon (Galy *et al.* 2007). Another method was proposed to replace the liquid acidification based on vapor acidification (Hedges & Stern 1984). According to Galy *et al.* (2007), the vapor acidification method appears to work for calcite-rich sediments, but it was not tested for sediments containing appreciable quantities of dolomite or siderite. Because the acid digestion of calcite is much faster than that of dolomite, it is unlikely that vapor acidification is an efficient method for the dissolution of dolomite.

Working on the determination of total organic carbon (TOC) content and $\delta^{13}C$ in carbonate-rich detrital sediments, Galy *et al.* (2007) proposed an analytical procedure based on liquid acidification that completely digests both calcite and dolomite, while the amount of C_{org} solubilized during acid leaching is estimated. After carbonate removal, the acid insoluble C_{org} present in the leached sediment is measured by combustion in an element analyzer. The amount of acid soluble C_{org} remaining in the leachate is determined using a TOC analyzer. Galy *et al.* (2007) studied river sediments from the Himalaya–Bengal Fan system, where the amount of calcite and dolomite (from weathering of ancient sedimentary Himalayan rocks) can be as high as 50% and the TOC is generally low (<1%). They also measured the C_{org} content, before and after acid leaching in sediments from the Amazon River, where carbonate is absent. Comparison of sediments from the Himalayan and Amazonian Rivers showed that the proportion of C_{org} solubilized during acid leaching was relatively

factor, $\rho = R_g/R_h$, was used to give an indication of how far the particles deviated from a spherical shape and R_g was used as a control to ensure undisturbed fractionation. Baalousha *et al.* (2005b) demonstrated that FlFFF–SLS coupling was a valuable method to fractionate and characterize particle size in river colloids. Ledin *et al.* (1995) found that seasonality affected size distribution of colloidal matter in a Swedish lake using DLS, with the smallest between 120 and 340 nm in spring, and between 280 and 700 nm during summer and fall. Li *et al.* (2007) verified that sediments were disturbed by different wind velocities, especially in shallow lakes. The three-dimensional fractal expression (D_f) of resuspended sediment particles was measured by light-scattering techniques because $I(\theta)\ \alpha\ q^{-Df}$. D_f was between 2.26 and 2.44 at different depths and under various wind velocities. Fractal geometry is a well-established means of describing the complicated structure of aggregates in colloidal suspensions.

3.3.2 Identification of minerals by X-ray diffraction

X-ray diffraction (XRD) is one of the main methods to identify minerals and has been in use since the 1960s. Owing to the discovery of X-ray emissions in 1895 and the discovery of their diffraction patterns through various materials in 1912, techniques and equipment to study minerals have been developed. It is fundamental to know the crystalline structure of minerals and their behavior when exposed to X-rays. Thus, mineral species can be identified in a heterogeneous sample from X-ray diffractograms after several tests, for instance the expansivity and collapse of interlayers in minerals (Dixon & Weed 1989; Bouchet *et al.* 2000).

This section briefly presents the background to identification of fine material such as clays, which at less than 2 µm are the main fraction studied as they are the most reactive.

Silicate minerals are predominantly found at the surface of the Earth, and contain silica (Si^{4+}) in the crystal lattice. However, other cations, as Al^{3+}, Mg^{2+}, Fe^{2+}, and Fe^{3+} are part of the lattice in minerals and coordinate oxygen atoms, O^{2-} or OH^- (Schulze 1989). The coordination number depends on the valency of the cation and its ionic radius, following Pauling's law (Pauling 1929, 1947). Assuming that ions are spherical, the ionic radii of cations are lower than O and OH, so the arrangement of ions within a crystalline structure is ruled by O and OH in compact or hexagonal planes with alternating cations. According to the arrangement of compact or hexagonal planes face to face, two polyhedral forms are found: the tetrahedron and octahedron.

The cation in the middle of the polyhedron can coordinate four oxygen ions in the tetrahedron, six or eight in an octahedron, and 12 when outside the polyhedron, in the interlayer. The ratio between the cationic radius, *x*, and the oxygen radius, *o*, (R_x/R_o) determines the cation inside the polyhedron, as reported in Dixon & Weed (1989). Therefore, ratios less than 0.41, such as Si^{4+} and Al^{3+}, determine that these cations can coordinate four molecules of oxygen inside the tetrahedron; ratios between 0.41 and 0.73, such as Al^{3+}, Mg^{2+}, Fe^{3+}, and Fe^{2+}, determine the possible cations inside tetrahedrons. Cations with values higher than 0.73, for example K, can only be located in the interlayer. When inside polyhedrons, cations with different valencies create a deficit of positive charge.

The formation of a mineral crystal structure depends on the organization of successive ionic layers or sheets, namely tetrahedrons or octahedrons. The combination of these layers results in 1:1 and 2:1 layers or even 2:1 layers with one more octahedron layer, resulting in chlorite, for example. The successive arrangement of these layers forms interlayer spaces of H^+ bonds with a large amount of energy. This bond occurs in the union between tetrahedron basal oxygen and the OH^- of the octahedron in phyllosilicates 1:1, resulting in a non-expanded interlayered rigid structure. On the other hand, when octahedron basal oxygen atoms are face to face bonded by van der Waals forces in a 2:1 structure, it can expand in response, for example, to water content. However, when non-hydrating K ions occupy the spaces between layers, or siloxane cavities, they are bonded to permanent charges in the tetrahedron layers, for example micas, and expansion does not occur.

The identification of the clay mineral species is based on the sequence of ionic planes that are part of the structure, and are represented by the Miller index (*hkl*) (Brindley & Brown 1980; Bouchet *et al.* 2000). Plane 001 refers to the basal distance of clay minerals in orientation (c).

Phyllosilicate mineral nomenclature is based on its structure, the presence of cations and interlayer expansion capability, with differentiation obtained mainly from basal distance (c) (Brindley & Brown 1980; Caillère *et al.* 1982). For instance, 1:1 clay mineral structure, such as kaolinite, has a 001 invariable basal distance of about 7 Å, despite many treatments (Churchman *et al.* 1984). For 2:1 clay minerals, as in micas, the 001 structure has an invariable basal distance of about 10 Å with solvation and heating treatments (Moore & Reynolds 1997); whereas vermiculites and smectites vary in their basal distance 001 at about 14 Å, depending on hydration in the presence of cations in the interlayers (Robert & Tessier 1974).

The formation of X-rays is based on the idea that they are produced when electrically charged particles are suddenly stopped (Schulze 1989). When these particles are electrons, they can be accelerated, and are halted by collision with other electrons (anticathodes), and produce X-rays. The electrons become excited because of the collision, and can pass from K to L layers or from K to M layers, after the emission of photons called K_α and K_β, respectively. The most commonly found anti-cathodes are Fe with wavelength K_α = 1,935 Å, Cu with K_α = 1,540 Å, and Co with K_α = 1,788 Å. The emission of K_β energy is filtered using monochromators.

When interacting with crystals, X-rays are subjected to diffraction, reflection, and refraction, among others. Bragg's law relates the position of diffracted peaks between atomic planes in the crystal (Schulze 1989). If the planes are coherent to X-ray diffraction, then high-intensity peaks in a region corresponding to the inclination angle of to the X-ray emitter will be produced. Bragg's law is expressed as:

$$d/n = \lambda/(2\sin\theta) \quad (9)$$

where λ is the wavelength of the X-ray beam (in ångströms); θ is the angle of incidence on atomic planes (in degrees); n is an integer determined by the order given; d is the spacing between the planes in the atomic lattice (in ångströms).

At a certain inclination angle, X-rays incident on coherent planes will reflect Bragg's law and vibrate in phase amplifying the resulting emissions, namely a DRX peak. Published reference X-ray diagrams of clay minerals with known basal distances enable a comparison to be made with the unknown sample and hence its identification (Brindley & Brown 1980). However, for samples with a particle diameter less than 50 μm, all ionic planes will be represented on the X-ray diagram and may result in interference. In this case, if the objectives of the study are phyllosilicates (plate form) the sample must be orientated so the other 00l planes are more likely to interact with the X-rays (Robert & Tessier 1974), for example by drying the clay suspension over a glass blade.

Thus, X-rays rely on sample pretreatments as shown in Fig. 3.2. Variation in basal distance by means of various treatments, such as saturation with ethylene glycol, formamide, and heating at 200, 300, and 550 °C, assists in the production of X-ray diagrams, and the identification of the mineral species required (Brindley & Brown 1980).

Figure 3.2 shows that basal distances can be altered owing to the opening or collapse of minerals interlayers. Such behavior is a diagnostic characteristic in the identification of clay mineral species (Brindley & Brown 1980). Fig. 3.3 presents an X-ray diagram of the less than 2 μm fraction of fluvial sediment from a watershed in Rio Grande do Sul, Brazil (Bortoluzzi 2004) after various pretreatments.

There are intense peaks between 2 and 16° of 2θ at room temperature (trace N), which indicates that smectite (15 Å) and kaolinite (7.2 Å) are present. After treatment at 200 and 550 °C, it is mainly smectite that reacts. The identification of clay minerals in this way is reasonably easy; however, there is no information on the ratio between species identified. Thus, posttreatment of X-ray diagrams with the aid of mathematical models, such as deconvolution analysis (Lanson 1997), or simulation of interstratified clay minerals can be used (Reynolds & Reynolds 1996). Posttreatment has been used in mineralogical studies for semi-quantitative analyses (Inoue *et al.* 1989; Lanson & Benson 1992; Moore & Reynolds 1997; Bortoluzzi 2004; Bortoluzzi *et al.* 2005), so that as well as being able to identify the clay minerals, their relative ratios can also be elucidated.

However, sediment particles not only have size and form but also varied mineralogy, according their source material (Hsieh 1984). Characterization of sediment using XRD requires careful sample preparation, knowledge of the minerals present and grain size distribution, as well as consideration of posttreatments.

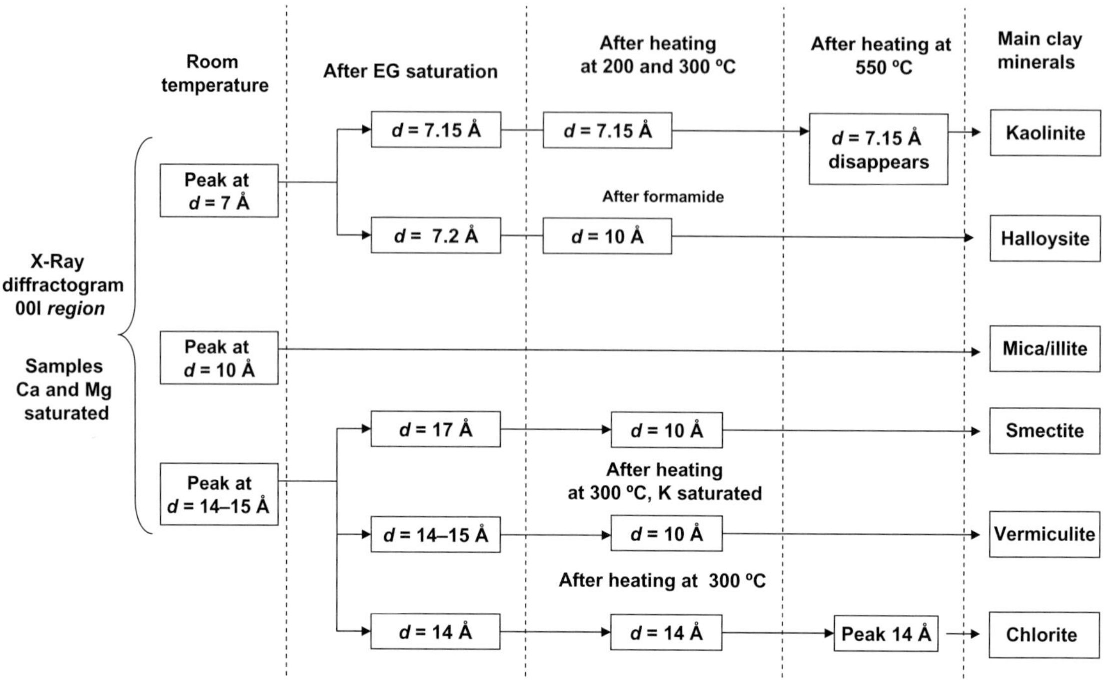

Fig. 3.2 Schema showing the position of peaks of the most common clay minerals in X-ray diagrams after pretreatments. From Bortoluzzi & Poleto (2006).

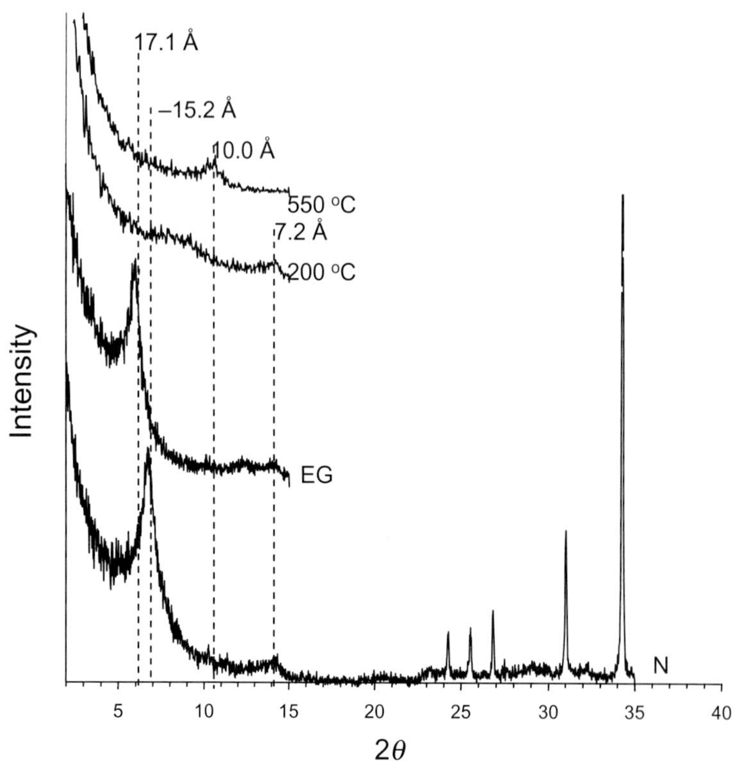

Fig. 3.3 X-ray diagrams of the less than 2 μm fraction in an oriented deposit under different treatments (N is at room temperature; EG is with ethylene glycol solution; heating at 200 and 550 °C). Samples of fluvial sediments in a watershed in southern Brazil. Adapted from Bortoluzzi (2004).

In a pretreatment, choosing the grain size to be analyzed is essential. Eliminating the organic fraction, Fe oxides and carbonates affects the quality of diagrams produced for sample orientation and the amount of background noise. Choosing the cations or organic molecules to saturate the particle charge or interlayers is fundamental to sample identification, in addition to treatment with heat which can differentiate the behavior of mineral species in a heterogeneous sample (Kunze & Dixon 1986). Figure 3.4 therefore proposes a sequence of pretreatments for studies of sediments in which treatments can be applied, according to the aim of the study and the condition of the sample.

The mineralogical characterization of sediment particles allows identification of the clay mineral species (Brindley & Brown 1980), the intrinsic chemical and physical properties of each species, as well as the relative ratio of the species in the sample (Hughes *et al.* 1994). This knowledge is fundamental in studies whose aim is to determine the chemical behavior of the sediments and potential pollutant transport.

3.3.3 Electron microscopy

Electron microscopy is used as a means of individual particle characterization (Elsass & Flores-Velez 1999) as well as of the elements associated with the particles (Lee & Fittrick 1984). The resolution of a scanning electron microscope (SEM) is 10 nm, whereas a transmission electron microscope (TEM) reaches 0.2 nm.

The applications of SEM include observations on mineral particles and living material (Castro 2002). In SEM, electron beams collide with the surface of the sample, which has been covered with carbon or gold, and the topography of the sample can be constructed by the interaction of the electrons with the particles in the sample. Depressions in the micro-

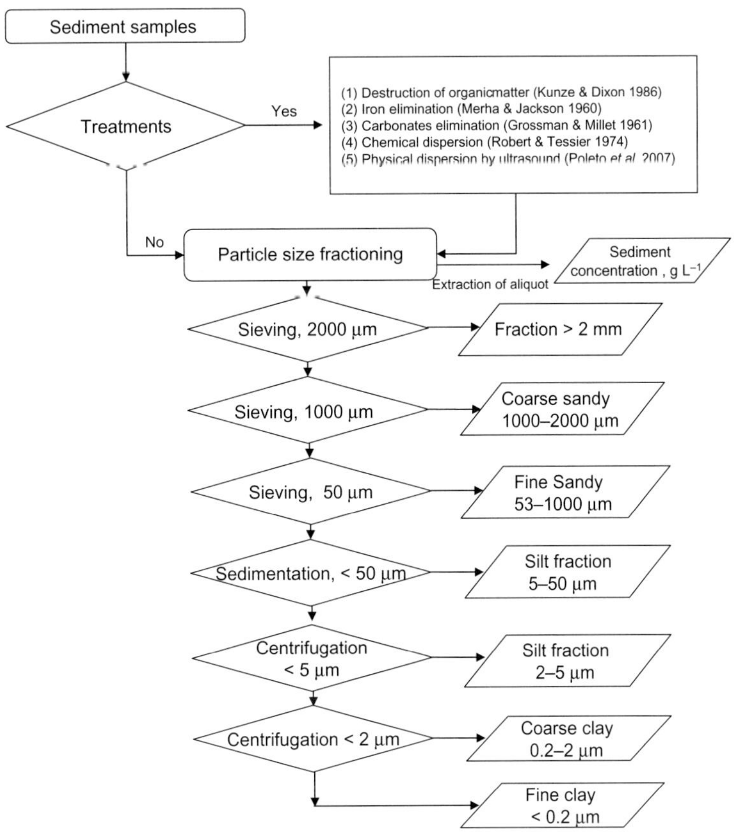

Fig. 3.4 Flowchart showing a sequence of pretreatments. From Bortoluzzi & Poleto (2006).

photography are in darker shades, whereas elevations are in lighter ones. The main advantage of using this method is that it enables the study of sample morphology; Figure 3.5 shows SEM images of two clays (<2 μm), illite and smectite, from fluvial sediment in a watershed in southern Brazil.

TEM observations are used for individual mineral particle and ionic species associated with fine fractions in sediments. Smaller particle sizes can be studied with TEM owing to its better resolution than SEM. The details of particle texture, as well as their crystalline structure, can be studied in detail with this technique. However, sample preparation is highly complex, including granulometric separation, ionic and molecular saturation, dilutions, dispersions, dehydrations, impregnations, inclusion in resin, drying, sectioning, and deposition on grids (Elsass *et al.* 2008). Basically, two forms of sample preparation are possible: ultrathin sectioning and deposits.

Fig. 3.5 Two clay mineral images obtained by scanning electron microscopy from the fine clay fraction of sediments in a watershed, southern Brazil (upper, illite; bottom, smectite). Photographs obtained by a JEOL® SEM apparatus. From Bortoluzzi *et al.* (2006).

In TEM, parallel electron beams pass through a set of objective lenses and illuminate the sample, the beams then spread out owing to interaction with it. The primary image approximates the inverse Fourier transform of the diffraction pattern and is subsequently magnified by additional lenses to form the final image.

Using both SEM and TEM, low- or high-resolution cameras capture digital images, which are subsequently treated using computer programs. This procedure assures a high turnaround of image analysis, so that information is obtained quickly, such as particle dimension, structure, and interlayer spaces. In Fig. 3.6, TEM micrographs of clays from subtropical soils are presented where the morphology of the clay mineral particles inside the resin can be seen as well as individual particles.

The chemical composition of particles can also be obtained using X-ray emission electronic microscopy done in association with TEM and SEM. In both techniques, one region of the image representing many or only one particle can be selected by microsound, where operating conditions, such as X-ray intensity and observing time, are also controlled (Elsass & Flores-Velez 1999). The image and elemental composition are obtained and analyzed simultaneously, and computer software further enables the relation between structural and absorbed elements to be explored (Dur *et al.* 2004). From a practical viewpoint, this information is valuable in studies of pollution and pollutant transport (Citeau *et al.* 2006).

3.4 Nuclear magnetic resonance

3.4.1 Basic theory of nuclear magnetic resonance

Although most of the chemical properties of the atoms are related only to the electrons that surround the nucleus, there are some characteristics of the

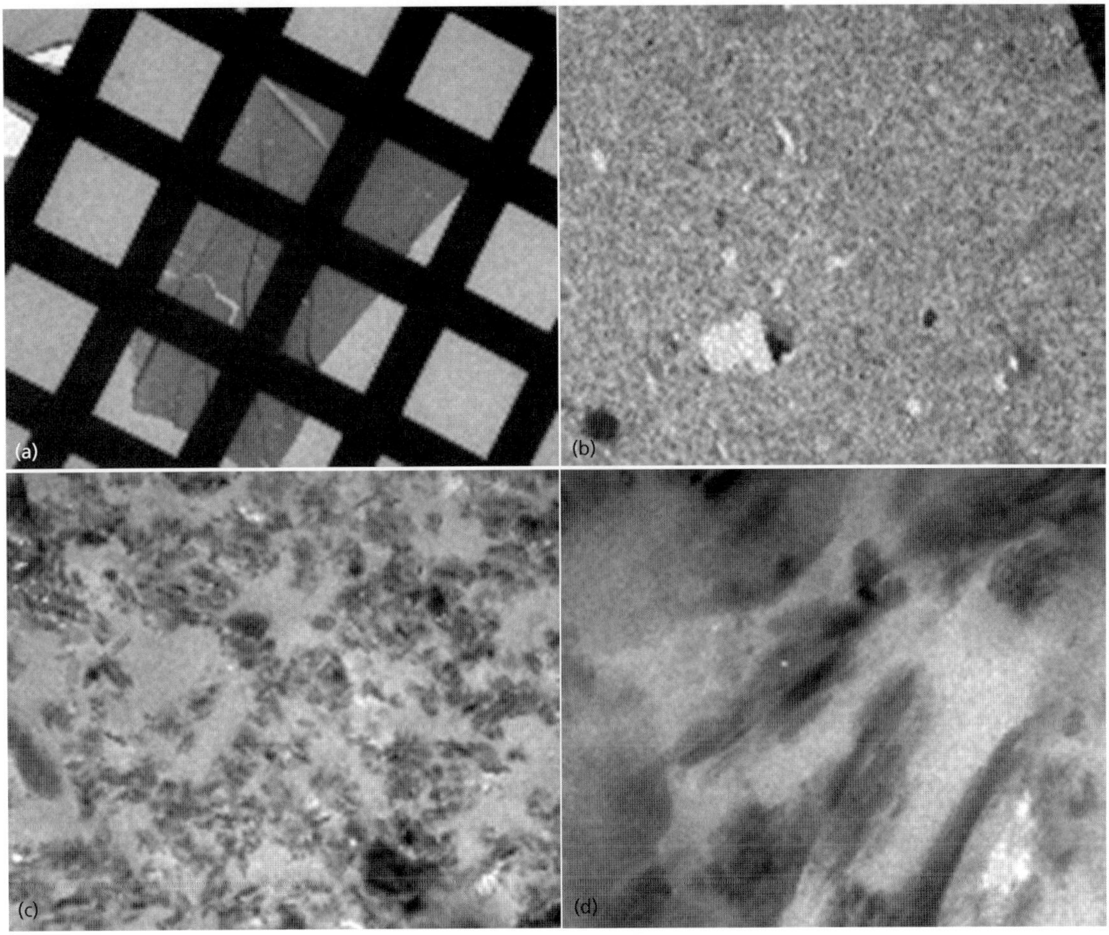

Fig. 3.6 Transmission electron micrographs of clay in soil samples, prepared with 70-nm thickness are shown at different magnifications: (a) ×3300, (b) ×10500; (c) ×32000; (e, d) ×110,000. From Bortoluzzi (2003), unpublished data.

nucleus itself which are important to chemistry, such as the magnetic properties of the nucleus, which are the basis for nuclear magnetic resonance (NMR) spectroscopy.

All atomic nuclei have charge, but not all of them are magnetically active and thus accessible to NMR spectroscopy. The magnetic activity of the nucleus is due to the charge of the nucleus flowing about a loop around a rotation axis, creating a magnetic dipole. Consequently, the nucleus behaves like a small magnet. The fundamental property of the atomic nucleus involved in NMR spectroscopy is nuclear spin. According to quantum mechanics, the angular momentum of the moving nuclear charge can be described in terms of a quantized spin number (I), which can have values of 0, 1/2, 1, 3/2, etc., depending on the nucleus under consideration.

In those nuclei that have no angular momentum (I = 0) it is not possible to induce an NMR signal; this is the case for ^{12}C, ^{16}O, and ^{32}S. Although these nuclei do not have spin (i.e. no associated magnetic moment) they are free to rotate in the classical sense, forming a current loop. However, the quantum mechanical concept of spin is different for classical rotation of charged nuclei. The particles that make up the nucleus (neutrons and protons) are called nucleons; as with electrons in atoms, nucleons possess an intrinsic spin. Nucleons of opposite spin can pair, in a similar manner as electrons do. However, only nucleons of the same type can be paired: that is, protons with protons, and neutrons with neutrons. Thus, if a nucleus contains even numbers of both protons and neutrons, all the spins are paired and then $I = 0$. The nuclear spin is non-zero when there are unpaired nucleons, which happens when there are odd numbers of either protons or neutrons, or when there are odd number of both protons and neutrons (Akitt 1983).

To be magnetically active, therefore (and accessible to NMR spectroscopy) a nucleus must have $I > 0$. That happens when the nucleus has either an odd number of nucleons (protons + neutrons = mass number) or an odd number of protons (atomic number). Fortunately, among the magnetically active nuclei, there are several nuclides important to the environmental chemistry of natural particles, including ^{1}H, ^{2}H, ^{13}C, ^{14}N, ^{15}N, ^{27}Al, ^{29}Si, and ^{31}P (Sposito 2004).

A nucleus with $I > 0$ has an associated magnetic moment μ, which is directly proportional to the spin number, I:

$$\mu = \gamma.I.h/(2\pi), \qquad (10)$$

where: h is the Planck constant and γ is the magnetogyric ratio which has a characteristic value for each magnetically active nucleus.

Because magnetically active nuclei act as small magnets, it can be expected that the application of a magnetic field will affect the behavior of molecules containing these kinds of nuclei. Indeed, in an external magnetic field, the nucleus undergoes precessional motion, which is the motion of a spinning body whose axis of rotation is constantly changing orientation. The spinning axis of the precessing nucleus describes a cone around the direction of the external magnetic field $\mathbf{B}_o$ (Fig. 3.7), but only certain orientations of the molecule with respect to this axis are allowed by quantum rules. The number of allowed orientations (magnetic energy states) is given by the formula: $2I + 1$. Thus, for nuclei such as ^{1}H, ^{13}C, ^{15}N, and ^{31}P, with a spin of ½, there are two possible states in the presence of an external magnetic field, each with a slightly different energy. In the absence of an external magnetic field, these two spin states are degenerate; That is, they have the same energy. When placed in the magnetic field $\mathbf{B}_o$, precessing nuclei with spin ½ may have their magnetic moments aligned with the field, in the lower energy state (α or +1/2), or aligned against the field

Fig. 3.7 Precessional orbit (dotted line) of the nucleus spin axis around the direction of the magnetic field $\mathbf{B}_o$.

at the higher energy level (β or −1/2) (McGregor 1997; Silverstein *et al.* 2005). The difference between these energy levels (ΔE) is proportional to the strength of $\mathbf{B}_o$ and the magnetogyric ration γ:

$$\Delta E = \gamma.h.B_o/(2\pi) \qquad (11)$$

To bring about the transition between these energy levels, radiation with energy equal to ΔE must be applied. For a given nuclear isotope, the transition occurs at a single frequency, because all the energy separations are equal and, by the selection rules of quantum mechanics, transitions are only allowed between adjacent levels. The frequency, υ, is obtained from the Planck relation, namely:

$$\Delta E = h.\upsilon \qquad (12)$$

where:

$$\upsilon = \gamma.B_o/(2\pi) \qquad (13)$$

Combining these two equations, one obtains the fundamental resonance condition for all NMR experiments:

$$\Delta E = h.\gamma.B_o/(2\pi) \qquad (14)$$

The transition between two energy states can be achieved for each element at a precise frequency of electromagnetic radiation, called the resonance frequency. For practical purposes, the difference in energy levels is small, corresponding to radiation in the radiofrequency region of the electromagnetic spectrum. NMR absorbs photon energy equal to the difference between these levels, causing a transition from a lower to a higher energy state. The resonance frequency depends only upon the applied magnetic field and the nature of the nucleus. NMR allows the identification of different elements in a sample because the resonance frequency differs for different nuclei (Abraham & Loftus 1985; Wilson 1987). The main application of NMR is as a technique for chemical analysis and structure determination known as NMR spectroscopy.

In an NMR experiment, nuclei are excited by a radio-frequency pulse and the excited nuclei undergo a relaxation process, whereupon they return to their ground state. While the excited nuclei relax back to equilibrium, the emitted energy is recorded as a peak after Fourier transformation. This excitation–relaxation cycle is repeated until a clear spectrum is obtained. There are two principal types of relaxation processes: spin–lattice relaxation and spin–spin relaxation (Pavia *et al.* 2001; McDowell *et al.* 2006).

Spin–lattice, or longitudinal, relaxation processes occur in the direction of the field. The spins transfer their energy to their surroundings – the lattice – as thermal energy. The rate of this process is related to the spin–lattice relaxation time, T_1. Intramolecular and intermolecular processes contribute to spin–lattice relaxation, but the principal contributor is *dipole-dipole interaction*, where the excited nuclei relax by exchanging energy with other magnetic nuclei that are in the same molecule or in nearby molecules. For carbon nuclei, this process is especially successful if there are hydrogen atoms nearby. The relaxation of the excited carbon nuclei is fastest if hydrogen atoms are directly bonded, as in CH, CH_2, and CH_3 groups. Spin–spin, or transverse, relaxation processes occur only between nuclei of the same type, in a plane perpendicular to the direction of the field. The rate of this process is related to the spin–spin relaxation time, T_2. Spin–spin relaxation is often described as an entropy process; it does not change the energy of the spin system (Pavia *et al.* 2001).

In solutions, spin–lattice processes dominate, whereas spin–spin relaxation is negligible (Pavia *et al.* 2001; McDowell *et al.* 2006).

3.4.2 The chemical shift

The nucleus is sensitive to the effects of small magnetic fields in its local molecular environment. The magnetic field generated by circulating neighbouring electrons affects the nucleus and may either oppose or enhance the much larger external field $\mathbf{B}_o$. When this local molecular magnetic field opposes $\mathbf{B}_o$, reducing its magnitude, the nucleus is shielded from the full effect of $\mathbf{B}_0$. A shielded nucleus feels a lower effective field strength and resonates at a lower frequency (Fig. 3.8). The opposite phenomenon, called deshielding, occurs, for example in the benzene molecule, where the moving electrons in the π orbitals give rise to a magnetic field in the hydrogen nuclei which reinforces the $\mathbf{B}_o$ field (McGregor 1997; Skoog *et al.* 1998).

In a molecule, the electron density around each nucleus changes for different types of nuclei and bonds. The opposing field and therefore the effective

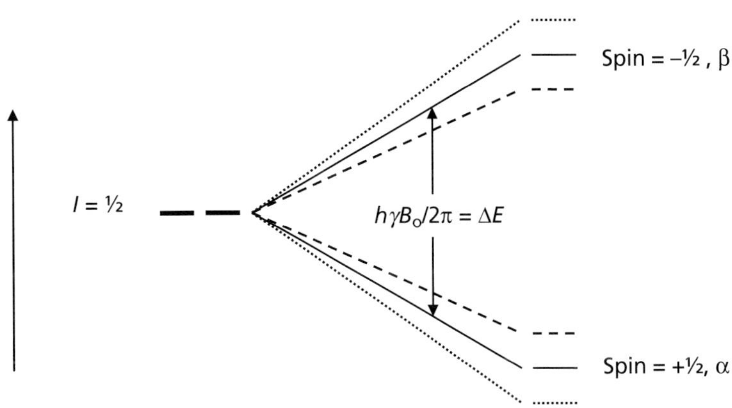

Fig. 3.8 Spin energy level diagram for a nucleus with spin = ½ which is brought into a magnetic field $\mathbf{B}_o$, showing the shielded (dashed line) and deshielded (dotted line) cases.

field for each nucleus will vary. As a result, each nucleus in an atom may have a slightly different resonance frequency. This is called the chemical shift phenomenon. Thus, the two kinds of carbon in an ethanol molecule (CH_3CH_2OH) differ because the CH_3 and CH_2 carbons have different chemical environments and therefore resonate at different frequencies. This frequency difference increases with increasing strength of the magnetic field;, consequently, it is difficult to compare NMR spectra taken on spectrometers operating at different field strengths. To overcome this problem, it is desirable to have a parameter that is independent of the magnetic field to be able to use different machines. This parameter is the chemical shift (δ), defined as the difference between the resonant frequency (υ) of a nucleus in one type of chemical environment and that of a reference nucleus (υ_{ref}), divided by the spectrometer frequency. This dimensionless quantity is expressed in parts per million (ppm) because the frequency of the spectrometer is usually in the megahertz range, whereas the chemical shift range is in the hertz or kilohertz range (McGregor 1997),

$$\delta\,(\text{ppm}) = (\upsilon - \upsilon_{ref}) \times 10^6 / \upsilon_{spectrometer}\,(\text{Hz}) \qquad (15)$$

The chemical shift is a molecular parameter that is dependent only on sample conditions (solvent, concentration, temperature) and not the spectrometer frequency.

In ^{1}H, ^{13}C, and ^{29}Si NMR spectroscopy, the reference standard is often tetramethysilane, $Si(CH_3)_4$, abbreviated to TMS (Wilson 1987). TMS is inert, readily soluble in most organic liquids, and its hydrogen atoms are more shielded than almost all other hydrogen atoms in organic compounds, providing agreed-upon chemical shift scales for all spectrometers. In addition, TMS is easily removed from samples by distillation (boiling point 27 °C). Moreover, TMS is a symmetric molecule, as all the protons are identical, so it produces only one sharp, strong absorption signal (Silverstein *et al.* 2005). Therefore, the standard for protons is the resonance frequency of ^{1}H and ^{13}C in $Si(CH_3)_4$ and for ^{31}P it is H_3PO_4 (aq) at 85% (Atkins 1994). For other nuclei, other standards are adopted. However, TMS is not soluble in water, so if using an aqueous solution, it is usually replaced by the sodium salt of 2,2-dimethyl-2-silapentane-5-sulfonic acid, $(CH_3)_3SiCH_3CH_2CH_2SO_3Na$, because the methyl protons of this salt produce a peak at virtually the same place in the spectrum as that of TMS (Skoog *et al.* 1998).

The determination of chemical shift is the principal application of NMR by which structural information is obtained in geochemistry, because the chemical shift is a very precise metric of the chemical environment around a nucleus. As an example, the hydrogen chemical shift in CH_3F is higher than that of CH_3Cl, because a more electronegative group (such as F) attached to the CH system is more effective at withdrawing electrons from the methyl protons, causing deshielding and consequently increasing δ. Shielding decreases with increasing electronegativity of adjacent groups, if other influences are not present.

The interaction of the magnetic field of a nucleus with the magnetic field of immediately adjacent nuclei gives rise to the splitting of chemical shift peaks. This effect is called spin–spin coupling and, in general, is observable if the distance between these two nuclei is less than or equal to three bond lengths. This coupling occurs by interactions between the nuclei and the bonding electrons. A nucleus without coupling produces a single sharp peak characteristic of isotropicity. If the same nucleus is in a molecule where it experiences spin–spin coupling with a neighboring nucleus, the NMR spectrum presents the splitting of this line as two absorption lines. The spin–spin coupling effect is a good tool for investigating stereochemical relations (McGregor 1997).

3.4.3 Solid-state nuclear magnetic resonance

Compared with liquid-state samples, solid-state samples present additional problems for NMR spectroscopy. With a powder sample a broad signal is observed, corresponding to all the possible orientations of the molecule with respect to the axis of the applied field $\mathbf{B}_o$ (and the chemical shifts related to each of these orientations). Therefore, the solid-state sample presents chemical shift anisotropy (CSA), namely a non-uniform chemical shift along the sample, owing to the directional dependence of electronic shielding in the molecule.

Another factor responsible for the broad lines observed with a solid-state sample is the dipole–dipole interaction, which arises from energy levels shifted slightly by local fields around the nucleus, derived from neighbouring nuclei (dipolar coupling).

NMR spectra of solids are more affected by dipole–dipole interactions as a result of the near-neighbor magnetic dipoles than liquids. In ^{13}C NMR spectroscopy of amorphous solid samples, static dipolar interactions between ^{13}C and ^{1}H result in a large amount of dipolar splitting (Skoog *et al.* 1998). The CSA and dipolar coupling are greatly or completely reduced in solution by rapid, random molecular tumbling (Brownian motion), because in this case what is observed is an average chemical shift (isotropic chemical shift), which is given by the average of all the possible orientations of the molecule in the applied field (McGregor 1997). In the solid state, molecular movement is restricted to small oscillations around fixed positions in the solid matrix and the sample presents both CSA and dipolar coupling phenomena, which are responsible for the broad lines in the NMR spectrum.

The broadening of the line-shape observed in NMR spectra of solid samples has posed challenges for those wishing to study solid samples, such as natural organic matter. Fortunately, broadening of the NMR line-shape due to CSA, as well as, at some level, due to dipole–dipole interactions can be alleviated to a large extent by an experimental technique, known as MAS (magic-angle-spinning). Both CSA and dipolar coupling effects have a $(1-3\cos^2\theta)$ term in their mathematical description, which is zero if the sample is rotated (about an axis making an angle θ with B_o) at a high enough speed to force the magnetic nuclei in the sample to experience the magic angle of 54.7°. Usually the very broad lines in the spectrum are sharpened significantly using MAS (McGregor 1997; Cook 2004; Sposito 2004).

Dipolar splitting in a ^{13}C spectrum can be removed by irradiating the sample with a second radio frequency corresponding to the peak of proton frequencies recorded when the spectrum was being obtained. This procedure, called dipolar decoupling uses a series of pulses to average the dipolar interactions by reorienting the spins. Dipolar coupling is used to increase the sensitivity of less-sensitive nuclei using a technique known as cross-polarization (CP), a complicated pulsed technique that causes the resonance frequencies of the ^{1}H and ^{13}C nuclei to become identical, and then promotes interactions between the magnetic fields of the two nuclei. Currently, there are instruments commercially available that incorporate dipolar decoupling, magic angle spinning and cross polarization, which can produce high-resolution ^{13}C spectra from solids (McGregor 1997; Skoog *et al.* 1998).

3.4.4 Nuclear magnetic resonance applied to sediments

The most used NMR method for the characterization of sediments is solid-state NMR, by cross polarization-magic-angle spinning (CP-MAS) ^{13}C NMR (Dickens *et al.* 2006).

NMR analysis of the natural organic matter in soils or sediment samples is highly complex owing to the intricate, heterogeneous nature of natural particles, but especially because of the naturally occurring paramagnetic centers, such as metal ions (especially iron) and organic stable radicals within the soil matrix. These paramagnetic centers are not evenly distributed throughout the sample and can lead to signal broadening ^{13}C NMR resonances, decreasing the signal-to-noise ratio and causing paramagnetic resonance shifts (Gélinas *et al.* 2001). Until now, the problem of signal loss due to stable organic radicals remains poorly understood and does not have a simple solution; in general, it is not addressed (Cook 2004). On the other hand, for solid samples such as soils, humin, or sediments, the removal of the inorganic paramagnetic centers can be done by repeated treatment with dilute HF through three or four cycles (Cook 2004; Hedges & Oades 1997). For liquid samples, such as humic or fulvic acids, the metal ions can be removed using cation exchange resin (Cook 2004).

Gélinas *et al.* (2001) studied marine sediments using CP-MAS ^{13}C NMR analysis. Before the NMR analysis, marine sediment samples were demineralized through acid treatment developed by the authors, which used HCl to dissolve carbonates and a mixture of diluted HCl/HF to dissolve silicates. This demineralization method removed minerals containing paramagnetic elements that otherwise could interfere with NMR analysis and allowed the authors to study recently deposited marine sediments with low organic carbon concentration, containing labile organic matter, with minimal alteration of organic structures.

Paramagnetic metal removal from marine sediments is not always necessary. Hedges & Oades (1997) performed CP/MAS ^{13}C NMR analysis of untreated surface sediments (8.4 wt% organic carbon) from the Peru Margin. The NMR spectrum showed resonance of alkyl, carbohydrate, aromatic, and carboxyl structures. This kind of marine sediment was accumulating in coastal regions, away from strong river discharges, and was predominantly composed of opal or carbonate. Iron metal was almost absent in this case, because there was no terrestrial input of sediments. Therefore the NMR analysis could be performed without pretreatment for paramagnetic impurity removal.

Dickens *et al.* (2006) used solid-state ^{13}C-NMR spectroscopy, along with elemental stable carbon isotopic (δ^{13}C) and lignin phenol analysis, to study the mechanisms controlling the preservation of organic carbon in ocean sediments. Sediment samples were demineralized in preparation for NMR analysis to remove paramagnetic cations that would interfere with NMR analysis, and to concentrate organic carbon. They studied two marine sediments, one containing a mixture of terrestrial and marine inputs, the other containing entirely marine organic carbon (from an anoxic region). Using solid-state CPMAS ^{13}C NMR spectroscopy, these authors identified and quantified functional groups such as alkyl C, unsaturated C, O-alkyl C, carbonyl and amide C, or ketone C for different size and density fractions for the two sediments. This information allowed inferences to be made about the molecular structure of the organic matter and investigated the mechanisms allowing preservation of organic carbon. Therefore, ^{13}C-NMR spectroscopy is a useful tool for determining how organic carbon is preserved in sediment.

In environmental samples, NMR can also be applied to the study of phosphorus, an essential nutrient used by all living organisms. Phosphorus (P) is easily detected by NMR spectroscopy, owing to the large magnetogyric ratio of ^{31}P and its natural abundance (Paytan *et al.* 2003). ^{31}P-NMR spectroscopy is a suitable method for identifying and quantifying phosphorus species in environmental samples. Phosphorus is present in aquatic systems in dissolved and particulate forms. In contrast to nitrogen, phosphorus has no gaseous phase; thus, its supply for living organisms in aquatic environments depends on external sources as well as internal recycling (Ahlgren *et al.* 2006b). Orthophosphate is the most important bioavailable form of phosphorus, although other inorganic and organic forms can also be used by organisms. Phosphorus availability in aquatic systems is regulated by the conversion of particulate phosphorus to dissolved forms and organic phosphorus to inorganic orthophosphate. There is relatively little information about concentrations, transport, and fate of particulate phosphorus in aquatic environments which primarily results from the current limitations in phosphorus analytical techniques. Most of the knowledge of phosphorus in sediment has been obtained from several sequential extraction procedures and is related to phosphorus in its inorganic form. In contrast, organic phosphorus concentrations in particulate and dissolved samples are determined indirectly by the difference between total phosphorus and soluble reactive phosphorus (SRP). SRP is the fraction that reacts to form a blue-colored phosphomolybdate complex under slightly acidic conditions (Cade-Menun *et al.* 2005). Although the degradation of organic phosphorus compounds in environmental samples, including sediments, may be an important source of bioavailable phosphorus, little is known about the chemical forms of organic phosphorus (Nanny & Minear 1997; Ahlgren *et al.* 2006a). Therefore, new analytical methods are required to study phosphorus in environmental systems. In this context, ^{31}P-NMR spectroscopy is a powerful tool that can identify inorganic phosphorus forms such as orthophosphate, pyrophosphate, or polyphosphate and organic forms such as orthophosphate monoesters, orthophosphate diesters, or phosphonates (Paytan *et al.* 2003). In addition, information on degradation and mineralization can be obtained using ^{31}P NMR in monitoring changes to P composition promoted by these processes (Ahlgren *et al.* 2006a).

^{31}P-NMR spectroscopy is widely used in the investigation of organic phosphorus speciation in terrestrial ecosystems; however, there are few studies of aquatic phosphorus using this technique (Cade-Menun *et al.* 2006). This is probably because, in environmental samples, the phosphorus concentration of the sample is below the lower limit of NMR detection, which therefore requires a concentration procedure (Nanny & Minear 1997). For solution ^{31}P-NMR, the sediment samples are usually concentrated by extraction with a NaOH-EDTA solution followed by lyophilization (Cade-Menun *et al.* 2005) or by rotatory evaporation (Ahlgren *et al.* 2006a) or just concentrating and fractionating with ultrafiltration/reverse osmosis membranes, without NaOH-EDTA extraction (Nanny & Minear 1997). Solution ^{31}P-NMR spectroscopy has been used to characterize phosphorus forms in sediments from oceans (Cade-Menun *et al.* 2005; Paytan *et al.* 2003; Ahlgren *et al.* 2006a), rivers (Cade-Menun *et al.* 2006) and lakes (Nanny & Minear 1997; Hupfer *et al.* 2004; Ahlgren *et al.* 2006b).

For the analysis of organic phosphorus and polyphosphate using solution ^{31}P-NMR spectroscopy, alkaline extraction procedures are usually chosen, because phosphorus in biological materials dissolves in alkaline extracts. EDTA is added to the alkaline extraction solution because its chelating ability increases the efficiency of NaOH by breaking phosphorus-containing organometal complexes. In addition, iron and other paramagnetic metals are minimized by the pre-extraction with EDTA and their interference in NMR spectra is reduced (Hupfer *et al.* 2004). However, extraction with NaOH-EDTA can introduce the possibility of base-catalyzed hydrolysis of organic phosphorus to orthophosphate, as well as by extracting only base-soluble phosphorus (Nanny & Minear 1997). According to Cade-Menun *et al.* (2005), most organic esters are removed by extraction with NaOH-EDTA but not all of the phosphonates are quantitatively removed; moreover, it is likely Ca-associated phosphates are preferentially extracted, rather than those associated with Fe and Al. Nanny & Minear (1997), have concentrated and fractionated water lake samples with ultrafiltration/reverse osmosis (UF/RO) membranes, avoiding extraction with NaOH-EDTA, before solution ^{31}P-NMR analysis. However, these authors concluded that this method possibly modifies the sample by incorporating soluble phosphorus into aggregates and that it was not possible to obtain 100% of soluble phosphorus recoveries by these UF/RO methods.

However, sample preparation for solid-state ^{31}P-NMR is minimal (with the exception of drying and grinding) and the technique allows the investigation of the abundance of phosphonate species in environmental samples, without the inconvenience of an extraction procedure, because the phosphonate peak is well separated from other phosphorus species peaks (Cade-Menun *et al.* 2005). The presence of a C–P bond, highly resistant to chemical hydrolysis, thermal decomposition, and photolysis, is probably responsible by the persistence of phosphonates relative to other phosphorus forms in marine samples. However, phosphonates may be mineralized under anoxic conditions (Cade-Menun 2005). Using solid-state ^{31}P-NMR spectroscopy to examine sediment samples across the oxic-anoxic boundary of a marine basin, Benitez-Nelson *et al.* (2004) obtained a clear signal of phosphonate, orthophosphate, and phosphorus esters at all depths. They observed that the phosphonate signal decreased relative to orthophosphate and phosphorus esters with depth, which was interpreted as being caused by rapid release of phosphonate from particles and/or preferential remineralization in anoxic environments.

Unfortunately, the characterization of orthophosphate monoesters and orthophosphate diesters in environmental samples is not possible using solid-state ^{31}P-NMR spectroscopy, because the peaks of these phosphorus species overlap in the spectrum. The reason for this is the chemical shift anisotropy of the solid sample and the presence of paramagnetic ions such as Fe and Mn, both of which are responsible for broadening the peaks and for reduced spectral resolution in the solid-state ^{31}P-NMR spectrum. In contrast, resolution of phosphorus forms in samples is much higher in solution ^{31}P-NMR spectroscopy, and compound groups such as pyrophosphates, orthophosphate monoesters, and orthophosphate diesters are clearly separated in the spectrum (Cade-Menun *et al.* 2005).

3.5 Infrared spectroscopy

3.5.1 Basic theory of infrared radiation

Electromagnetic radiation can be characterized by its wavelength, λ, its frequency, υ, and its wavenumber, $\hat{u}$. Infrared absorption positions are generally presented as either wavelengths or wavenumbers. The latter is usually expressed in cm^{-1} and corresponds to the number of waves in 1 cm.

Wavenumbers and wavelengths are related by the following equation:

$$\hat{u}(cm^{-1}) = 10^4/[\lambda(\mu m)] \quad (16)$$

Infrared radiation (IR) encompasses a section of the electromagnetic spectrum with wavenumbers ranging from about 12,800 to 10 cm^{-1} or wavelengths from 0.78 to 1000 μm. It is limited at high frequencies by the red end of the visible region and at low frequencies by the microwave region (Hsu 1997; Skoog *et al.* 1998).

A linear wavenumber scale is usually preferred in infrared spectroscopy because wavenumber is directly proportional to both frequency and energy of infrared absorption (Pavia *et al.* 2001).

The energy of a molecule encompasses translational, rotational, vibrational, and electronic energies. Electronic transitions occur when the molecule absorbs radiation in the ultraviolet (UV)–visible region of the electromagnetic spectrum. Infrared radiation is less energetic than visible radiation and therefore cannot bring about electronic transitions. However, the energy differences between various vibrational and rotational states of molecular species are in the infrared region of the electromagnetic spectrum (Colthup *et al.* 1975).

The translational, rotational, and vibrational energies of the molecule are related to each atom in their structure. Above absolute zero temperature, all the atoms in the molecules are continually vibrating with respect to each other. Each atom can be described by its own cartesian coordinate system with the origin defined by the equilibrium position of the atom. An atom can move along any of the three coordinate axes (x, y, z) and each coordinate corresponds to one degree of freedom. Thus, a polyatomic molecule of n atoms has $3n$ total degrees of freedom. However, the motion of the entire molecule through space (translation) corresponds to three degrees of freedom; another three degrees of freedom are needed to describe the rotation of the entire molecule around its center of gravity. Therefore, for a nonlinear molecule, the true (fundamental) vibrations are the remaining $3n–6$ degrees of freedom. To describe rotation of linear molecules two degrees of freedom are sufficient, because rotation about the bond axis is not possible. Thus, the number of fundamental vibrations for a linear molecule is given by $3n–5$. These fundamental vibrations are also called normal modes of vibration (Hsu 1997; Skoog *et al.* 1998).

However, not all the fundamental modes of the molecule have infrared activity. Only those vibrations that promote a net change in the dipole moment of the molecule may give rise to infrared absorption by the molecule. In the case of a simple dipole (such as the HCl molecule), the dipole moment is defined as the magnitude of either charge in the dipole (positive for the H atom and negative for the Cl atom) multiplied by the charge spacing (Colthup *et al.* 1975). Therefore, only those bonds that have a dipole moment that changes as a function of time are able to absorb infrared radiation (Pavia *et al.* 2001).

In summary, the origin of the infrared absorption by molecules is related to their vibrational and rotational motions. The molecule can absorb incident infrared radiation if it has a frequency equal to that of a specific molecular vibration and if it results in a change in the dipole moment of the molecule (Ewing 1985; Hsu 1997). Nearly all molecules, whether organic and inorganic, absorb various frequencies of radiation in the infrared region of the electromagnetic spectrum. The only exceptions are diatomic homonuclear molecules such as H_2, N_2, and O_2, because only in these can no vibration or rotation be found that will produce a dipole moment (Ewing 1985).

Generally, the total number of fundamental vibrations does not coincide with the total number of observed absorption bands. The reason for this difference is the fact (already mentioned) that when the vibration does not causes a net change in the dipole moment of the molecule, its fundamental mode is infrared inactive. On the other hand, additional bands are generated by the appearance of overtones at frequencies approximately two or three times that of the fundamental line. Overtones result from excitation from the ground state to higher energy states, which correspond to integral multiples of the frequency of the fundamental mode. Another phenomenon, called combination bands, can occur when the energy of a photon is absorbed by two bonds rather than one, exciting two vibrational modes simultaneously. Combination bands are the consequence of a coupling of these two vibrational frequencies in a molecule, which gives rise to the vibration of a new infrared active frequency within the molecule. A combination band usually occurs at a frequency that corresponds to approximately the sum of the two fundamental frequencies. Difference bands are analogous to combination bands, but the observed frequency in this case is the difference between the two coupling bands. The intensities of overtone, combination, and difference bands are less than those of fundamental bands (Hsu 1997; Skoog *et al.* 1998; Pavia *et al.* 2001).

Owing to the coupling of vibrations, the position of an absorption peak related to a given organic functional group cannot always be specified exactly; usually some range of wavenumbers are associated with each functional group (Skoog *et al.* 1998).

Vibrations can be classified into two basic categories: stretching and bending. When the infrared radiation is absorbed, the associated energy is converted into these types of motions. A stretching vibration is characterized by a continuous change in the interatomic distance along the axis of the bond between two atoms. Bending vibrations involves a change in the angle between two bonds. There are of four types of bending: scissoring, rocking, wagging, and twisting (Skoog *et al.* 1998; Pavia *et al.* 2001).

Vibrational infrared absorption involves discrete, quantized energy levels. Although rotational frequencies of the entire molecule are not infrared active, they frequently couple with the vibration modes in the molecule to give additional fine structure to these absorptions. These combinations lead to the commonly observed broad bands rather than discrete lines in the infrared spectrum (Hsu 1997; Pavia *et al.* 2001).

3.5.2 Infrared applied to sediments

To give a better description of application and instrumentation, the infrared spectrum is conveniently divided into near-, mid-, and far-infrared radiation. Table 3.1 shows the rough limits of each infrared region.

The far-infrared region of the spectrum is particularly useful for studies of vibration absorptions of inorganic solids (especially semiconductors) and for investigation of pure rotational absorption by molecules that present permanent dipole moments, such as O_3 and H_2O in the gaseous state (Skoog *et al.* 1998).

The near-infrared region has several applications in the study of sediments. Some of the fundamental stretching vibrational bands that occur in the middle-infrared region of 3000–1700 cm^{-1} give rise to overtones or combinations, which are the absorption bands observed in the near-infrared region. The bonds usually involved are C–H, N–H, and O–H. What appears in the near-infrared region of the spectrum is the result of vibrations of light atoms that have strong molecular bonds. Weak chemical bonds or bonds involving heavy atoms have a low vibrational frequency; thus, their overtones will not be detectable in the near-infrared. Consequently, the most observable overtones and combination bands in the near-infrared are the result of chemical bonds containing hydrogen attached to atoms such as nitrogen, oxygen, or carbon; that is, the chemical structures that are common in many organic compounds. Moreover, these weak overtone bands are more sensible to their environment than the fundamental mode of the same vibration. A slight perturbation in the bonding produces small changes in the fundamental mode, but great frequency shifts and amplitude changes in the near-infrared (Wetzel 1983). The near-infrared spectrum has several regions that are sensitive to the environment of the absorbing molecules and to the number of molecules present, allowing for quantitative measurements. However, the combination bands and spectral overtones do not occur in distinct absorption peaks; instead, several overlapping peaks are observed. Thus, to extract information on chemical compounds from the spectral response of a sample it is necessary to perform a calibration. The near-infrared spectral data of the samples must be correlated to other chemical data, obtained by different methods, by an appropriate statistical relation to make it possible to predict the chemical constituent of interest from the near-infrared spectra of unknown samples (Wetzel 1983; Korsman *et al.* 1999).

Near infrared spectroscopy (NIRS) needs minimal or no sample preparation. The near-infrared region has been extensively used in quantitative analysis of

Table 3.1 Infrared spectral regions.

	Near infrared	Middle infrared	Far infrared
Wavenumber ($\hat{u}$)	12.800–4000 cm^{-1}	4000–200 cm^{-1}	200–10 cm^{-1}
Wavelength (λ)	0.78–2.5 μm	2.5–50 μm	50–1000 μm

organic matter in industrial and agricultural materials and for process control (Hsu 1997; Skoog *et al.* 1998). The application of NIRS to the analysis of environmental samples started in the 1990s and since then has been increasing. NIRS has been used for prediction of heavy metal concentration in freshwater sediments (Malley 1997), analysis of spatial variability in surface lake sediments (Korsman *et al.* 1999), analysis of C, CO_3^{-2}, N and P in freshwater sediments (Malley 1997), and the determination of carbon in marine sediments (Chang *et al.* 2005).

In the infrared spectra, the detected changes in transmittance (or absorption) intensity are presented as a function of frequency. The separation and measurement of infrared radiation in most commercial instruments is performed using dispersive spectrometers (based on diffraction gratings) or Fourier transform spectrometers (based on interferometer filters). The photometers and spectrophotometers used to perform measurements in the near-infrared region are dispersive spectrometers, similar in design and components to those used in ultraviolet/visible absorption spectrometry (Skoog *et al.* 1998). In Fourier transform infrared spectroscopy, all frequencies are analyzed simultaneously, rather than examining each component frequency sequentially, as in the dispersive infrared spectrometer. Interferometric instruments have high resolutions and are very accurate with reproducible frequency determinations. Moreover, their signal-to-noise ratios are better than those of a good-quality dispersive instrument by more than an order of magnitude (Hsu 1997; Skoog *et al.* 1998).

Until the early 1980s, the use of the mid-infrared region was limited to qualitative organic analysis and structure determination based on absorption spectra, because the only available instruments were of the dispersive type. The multiple layers of information featured in a mid-infrared spectrum were a major challenge for the interpretation and quantification of the data. Since then, however, the appearance of Fourier transform spectrometers, based on interference filters, has brought a dramatic increase in the number and type of applications of mid-infrared radiation. This technique is known as Fourier transform infrared spectroscopy (FTIRS). It offers some important advantages in sediment analysis, such as the ability to assess both mineral and organic structures in particles, and good sensitivity (Gallé *et al.* 2004). FTIRS was used by Wirrmann *et al.* (2001) to quantify the mineral abundance in dated lake sediments, with the aim of investigating hydrologic records during the Late Holocene. The sediment samples were ground and diluted in KBr pellets before measurement of the mid-infrared spectra. The infrared absorbance was linearly correlated to the composition and mass of constituents in the KBr pellet. The spectra of sediment samples were compared with spectra of pure mineral phases similar to the ones found in the sediments, to perform a calibration. The authors observed a good correlation between their quantitative results from FTIR spectroscopy and chemical analysis.

However, the technique usually used in the mid-infrared region by those studying sediments is diffuse reflectance infrared Fourier transform spectroscopy (DRIFTS or DRIFT). DRIFT is a surface characterization method, involving a reflection experiment where the typical depths of penetration of the infrared beam into the surface are 1–10 μm, sufficient depth to characterize the organic layer on mineral surfaces. However, obtaining reproducible quantitative DRIFT measurements requires strict attention to experimental details, especially to particle size distribution and packing density of the sample (Belton & Wilson 1990; Gallé *et al.* 2004).

To address the question of whether the TOC content of lake water follows changes in climate and vegetation on a millennia timescale, Rosén & Persson (2006) tested the hypothesis that DRIFTS of lake sediments can be used to infer past changes in tree-line position and TOC content of lake water. The statistical method of principal component analysis was used to get an overview of the spectral variability of the lakes. Partial least square regression was used to develop a transfer function between DRIFT spectra of surface sediment (0–1 cm) and TOC. Both quality and quantity of organic material can be measured by DRIFTS. The relation between FTIR spectra of sediment and the TOC content in the lake water was probably because the sediments in lakes with high and low TOC levels, respectively, have quantitatively and qualitatively different composition, owing to different types of vegetation, algae, input and degradation of organic material in the water column. The authors succeeded in using the transfer function developed between FTIR data of the sediments and TOC to obtain information about past changes in tree-line position and TOC of another lake. However, they emphasized that this work was just a first step in developing DRIFTS into a new paleolimnological tool and that future research was needed to include many more lakes to assess further the uniformity of reconstructions among different types of lake.

Gallé *et al.* (2004) applied DRIFTS to follow the sediment composition of a mountainous river during changes in its hydrological life cycle for one and a half years. A set of 57 sediment samples collected on a weekly basis were wet-sieved down to less than 63 µm, freeze-dried, and homogenized before analysis. All samples were ground and mixed with KBr before DRIFS analysis. Usually, a drawback is the presence of inorganic carbonate in the sediment samples submitted to FTIRS analysis, because it gives rise to a broad signal around 1650 cm^{-1} that can mask the asymmetric COO^{-}/C–C stretches bands. However, the sediment samples studied by Gallé *et al.* (2004) were practically free from inorganic carbonate. Therefore, the refractory organic matter contribution to the overall C_{org} was easily detectable in the DRIFTS spectra, without the removal of inorganic carbonates by chemical methods.

Gallé *et al.* (2004), observed that during or shortly after flood events particulate organic matter content in sediments was reduced and sediments poor in C_{org} content gave rise to DRIFTS spectra enriched in carboxylic and aromatic signals. These signals were considered characteristic of terrestrial oxidized vascular plant debris (humic substances). The reduction in organic matter content during flooding was attributed to the fact that, at higher flow velocities, the loosely organized upper parts of biofilms were removed from the particle, with only diatoms or cyanobacteria remaining attached directly to the surface. Although often damaged, these algae and bacteria can serve as inocula for the recolonization of the particle surface. DRIFT spectra obtained during low-flow conditions showed growth of bands corresponding to –OH, $-CH_3$, $-CH_2$, and secondary amide –C O stretches and the –NH band. According the authors, these bands are considered to be the most prominent features of microbial (bacterial) spectra. Therefore, the DRIFTS results seem to indicate that low-flow conditions allow the recolonization of the sediment particle by bacteria, which are rich in amide, aliphatic, and polysaccharide moieties.

3.4 Final considerations

Mineralogical characterization of particles obtained using electronic microscopy and X-ray diffraction techniques in combination with information of particle size distribution from laser diffraction and functional groups enables advances in understanding of sediments. Understanding the reactivity of particles that are found in sediment and their capacity to interact with the environment, either in water or organisms, is the greatest benefit of such studies of sediment characterization.

References

Abraham, R. J. & Loftus, P. (1985) *Proton and Carbon-13 NMR Spectroscopy: An Integrated Approach*. New York: John Wiley, 230pp.

Accioly, A. M. A. & Siqueira, J. O. (2000) Contaminação química e biorremediação de solo. In: *Tópicos Especiais em Ciência do Solo, SBCS, Viçosa*, vol. **1**, R. F. Novais, V. V. H. Alvarez & C. E. G. R. Schaefer (eds), 299–352.

Akitt, J. W. (1983) *NMR and Chemistry: An Introduction to the Fourier Transform Multinuclear Era*. 2nd edition. New York: Chapman and Hall, 263pp.

Ahlgren, J., Reitzel, K., Tranvik, L., Gogoll, A. & Rydin, E. (2006a) Degradation of organic phosphorus compounds in anoxic Baltic Sea sediments: a ^{31}P nuclear magnetic resonance study. *Limnology and Oceanography*, **51**, 2341–48.

Ahlgren, J., Reitzel, K., Danielsson, R., Gogoll, A. & Rydin, E. (2006b) Biogenic phosphorus in oligotrophic mountain lake sediments: differences in composition measured with NMR spectroscopy. *Water Research*, **40**, 3705–12.

Atkins, P. (1994) *Physical Chemistry*. 5thedition. Oxford: Oxford University Press, 1031pp.

Baalousha, M., Kammer, F. V. D., Motelica-Heino, M. & Le Cooustumer, P. (2005a) 3D characterization of natural colloids by FIFFF-MALLS-TEM. *Analytical and Bioanalytical Chemistry*, **383**, 549–56.

Baalousha, M., Kammer, F. V. D., Motelica-Heino, M. & Le Coustumer, P. (2005b) Natural sample fractionation by FIFFF-MALLS-TEM: sample stabilization, preparation, pre-concentration and fractionation. *Journal of Chromatography A*, **1093**, 156–66.

Baalousha, M., Kammer, F. V. D., Motelica-Heino, M. M., Hilal, H. S. & Coustumer, P. L. (2006) Size fractionation and characterization of natural colloids by flow-field flow fractionation coupled to multi-angle laser light scattering. *Journal of Chromatography A*, **1104**, 272–281.

Beckett, R., Jue, Z. & Giddings, C. (1987) Determination of molecular weight distributions of fulvic and humic acids using flow field-flow fractionation. *Environmental Science and Technology*, **21**, 289–95.

Belton, P. S. & Wilson, R. H. (1990) Infrared sampling methods. In: *Perspectives In Modern Chemistry Spectroscopy*, D. L. Andrews (ed.), Chapter 3, 67–86, . Berlin: Springer-Verlag, 325pp.

Benitez-Nelson, C. R., O'Neill, L., Kolowith, L. C., Pellechia, P. & Thunell, R. (2004) Phosphonates and particulate organic phosphorus cycling in an anoxic marine basin. *Limnology and Oceanography*, **49**, 1593–604.

Blume, T., Weisbrod, N. & Selker, J. S. (2005) On the critical salt concentrations for particle detachment in homogeneous sand and heterogeneous Hanford sediments. *Geoderma*, **124**, 121–32.

Bortoluzzi, E. C. (2004) Caracterização quali-quantitativa de sedimento fluvial oriundo da microbacia hidrográfica fumageira de Agudo, Rio Grande do Sul Brasil. Relatório Técnico, Brasília: CNPq. 75pp.

Bortoluzzi, E. C. & Petry, C. (2008) Partículas Minerais: da rocha ao sedimento. In: *Ambiente e Sedimentos*. Poleto, C. (org.). 1ed. ABRH: Porto Alegre, 404pp.

Bortoluzzi, E. C. & Poleto, C. (2006) Metodologias para estudos de sedimentos:ênfase na proporção e na natureza mineralógica das partículas. In: *Qualidade dos Sedimentos*. Poleto C., Merten G. H. (org.). ABRH: Porto Alegre. 397pp.

Bortoluzzi, E. C., Rheinheimer, D. S., Kaminski, J., Gatiboni, L. C. & Tessier, D. (2005) Alterações na mineralogia de um Argissolo do Rio Grande do Sul submetido à fertilização potássica. *Revista Brasileira de Ciência do Solo*, **29**, 327–35.

Bortoluzzi, E. C., Rheinheimer, D. S., Pellegrini, J. B. R., Pernes, M. & Dur, J. C. (2006) Qualidade de sedimento e processo de eutroficação de águas fluviais em uma microbacia hidrográfica do Rio Grande do Sul. In: *V simpósio internacional de qualidade ambiental: gestão integrada do ambiente*. PUCRS.

Bouchet, A., Meunier, A. & Sardini, P. (2000) Minéraux argileux: structure cristaline, identification par diffration de rayons-X. *Bulletin Centre Rech. Elf. Explor. Prod., Mém.* **23**, 136pp.

Brindley, G. W. & Brown, G. (1980) Crystal structures of clays minerals and their x-ray identification. *Mineralogical Society Monograph* **5**, London, 495pp.

Buffle, J. (2006) The key role of environmental colloids/nanoparticles for the sustainability of life. *Environmental Chemistry*, **3**, 155–58.

Byers, S. C., Mills, E. L. & Stewart, P. L. (1978) A comparison of methods of determining organic carbon in marine sediments, with suggestions for a standard method. *Hydrobiologia*, **58**, 43–47.

Cade-Menun, B. J. (2005) Characterizing phosphorus in environmental and agricultural samples by ^{31}P nuclear magnetic resonance spectroscopy. *Talanta*, **66**, 359–71.

Cade-Menun, B. J., Benitez-Nelson, C. R., Pellechia, P. & Paytan, A. (2005) Refining ^{31}P nuclear magnetic resonance spectroscopy for marine particulate samples: Storage conditions and extraction recovery. *Marine Chemistry*, **97**, 293–306.

Cade-Menun, B. J., Navaratnam, J. A. & Walbridge, M. R. (2006) Characterizing dissolved and particulate phosphorus in water with ^{31}P nuclear magnetic resonance spectroscopy. *Environmental Science and Technology*, **40**, 7874–80.

Caillère, S., Hénin, S. & Rautureau, M. (1982) Minéralogie des argiles: structure et propriétés physico-chimiques. 2nd edition. *Actualités Scientifiques et Agronomiques/INRA*, **8**, 184pp.

Castro, L. A. S. (2002) Processamento de amostras para microscopia de varredura. EMBRAPA: Pelotas: Embrapa Clima temperado. 2002. 37pp. (Doc. 93).

Chang, C. W., You, C. F., Huang, C. Y. & Lee, T. Q. (2005) Rapid determination of chemical and physical properties in marine sediments using a near-infrared reflectance spectroscopic technique. *Applied Geochemistry*, **20**, 1637–47.

Chantiwas, R., Beckett, R., Jakmunee, J., McKelvie, I. D. & Grudpan, K. (2002) Gravitational field-flow fractionation in combination with flow injection analysis or electrothermal AAS for size based iron speciation of particles. *Talanta*, **58**, 1375–83.

Chenu, C. (2001) Le complexe argilo-humique des sols: etat des connaissances actuelles. *Comptes Rendus de l'Academie d'Agriculture de France*, **87**(3), 3–12.

Chenu, C. & Plante, A. (2006) Clay-sized organo-mineral complexes in a cultivation chronosequence: revisiting the concept of the primary organo-mineral complex. *European Journal of Soil Science*, **57**, 596–607.

Churchman, G. J., Whitton, J. S., Claridge, G. G. C. & Theng B. K. G. (1984) Intercalation method using formamide for differentiating halloysite from kaolinite. *Clays and Clay Minerals*, **32**, 241–48.

Citeau, L., Gaboriaud, F., Elsass, F., Thomas, F. & Lamy, I. (2006) investigation of physico-chemical features of soil colloidal suspensions. *Colloids and Surfaces A: Physicochemical and Engineering Aspects*, **217**, 11–19.

Colthup, N. B., Daly, L. H. & Wiberley, S. E. (1975) *Introduction to Infrared and Raman Spectroscopy*. 2nd edition. New York: Academic Press. 523pp.

Cook, R. L. (2004) Coupling NMR to NOM. *Analytical and Bioanalytical Chemistry*, **378**, 1484–503.

Coplen, T. B. (1996) New guidelines for reporting stable hydrogen, carbon, and oxygen isotope-ratio data. *Geochimica Cosmochimica Acta*, **60**, 3359–60.

Czigány, S., Flury, M. & Harsh, J. (2005) Colloid stability in vadose zone Hanford sediments, *Environmental Science and Technology*, **39**, 1506–12.

Dickens, A. F., Baldock, J. A., Smernik, R. J., *et al.*. (2006) Solid-state ^{13}C NMR analysis of size and density fractions of marine sediments: insight into organic carbon sources and preservation mechanisms. *Geochimica Cosmochimica Acta*, **70**, 666–86.

Dixon, J. B. & Weed, S. B (1989) *Minerals in Soil Environments*. 2nd edition. Madison, USA: Soil Science Society of America. 1244pp.

Dur J. C., Elsass F., Chaplain V. & Tessier D. (2004) The relationship between particle-size distribution by laser granulometry and image analysis by transmission electron microscopy in a soil clay fraction. *European Journal of Soil Science*, **55**, 265–70.

Effler, S. W., Matthews, D. A., Kaser, J. W., Prestigiacomo, A. R. & Smith, D. G. (2006) Runoff event impacts on a water supply reservoir: suspended sediment loading, turbid plume behavior, and sediment deposition. *Journal of the American Water Resources Association*, **42**, 1697–710.

Elsass, F. & Flores-Velez, L. M. (1999) *Spéciation physique de métaux traces dans les sols – analyse statistique des données de MET*. In: *Structure et ultrastructure des sols et des organisms vivants*, F. Elsass, & A. M. Jaunet (eds) Versailles. Paris: INRA, 231pp.

Elsass, F., Chenu, C. & Tessier, D. (2008) Transmission electron microscopy for soil samples: preparation methods and use. Chapter 9. In Methods of Soil Analysis. Part 5. Mineralogical methods. Soil Science Society of America, Book Series, no. 5, 235–68.

Entwistle, J., Williamson, B. & Hudson-Edwards, K. (2003) Mineral particles and the environment. *Mineralogical Magazine*, **67**, 127pp.

Ewing, G. W. (1985) *Instrumental Methods of Chemical Analysis*. 5th edition. New York: McGraw-Hill.

Filella, M., Zhang, J., Newman, M. E. & Buffle, J. (1997) Analytical applications of photon correlation spectroscopy for size distribution measurements of natural colloidal suspensions: capabilities and limitations. *Colloids and Surfaces A: Physicochemical and Engineering Aspects*, **120**, 27–46.

FAO–WRB. (2006) IUSS working Group WRB. World reference base for soil resources, 2006. 2nd edition. World Soil Resources Reports No. 103. FAO, Rome. 128pp.

Gallé, T., van Lagen, B., Kurtenbach, A. & Bierl, R. (2004) An FTIR-DRIFT study on river sediment particle structure: implications for biofilm dynamics and pollutant binding. *Environmental Science and Technology*, **38**, 4496–502.

Galy, V., Bouchez, J. & France-Lanord, C. (2007) Determination of total organic carbo content and $\delta^{13}C$ in carbonate-rich detrital sediments. *Geostandards and Geoanalytical Research*, **31**, 199–207.

Gélinas, Y., Baldock, J. A. & Hedges, J. I. (2001) Demineralization of marine freshwater sediments for CP/MAS ^{13}C NMR analysis. *Organic Geochemistry*, **32**, 677–93.

Gonçalves, C. S., Rheinheimer, D. S., Pellegrini, J. B. R. & Kist, S. (2005) Qualidade da água numa microbacia hidrográfica de cabeceira situada em região produtora de fumo. *Revista Brasileira de Engenharia Agrícola e Ambiental*, **9**, 391–99.

Graham, T. (1861) Liquid diffusion applied to analysis. *Philosophical Transactions of the Royal Society of London*, **151**, 183–224.

Grossman, R. B. & Millet, J. C. (1961) Carbonate removal from soils by a modification of acetate buffer method. *Proceedings of the Soil Science Society of America.*, **25**, 325–26.

Hedges, J. I. & Oades, J. M. (1997) Comparative organic geochemistries of soils and marine sediments. *Organic Geochemistry*, **27**, 319–61.

Hedges, J. I. & Stern, J. H. (1984) Carbon and nitrogen determinations of carbonate-containing solids. *Limnology and Oceanography*, **29**, 657–63.

Horowitz, A. J. (1991) *A Primer on Sediment-Trace Element Chemistry*. Chelsea, EUA: Lewis Publishers.

Hsieh, Y. P. (1984) Using clay mineralogy to infer sources of suspended clay and silt in a watershed: quantitative approach. *Journal of the Soil Science Society of America*, **48**, 1446–50.

Hsu, C. P. (1997) Infrared spectroscopy. In: *Handbook of Instrumental Techniques for Analytical Chemistry*, F. A. Settle (ed.), Chapter 15, 247–83. Upper Saddle River, USA: Prentice Hall PTR.

Hughes, R. E., Moore D. M. & Glass, H. D. (1994) Qualitative and quantitative analysis of clay minerals in soils. 330–In: *Quantitative Methods in Soil Mineralogy*, J. M. Baterls (ed.). Madison, Wisconsin, USA: Soil Science Society of America.

Hupfer, M., Rübe, B. & Schmieder, P. (2004) Origin and diagenesis of polyphosphate in lake sediments: A ^{31}P-NMR study. *Limnology and Oceanography*, **49**, 1–10.

Inoue, A., Bouchet, A., Velde, B. & Meunier, A. (1989) Convenient technique for stimating smectite layer percentage in randomly intertratified illite/smectite minerals. *Clays and Clay Minerals*, **37**, 227–34.

Kammer, F. & Forstner, U. (1998) Natural colloid characterization using flow-field-flow-fractionation followed by multi-detector analysis. *Water Science and Technology*, **37** (6–7), 173–80.

Kammer, F. V. D., Baborowski, M. & Friese, K. (2005) Field-flow fractionation coupled to multi-angle laser light scattering detectors: applicability and analytical benefits for the analysis of environmental colloids. *Analytica Chimica Acta* **552**, 166–74.

Korsman, T., Nilsson, M. B., Landgren, K. & Renberg, I. (1999) Spatial variability in surface sediment composition characterized by near-infrared (NIR) reflectance spectroscopy. *Journal of Paleolimnology*, **21**, 61–71.

Kunze, G. W. & Dixon, J. B. (1986) Pretreatment for mineralogical analysis. In *Method of Soil Analysis: Part 1 Physical and Mineralogical Methods*, 2nd edition, A. Klute (ed.), 91–100. Madison, USA: SSSA.

Lanson, B. (1997) Decomposition of experimental X-ray diffraction patterns (profile fitting): a convenient way to study clay minerals. *Clays and Clay Minerals*, **45**, 132–146.

Lanson, B. & Besson, G. (1992) Characterization of the smectite and of smectite-to-illite transformation: decomposition of X-ray patterns. *Clays and Clay Minerals*, **40**, 40–52.

Ledin, A., Karlsson, S., Duker, A. & Allard, B. (1995) Characterization of the submicrometer phase in surface waters. *The Analyst*, **120**, 603–08.

Lee, F. Y. & Fittrick, J. A. (1984) Electron microprobe analysis of elements associated with zinc and copper in an oxidizing and an aerobic soil environmental. *Journal of the Soil Science Society of America*, **48**, 548–54.

Li, T., Wang, D., Zhang, B., Liu, H. & Tang, H. (2007) Morphological characterization of suspended particles under wind-induced disturbance in Taihu lake, China. *Environmental Monitoring and Assessment*, **127**, 79–86.

Lin, S., Hsieh, I. J., Huang, K. M. & Wang, C. H. (2002) Influence of the Yangtze River and grain size on the spatial variations of heavy metals and organic carbon in the East China Sea continental shelf sediments. *Chemical Geology*, **182**, 377–94.

Loeppert, R. H. & Suarez, D. L. (1996) Carbonate and gypsum. In: *Methods of Soil Analysis, Part 3 Chemical Methods*, J. M. Bartels (ed.), chapter 15, 437–74. Madison, USA: SSSA.

Lorrain, A., Savoye, N., Chauvaud, L., Paulet, Y. M. & Naulet, N. (2003) Decarbonation and preservation method for the analysis of organic C and N contents and stable isotope ratios of low-carbonated suspended particulate material. *Analytica Chimica Acta*, **491**, 125–33.

Luczak, C., Janquin, M. A. & Kupka, A. (1997) Simple standard procedure for the routine determination of organic matter in marine sediment. *Hydrobiologia*, **345**, 87–94.

Malley, D. F. (1997) Use of near-infrared reflectance spectroscopy in prediction of heavy metals in freshwater sediment by their association with organic matter. *Environmental Science and Technology*, **31**, 3461–67.

McDowell, R. W., Stewart, I. & Cade-Menun, B. J. (2006) An examination of spin-lattice relaxation times for analysis of soil and manure extracts by liquid state phosphorus-31 nuclear magnetic resonance spectroscopy. *Journal of Environmental Quality*, **35**, 293–302.

McGregor, M. (1997) Nuclear magnetic resonance spectroscopy. In: *Handbook of Instrumental Techniques for Analytical Chemistry*, F. A. Settle (ed.), chapter 17, 309–37. Upper Saddle River: Prentice Hall PTR. 995pp.

Mehra, O. P. & Jackson, M. L. (1960) Iron removal from soils and clays by a dithionite citrate system buffered with sodium bicarbonate. *Clays and Clay Minerals*, **7**, 312–27.

Minella, J. P. G., Merten, G. H., Reichert, M. & Rheinheimer, D. S. (2007) Identificação e implicações para a conservação do solo das fontes de sedimentos em bacias hidrográficas. *Revista Brasileira de Ciência do Solo*, **31**, 1637–46.

Moore, D. M. & Reynolds, R. C. (1997) *X-ray Diffraction and the Identification and Analysis of Clay Minerals*, 100–200. Oxford University Press.

Nanny, M. A. & Minear, R. A. (1997) Characterization of soluble unreactive phosphorus using ^{31}P nuclear magnetic resonance spectroscopy. *Marine Geology*, **139**, 77–94.

Nelson, D. W. & Sommers, L. E. (1996) Total carbon, organic carbon and organic matter. In: *Methods of Soil Analysis, Part 3, Chemical Methods*, J. M. Bartels (ed.), chapter 34, 961–1010. Madison, USA: SSSA.

Nowicki, W. & Nowicka, G. (1991) Coagulation kinetics. *Journal of Chemical Education* **66**, 523–25.

Pauling, L. (1929) The principles determining the structure of complex ionic crystals. *Journal of the American Chemical Society*, **51**, 1010–26.

Pauling, L. (1947) Atomic radii and interatomic distances in metals. *Journal of the American Chemical Society*, **69**, 542–53.

Pavia, D. L., Lampman, G. M. & Kriz, G. S. (2001) *Introduction to Spectroscopy: a Guide for Students of Organic Chemistry*, 3rd edition. South Melbourne: Thomson Learning, Brooks/Cole. 579pp.

Paytan, A., Cade-Menun, B. J., McLaughlin, K. & Faul, K. L. (2003) Selective phosphorus regeneration of sinking marine particles: evidence from ^{31}P-NMR. *Marine Chemistry*, **82**, 55–70.

Pecora, R. (1985) *Dynamic Light Scattering Applications of Photon Correlation Spectroscopy*. New York: Plenum Press, 436pp.

Poleto, C. (2007) Fontes potenciais e qualidade dos sedimentos fluviais em suspensão em ambiente urbano. UFGRS-IPH. 137pp. Doctoral thesis.

Poleto, C., Bortoluzzi, E. C. & Merten, G. H. (2007) Uso de ultra-son como pré-tratamento de amostras de sediemento fluvial em difratometria laser. In. *Sedimentos: o desafio da multidisciplinaridade*. G. H. Merten, C. Poleto, & A. L. O. Borges (eds), 51–66. Porto Alegre: ABRH.

Provencher, S. W. (1982a) A constrained regularization method for inverting data represented by linear algebraic or integral equations. *Computer Physics Communications*, **27**, 213–27.

Provencher, S. W. (1982b) CONTIN: a general purpose constrained regularization program for inverting noisy linear algebraic and integral equations. *Computer Physics Communications*, **27**, 229–42.

Reynolds, R. C. Jr & Reynolds, R. C. (1996) NEWMOD for Windows©: the calculation of on-dimensional X-ray diffraction patterns of mixed-layered clay minerals. 8 Brook Road, Hanover, New Hampshire 03755, USA. 25pp.

Robert, M. & Tessier, D. (1974) Méthode de préparation des argiles de sols pour les études minéralogiques. *Annalles Agronomique*, **25**, 859–82.

Rosén, P. & Persson, P. (2006) Fourier-transform infrared spectroscopy (FTIRS), a new method to infer past changes in tree-line position and TOC using lake sediment. *Journal of Paleolimnology*, **35**, 913–23.

Schmitz, K. S. (1990) *An Introduction to Dynamic Light Scattering by Macromolecules*. New York: Academic Press, 895pp.

Schurtenberger, P. & Newman, M. E. (1993) Characterization of biological and environmental particles using static and dynamic light scattering. In: *Environmental Particles, 37–115*, vol. 2. Boca Raton, FL: Lewis.

Schubert, C. J. & Nielsen, B. (2000) Effects of decarbonation treatments on $\delta^{13}C$ values in marine sediments. *Marine Chemistry*, **72**, 55–59.

Schulze, D. G. (1989) An introduction to soil mineralogy. In *Minerals on Soil Environments*, 2nd edition, J. B. Dixon & S. B. Weed (eds). Madison, USA: Soil Science Society of America, 1244pp.

Seta, A. K. & Karathanasis, A. D. (1996) Water dispersible colloids and factors influencing their dispersibility from soil aggregates. *Geoderma*, **74**, 255–66.

Silverstein, R. M., Webster, F. & Kiemle, D. J. (2005) *Spectrometric Identification of Organic Compounds*. 7th edition. New York: John Wiley, 512pp.

Skoog, D. A., Holler, F. J. & Nieman, T. A. (1998) *Principles of Instrumental Analysis*. 5th edition. Philadelphia: Harcourt Brace College Publishing, 942pp.

Sposito, G. (2004) *The Surface Chemistry of Natural Particles*. New York: Oxford University Press, 242pp.

Stock, R. S. & Ray, W. H. (1985) Interpretation of photon correlation spectroscopy data: a comparison of analysis methods. *Journal of Polymer Science: Polymer Physics*, **23**, 1393–47.

Stumm, W. (1993) Aquatic colloids as chemical reactants: surface structures and reactivity. *Colloids and Surfaces A: Physicochemical and Engineering Aspects*, **73**, 1–18.

Wetzel, D. L. (1983) Near-infrared reflectance analysis: sleeper among spectroscopic techniques. *Analytical Chemistry*, **55**, 1165A–76A.

Wilson, M. A. (1987) *NMR Techniques and Applications in Geochemistry and Soil Chemistry*. Oxford: Pergamon Press, 353pp.

Wirrmann, D., Bertaux, J. & Kossoni, A. (2001) Late Holocene paleoclimatic changes in Western Central Africa inferred from mineral abundance in dated sediments from Lake Ossa (Southwest Cameroon). *Quaternary Research*, **56**, 275–87.

Wyatt, P. J. (1998) Submicrometer particle sizing by multiangle light scattering following fractionation. *Journal of Colloid and Interface Science*, **197**, 9–20.

van Olphen, H. (1977) *Clay Colloid Chemistry*. New York: John Wiley, New York, 318pp.

4 Trace elements in urban environments: a review

Susanne Charlesworth[1], Eduardo De Miguel[2] & Almudena Ordóñez[3]

[1]*Department of Geography, Environment and Disaster Management, Coventry University, UK*
[2]*Environmental Geochemistry Group, Madrid School of Mines, Spain*
[3]*Oviedo School of Mines, University of Oviedo, Spain*

4.1 Introduction

Several facts reveal clearly the need to gain a better understanding of the behavior of urban environments and the consequences of living within or close to a city's boundaries. It was estimated by Mock (2000), that urban and built areas occupied more than 471 million hectares (1 hectare = $10^4\,m^2$), which amounted to about 4% of land area. According to the United Nations Department of Economic and Social Affairs (UNESA 2008), the area covered by cities in the United States roughly doubled from 1950 to 1990 and that whereas 3.3 billion people lived in cities in 2007, it is projected that this figure will nearly double by 2050 to 6.4 billion.

It is not only the gross increase in urban population, rather its relative growth, that highlights the increasing relevance of urban environments as human habitats. It is estimated by UNESA (2008) that nearly all the population growth in the next 30 years will be concentrated in the urban areas of the world. The percentage of urban population will accordingly increase from 50% of the total world population in the year 2007 to 69.6% in 2050. It is interesting to note that in 1950 that percentage was a mere 29.7%. If the statistical analysis is restricted to the more developed regions, it is projected that the percentage of urban population in 2050 will reach 86%, up from 76% in 2000.

Given that cities have become the habitat where most the world population is housed, it is no surprise that urban geochemists have from the beginning concentrated on the environmental aspects of the geochemical problems they have researched. A clear emphasis has always been placed on toxic elements and compounds, on the materials the population would more readily be exposed to (house dust, playgrounds dirt, etc.) and on the chemical forms of a given element that would result in more serious adverse health effects (speciation).

4.2 Urban particulate materials

The earliest efforts in urban geochemistry and the largest body of results it has produced, are concerned with the levels of trace elements in the solid materials humans may be exposed to in urban environments: atmospheric suspended particles, street dust, house dust, and soil. Much research has been devoted to the identification and characterization of the urban and non-urban sources of those trace elements, both in qualitative terms (which elements are associated to which sources) and quantitative terms, namely relative contributions of different emission sources to the total amount of a given element in a given urban material. With the refinement in analytical techniques and the growing awareness of the potential health effects of trace elements in particulate form, the range of interests of urban geochemists have widened to include the problems of speciation, modes and rates of transfer between different urban media, estimates of exposure through different routes (inhalation, ingestion, dermal adsorption (Bowman *et al.* 2003)), potential adverse health effects, etc.

Lead has been by far the most extensively researched trace element in urban environments because of its potential toxicity, widespread occurrence in urban particulate materials, and well-established main urban source, namely traffic. The principal developments in urban geochemistry listed

Sedimentology of Aqueous Systems, 1st edition.
Edited by Cristiano Poleto and Susanne Charlesworth.

in the previous paragraph have generally taken place first in investigations on lead, and since the now widespread introduction of unleaded fuels, have subsequently been applied to other trace elements. However, it comes as no surprise that historically, the largest body of published geochemical research on urban particulate materials is concerned with lead, a fact that is reflected in the following sections.

4.2.1 Urban aerosols

Concern over the quality of urban air has driven the need for targets to improve emissions with legislative controls implementing the agreed improvements (Williams 2004). These targets have been based on the sources of emissions, beginning in the UK in the 1950s as a reaction to the "London smogs" of 1952 and concentrating mainly on smoke. However, suspended particles in an urban aerosol can have their origin outside the city limits or in typically urban sources (i.e. vehicular traffic, domestic heating systems, etc.) and much research has concentrated on attempts to differentiate natural from anthropogenic contributions (discussed later in section 4.2.2). There is a general agreement (cf Van Dingenen *et al.* 2004) that the size of urban suspended particles follows a bimodal distribution, in which particulate matter of a "natural" origin (resuspended soil and mineral particles) constitutes the coarsest fraction of the urban aerosol, while particles emitted from anthropogenic sources (combustion processes, in most cases) are smaller, with a diameter usually below 2 μm. Investigations in different cities have concluded that the dominant particle size in urban environments lies in the sub-micron size range (Oberdörster *et al.* 1995; Kasparian *et al.* 1998).

Size and chemical composition determine the potential health effects of atmospheric particles. Particulate matter with a diameter below 10 μm (PM_{10}) is considered "inhalable", whereas atmospheric particles with a diameter less than 2.5 μm ($PM_{2.5}$) are regarded as "respirable". The $PM_{2.5}$ fraction has been found to be associated with adverse health effects, such as mortality and asthma (Kappos *et al.* 2004), as well as with ambient air quality problems, including visibility reduction (Larson *et al.* 1989; Lin & Tai 2001). Consequently, research efforts on the geochemistry of the urban aerosol have mostly concentrated on the finest fraction, although some researchers (e.g. Kappos *et al.* 2004) have reported difficulties in assessing the health effects of ultra fine particles i.e. those less than 0.5 μm. Some of the trace elements of toxicological concern in the atmospheric aerosol include As, Cd, Cr, Hg, Mn, Ni, Pb, and V. Exposure to airborne compounds of these elements in occupational environments has been suspected of causing effects ranging from sinusitis, asthma and chronic bronchitis to pneumonia, lung hemorrhage, lung cancer, and brain hemorrhage (Doadrio 1984; Sadiq & Mian 1994; Crosby 1998).

Although the possible health effects of exposure to those elements and compounds in open, urban atmospheres are difficult to ascertain, their potential toxicity has nonetheless encouraged much scientific research into the sources and levels of particulate trace elements in urban aerosols. Among the most relevant emission sources of urban suspended particles, the following can be cited.

4.2.1.1 Traffic

The emission source most thoroughly researched in urban environments is automotive traffic. The particle size distribution of exhaust aerosols is strongly affected by driving patterns; for example, freeway exhaust particles usually exhibit median diameters close to 0.1 μm, whereas urban driving causes a distinct shift towards coarser particle sizes, probably around 5 μm and larger. Urban traffic has in the past contributed large amounts of lead to the atmospheric aerosol as a consequence of the use of leaded petrol in internal combustion engines. Kowalczyk *et al.* (1978) concluded that the absolute concentration of Pb associated with motor vehicle particles could range from about 40% if there is little contribution from diesel traffic, to 4% when the contribution of diesel traffic is significant (the large amount of carbonaceous particles emitted by diesel vehicles exerts a noticeable diluting effect). However, the gradual shift from leaded to unleaded petrol has drastically reduced vehicular emissions of this element, to the extent in fact that some countries have dropped lead from their atmospheric monitoring programs, concentrating instead on Zn and Cu from the original five metals of concern which included Cd, Pb, and Ni (Foster & Charlesworth 1996). Although studies have shown a reduction in the lead concentration in atmospheric particulates, others monitoring lead concentrations in urban soils and deposited dust have reported still significant levels of the element in solid material (e.g. Charlesworth *et al.* 2003), which reflects the storage of historical lead levels. This is discussed further in the sections on street dusts and urban soil. As well as the Zn and Cu mentioned above, traffic also contributes significant amounts of Ba, Cd, and Ni, a detailed account of the origin of which is provided in the next sections, devoted to street and house dust. The relative contribution of traffic to the trace element load in urban particulate materials has been evaluated alternatively by the ratios Ba/Pb, Br/Pb, and, using factor analysis, by the scores on a factor that includes Pb, Cu, Ba, and Zn (Kowalczyk *et al.* 1982; Sturges & Harrison 1986; Boni *et al.* 1988; Cornille *et al.* 1990; Paterson *et al.* 1996; De Miguel *et al.* 1999; Viana *et al.* 2006). Recent modifications to Factor Analysis, Positive Matrix Factorization, has been used with multiple sources to give "significant information on anthropogenic sources" (Mazzei *et al.* 2008, p. 87). However, studies using radioactive isotope ratios, for instance lead and carbon (Widory *et al.* 2004; Chen *et al.* 2005) have enabled the possibility of finer discrimination with Widory *et al.* (2004) using carbon isotopes to differentiate between diesel emissions and those of fuel oil, although they do admit that these conclusions are "subject to debate" (p. 959).

With the introduction of catalytic converters in the mid-1970s in the USA and mid-1980s in Europe, and increasing use of multi-element analytical techniques such as that afforded by inductively coupled plasma-atomic emission spectroscopy (ICP–AES), it was realized that the so-called platinum group elements (PGEs) or platinum group metals (PGMs), which include Pt, Pd, Rh, Ru, Ir, and Os (Ravindra *et al.* 2004) had begun to accumulate in the environment. In fact, Barbante *et al.* (2001) estimated that Pt from vehicle catalytic converters alone could release up to 1.4 tonnes of Pt per year globally, and Schäfer *et al.* (1999) found that the daily deposition rate for Pt in a typical urban site could reach $23\,\mathrm{ng\,m^{-2}}$. It was found (Palacios *et al.* 2000) that these elements bioaccumulate and are transported in the ultrafine particle sizes, generally less than $0.39\,\mu\mathrm{m}$, at sizes considered inhalable and therefore of most concern to human health (Kappos *et al.* 2004). Ravindra *et al.* (2004) provide a detailed review of PGE levels in environmental materials, including the urban aerosol and their subsequent health effects which is beyond the scope of this chapter, but other studies since then have concentrated on the distribution of PGEs in deposited road and street dust which will be covered in the sections which follow.

4.2.1.2 Domestic heating, coal and oil combustion

Depending on the fuel burnt for domestic heating, its emission profile can vary noticeably, for instance, coal combustion is one of the main sources of Mn, Cr, Cu, Co, As, and Se but its exact emission profile depends on the type of coal burnt. Estimates of the relative contribution of coal combustion to the trace element load of the urban aerosol are hindered by the fact that its emission profile coincides largely with that of soil resuspension, owing to the similarity between the alumino-silicate matrix of soil particles and that of coal fly ash (Kowalczyk *et al.* 1978; Tomza 1984). However, a differentiating factor between both sources of particulate material, according to these authors, would be the relative enrichment in As and Se and depletion in Mn of coal fly ash. Other elemental markers used to trace the influence of coal combustion include Al, Si, and Ti (Kowalczyk *et al.* 1978; Pacyna 1991; Rose *et al.* 1994).

Vanadium and, to a lesser extent, Ni and S have been almost universally used as tracers of oil combustion (Kowalczyk *et al.* 1978, 1982; Boni *et al.* 1988; Cornille *et al.* 1990; Sadiq & Mian 1994), although some authors have assigned up to 40% of all the vanadium in the aerosol of an arid area to shale-like soil resuspension (Cornille *et al.* 1990). The exact contribution of oil combustion to the urban aerosol is difficult to ascertain because its emission profile depends greatly on the origin of the oil (Kowalczyk *et al.* 1978).

4.2.1.3 Resuspension of soil and street dust particles

Soil and street dust particles can be lifted by wind currents and incorporated into the urban aerosol, where they represent a significant proportion of its coarse fraction (Harrison *et al.* 1974). Although wind is clearly one of the main resuspension agents, vehicular and pedestrian traffic (see, for example,

Kupiainen 2007; Patra *et al.* 2008), agricultural activities, street sweeping (Yuan *et al.* 2003), and construction operations also contribute to this process. Soil resuspension is probably the main source of K, Mg, and Mn in an urban aerosol, and together with coal combustion (whose emission profile, as discussed above, is very similar and difficult to individualize) should provide a significant amount of Al, Ca, Ce, Cr, Fe, La, Sc, Sr, Ti, and Th (Kowalczyk *et al.* 1978; Boni *et al.* 1988). Street dust and the fine soil fraction are enriched in anthropogenic trace elements relative to coarser soil particles. If resuspended, they can make a notable contribution to the trace element load of the inhalable fraction of an urban aerosol, for instance a study by Laidlaw & Filippelli (2008) found significant health risks to young people from resuspended soil contaminated with Pb both outside and in the home, leading to blood lead levels (BLLs) in children in excess of $10\,\mu m\,dL^{-1}$ in some US cities. In Cairo, Egypt, Sharaf *et al.* (2008) found BLLs up to $14.3\,\mu m\,dL^{-1}$ in children living by heavily trafficked roads, and asserted than the CDC (2007) $10\,\mu m\,dL^{-1}$ advisory level is too high. However, soil particles that enter the urban aerosol can have their origin outside the city limits. The finest fraction of these "natural" particles that result from crustal erosion can travel long distances and their chemical makeup reflects the mineral composition of the original soil (Cornille *et al.* 1990).

4.2.1.4 Other urban sources

Traffic, domestic heating, and soil resuspension do not account for all the particulate matter that is emitted to the atmosphere in an urban environment. Other sources include specific industrial sources, incineration, construction activities, road weathering and maintenance, etc. The emission profile of refuse incineration depends on several factors (refuse composition, design of combustion chamber, efficiency of filters, and other particle collection equipment), but it has been reported that incineration is a major source of Zn, Cd, and Sb in the urban aerosol (Kowalczyk *et al.* 1978, 1982; Pacyna 1983). Wadge *et al.* (1986) found high levels of Pb and Cd in the finest fraction of refuse incineration fly-ash. The effects of building construction and renovation, and weathering of building materials are discussed in detail in the following section. It should be noted here, however, that Gatz (1975) has used Ca as a tracer of suspended cement dust. According to Kowalczyk *et al.* (1982), airborne cement particles should exhibit concentration ratios K/Ca of approximately 0.006 and Mg/Ca of approximately 0.16. Possibly associated with pulses of construction activity, high activities of radioactive nuclides have been found in some urban dusts in Coventry, UK (Charlesworth & Foster 2005). The highest activities were found in road gutter and street dusts where some samples approached, and even exceeded, the ICRP (1991) guidelines of $1\,mSv\,yr^{-1}$ for members of the public.

In northern countries, the use of spiked tires in the cold season results in the abrasion of the road surface. Dust particles thus generated are a major source of atmospheric particulate matter in clear winter days. Lastly, depending on the location and urban characteristics of a city, other specific sources of airborne particles might exert a significant influence. As an example, sea spray supplies considerable amounts of Na to the atmosphere of coastal cities (Kowalczyk *et al.* 1978) and studies such as that by Pryor *et al.* (2008) suggest that neglecting the interactions of sea spray in considerations of urban air quality may lead to misleading conclusions being drawn.

4.2.2 Source apportionment

The terms "source apportionment" and "receptor modelling" are used to describe the attempt to apportion the aerosol measured at a receptor site to its likely sources, making use of various mathematical models. The two most widely used categories of mathematical models are chemical mass balance (CMB) and multivariate models. The latter use multivariate analyses techniques (i.e. factor analysis, target transformation factor analysis, Q-mode factor analysis) to predict the number of relevant emission sources in the area and their individual contributions to a series of aerosol measurements (Harrison *et al.* 1997). Receptor models based on factor analysis, however, are not well suited for source apportionment when two or more emission sources in the study area have similar "signatures" or elemental emission profiles. Furthermore, an infinite number of models can be produced that will satisfy a given aerosol composition and all natural constraints, i.e.

predicted source compositions must be non-negative which means the sum of the predicted elemental mass fractions for each source must be less than or equal to 1 (Henry 1987).

Source apportionment based on CMB does not suffer from the same problems. CMB models assume that, in addition to the composition of the aerosol at the receptor site, the elemental composition of the emissions from the different sources is known. The individual contribution of each source at a particular receptor site can then be estimated by solving a system of linear equations (Gatz 1975; Kowalczyk *et al.* 1978, 1982; Batterman *et al.* 1988; Cornille *et al.* 1990; Adgate *et al.* 1998). Although CMB source apportionment has certain advantages over multivariate methods, its application is restricted to sites where the emission profiles of the main sources in the area are known, and the reliability of the results is limited by the accuracy in the estimates of the emission profiles.

Depending on its physical characteristics, the airborne particulate material may settle onto a surface of some kind in the urban area. The next sections follow these particles onto the street and thence indoors and assess their eventual risk to the environment as a whole and to those who live in it in particular.

4.2.3 Street dust

Whether a solid particle remains airborne or settles down onto an urban surface (pavement, soil, roof, window ledge, playground area, etc.) is related to its aerodynamic diameter and to weather conditions, the finest materials staying suspended for longer periods. Fine particles are preferentially removed from the urban aerosol by wet deposition, whereas coarse particles are sedimented by dry deposition (Jaffé *et al.* 1993). Solid particles that accumulate on outdoor, impervious materials are collectively referred to as "street dust", whereas particles found inside urban dwellings are commonly termed "house or indoor dust" (see section 4.2.4), which suggests that all this material is extremely fine (i.e. less than 10 μm) and therefore inhalable. However, studies of various urban environments (e.g. Sansalone *et al.* 1998) have found that most urban sediments are greater than 400 μm by mass, and therefore the term "dust" may be inappropriate. The largest particles of the urban aerosol are not the only source of street dust, which also incorporates a large amount of displaced urban soil as well as particles that never become fully suspended after they are emitted.

Studies of particle size are important, however, in identifying whether they pose a health hazard to the city's population. According to Horowitz (1991), there is a strong positive correlation between the decrease in the size of particles and the increase in the concentration of trace elements, depending on the greater surface area of the particle and the increase of the cation-exchange capacity (CEC). Particles less than 100 μm can reach the respiratory system by inhalation through the mouth or nose, but of that, only the fraction less than 10 μm can reach the alveoli of the lungs where they can cause irritation and disease. This is further explored in section 4.3.

Street dust does not remain deposited in place for a long time. In fact a study by Allott *et al.* (1990) in a coastal town in northwest England, using ^{137}Cs, found that the half-life of street dust was between 190 and 370 days. It is easily resuspended back into the atmospheric aerosol, to which it contributes a significant amount of trace elements (Maxwell & Nelson 1978), or precipitation washes it away becoming an important component of the suspended and dissolved solids in street run-off (Vermette *et al.* 1991 and references therein). Consequently, the temporal variability of the concentration of trace elements in street dust is high (Duggan 1984), and most studies do not monitor for long enough to evaluate it. Street dust also presents a pronounced small-scale heterogeneity (Duggan 1984; Leharne *et al.* 1992), a reflection not only of the mobility and of rapid environmental alteration of street dust, but also of the heterogeneity in the distribution of its urban sources.

As was mentioned in section 4.2.1, the two main sources of street dust, and consequently of the trace elements found therein, are deposition of previously suspended particles (atmospheric aerosol) and urban soil. However, there are several point sources whose emissions contribute directly to the street-dust load in their proximity (Harrison 1979; Hopke *et al.* 1980; Schwar *et al.* 1988). The most relevant among them is vehicular traffic.

Car exhaust emissions are responsible for elevated concentrations of Pb, Zn, Cd, Cu, and Ba in the vicinity of roads. Lead is obviously associated with the use of leaded petrol in internal combustion engines (Archer & Barrat 1976), although, as has been stated, as a consequence of the gradual shift to unleaded petrol, the contribution of traffic to the load of Pb in the street dust near busy streets and roads has been significantly reduced (De Miguel *et al.* 1997). High concentrations of Zn and Cd have been traditionally related to tire wear (Stigliani & Anderberg 1992; Fergusson & Kim 1991). Zn compounds are also used as antioxidants and as detergent/dispersant improvers in lubricating oils (Drew 1975), contributing to the influence of traffic on the Zn load in street dust. Some authors in the past, however, played down the role of traffic as a source of Zn, noting that this element is a negligible component of the granulated material associated with vehicular emissions (Pierson & Brachaczek 1976). However, when lead in petrol was reduced from 0.4 to $0.15\,g\,L^{-1}$ in 1985 throughout the UK and across countries in the European Union, the atmospheric levels of lead in those countries "more than halved in a matter of weeks" (Williams 2004, p. 19). In fact, it was found by Charlesworth *et al.* (2003), that Pb concentrations in street dusts from Birmingham, West Midlands, UK, had declined over a nearly 30 year period from an average of $1300\,mg\,kg^{-1}$ in residential streets in 1976 (Davies *et al.* 1987) to $48\,mg\,kg^{-1}$ by 2003. As a result, the focus of research settled on both Zn and Cu, with the suggestion being made by Wong *et al.* (2006) that an inventory of their isotopic signatures be made, similar to that for Pb in order to assist in identifying and eventually quantifying their sources. Dispersions of Ba are widely used as detergents/dispersants and oxidation and corrosion inhibitors in lubricating oils for diesel and other combustion engines, and as smoke suppressant additives in diesel fuels. This fact explains the association of high Ba concentrations in street dust and circulation of diesel vehicles (Kowalczyk *et al.* 1982). Oxidation of lubricating oils upon exposure to air at high temperatures results in the formation of organic acids, alcohols, ketones, aldehydes, and other organic compounds that are corrosive to metal. This corrosive action causes wear of those metal parts that come into contact with the oil and which in many cases consist of zinc-, copper-, and cadmium-bearing alloys (Drew 1975) or, as in the case of sinterized materials used in automobiles' oil pumps, of Ni, Cu, and Mo. This process results ultimately in the release of those metals to the urban environment and their accumulation in street dust (De Miguel *et al.* 1997).

Although particles emitted directly from combustion engines usually lie in the range less than $1\,\mu m$ under normal driving conditions, at least two facts alter this size distribution. Stop–go activities, and acceleration–deceleration, situations commonly met in urban traffic (Ellis & Revitt 1982; Kim *et al.* 1998), cause resuspension and emission of larger particles that accumulate on exhaust systems. Also, fine atmospheric and street dust particles undergo an intense process of condensation growth and agglomeration that results in aggregates of larger size, as revealed under electronic microscope inspection of these materials (Dongarrà *et al.* 2003). This increase in particle size explains the fact that a large percentage of the trace elements emitted from automobile exhausts does not travel far, but is deposited on the soil or impervious surfaces (street dust) close to the road, street, or motorway where it originated (Raunemaa *et al.* 1986; Warren & Birch 1987).

Another localized source of trace elements in street dust is the weathering, construction, renovation, and redecoration of buildings and building materials. Corrosion of galvanized-metal structures (roofs, balconies, window ledges, etc.) contributes large amounts of Zn and Cd to street dust (Fergusson & Kim 1991). This process can raise the concentration of both elements in street dust to values of $44{,}000\,\mu g\,g^{-1}$ of Zn and $20\,\mu g\,g^{-1}$ of Cd in the particulate material collected from under the metal ledges and balconies of old buildings (De Miguel *et al.* 1997). High levels of Ca can be related to the presence of cement dust, especially if the ratio Mg/Ca is close to 0.14 (Kowalczyk *et al.* 1982). The most widely researched and clearly asserted influence of these activities and processes on the trace element load of street dust refers to the elevated levels of Pb, and to a lesser extent Cd, associated with the accumulation of paint flakes from deteriorating old facades or recently redecorated walls (Rundle & Duggan 1986; Davies *et al.* 1987; Schwar *et al.* 1988; Fergusson & Kim 1991). The interest in this source of lead in street dust arose from evidence that children were readily exposed to these particles, both at home and at school (Duggan *et al.* 1985), and that this exposure resulted in toxic effects (Harvey *et al.* 1985; Mielke *et al.* 1999).

4.2.4 Indoor dust

The main sources of indoor house dust include soil and street dust particles; these are carried indoors adhered to clothes and shoes, swept indoors by wind drafts, or even brought inside on the fur of domestic animals (Tong 1998). The relative contribution of the urban aerosol and of indoor sources of trace elements (cooking and other combustion processes, rubber, wall paint, fabrics, pigments) has not been conclusively evaluated (Fergusson & Kim 1991; Adgate *et al.* 1998; Edwards *et al.* 1998). As in the case of street dust, concern over inhalation, ingestion, and dermal exposure to house dust has fuelled research on this material. House dust has been cited as one of the major sources of exposure to pesticides and metals, particularly lead, in children (Edwards *et al.* 1998 and references therein). Turner & Simmonds (2006) reported that, in common with many other studies worldwide, enrichment of Cd, Cu, Pb, Sn, and Zn in dusts from four regions across the UK were of concern. However, Tong & Lam (2000) found that activities such as floor sweeping and dusting reduced the levels of metals in houses in Hong Kong, although the type of paint used to decorate the house and its age were of significance when determining indoor metal levels. Chattopadhyay *et al.* (2003) found that, whereas atmospheric concentrations of Pb have reduced since the introduction of unleaded petrol (section 4.2), that of household dusts in Sydney, Australia, have remained essentially unchanged. This, they assert, is due not only to the accumulation of Pb inside the house from the use of old leaded paints, but also the historical accumulation of more than 80 years of leaded petrol deposition in the urban area. As was mentioned above, one of the vectors for the transport of contaminants indoors is soil. It has been found that soil can act as a repository for historical contamination not only caused by traffic, but also industry (see, for example, Charlesworth *et al.* 2003). The next section considers levels and sources of soil contamination in urban areas.

4.2.5 Urban soil

The term "urban soil" can be understood to encompass all types of non-paved land within the city limits: public and private green areas (parks and gardens), undeveloped land, building lots, etc. Whereas the characterization of street dust offers an instantaneous "snapshot" of an urban environment's condition, urban soil rather acts as a pollutant sink and, if undisturbed, preserves the cumulative history of trace elements inputs into it (although not in the orderly, sequential fashion of an urban lake sediment).

However, soil particles do not necessarily remain in place, but can become part of street dust or even of the urban aerosol. Particles smaller than 100 μm move in "suspension" and the finest among them may remain airborne for prolonged periods of time. The process of suspension is all the more intense if the small particles are accompanied by particles moving by "saltation", which upon landing back on the surface will help to lift the finest material (Sehmel 1980; Nicholson 1988). Consequently, exposure to trace elements in urban soil does not occur solely by ingestion or dermal contact but also through inhalation of resuspended soil particles. However, the most immediate route of exposure for children, the most sensitive segment of the population, is hand-to-mouth activity during games and the habit of "pica", i.e. mouthing of non-food objects. As has been outlined in section 4.2.3, several investigations have suggested that urban soil and dirt represent a significant intake of trace elements for children living in urban areas (see Biggins & Harrison 1980 and references therein; Rundle *et al.* 1985; Watt *et al.* 1993; Abrahams 2002).

Although generally lower than those found in street dust, urban soil can contain enriched levels of trace elements relative to natural background levels (see, for example, Charlesworth *et al.* 2003; Charlesworth & Foster 2005; Biasioli *et al.* 2007). The main sources of these trace elements include the atmospheric deposition of particles generated by automotive traffic, heating systems, and resuspended street dust, the uncontrolled disposal of urban and commercial wastes, and the addition of fertilizers and composted sewage sludge to the soil (Carey *et al.* 1980; Haines 1984; Warren & Birch 1987; Fergusson 1990; Kabata-Pendias & Pendias 1992; Tiller 1992; Strnad *et al.* 1993; Sánchez-Camazano *et al.* 1994; De Miguel *et al.* 1998; Imperato *et al.* 2003; Shi *et al.* 2008; Yesilonis *et al.* 2008).

The exact contribution of each single source to the load of trace elements in urban soils is difficult to quantify, because all the various inputs are integrated in the soil over time, and urban soils are periodically disturbed by landscaping, construction, irrigation, and partial or total replacement, to name a few. Nevertheless, certain general conclusions can be drawn. Firstly, the influence of atmospheric deposition is fairly uniform across the city and gives rise to "urban background" levels of trace elements, which are higher than those in natural soils. De Miguel *et al.* (1998) cite enrichment factors of 2.3, 2.6, and 4.0 for Zn, Cu, and Pb, respectively, in the urban soil of Madrid relative to natural background levels. If the atmospheric aerosol has received contributions from industrial sources, the increase in trace element soil concentrations relative to natural levels is much more pronounced. Ordóñez *et al.* (2003) found concentrations of Zn and Cd as high as $2000\,\mu g\,g^{-1}$ and $8\,\mu g\,g^{-1}$, respectively, downwind from a Zn smelter. Moreover, the influence of atmospheric deposition is not restricted to soils within the city limits. The effect of the atmospheric fallout from the city of Madrid can be significantly noticed in soils up to a distance of 15 km from the city centre for Pb, Cu, and Zn (Fig. 4.1), decreasing abruptly or disap-

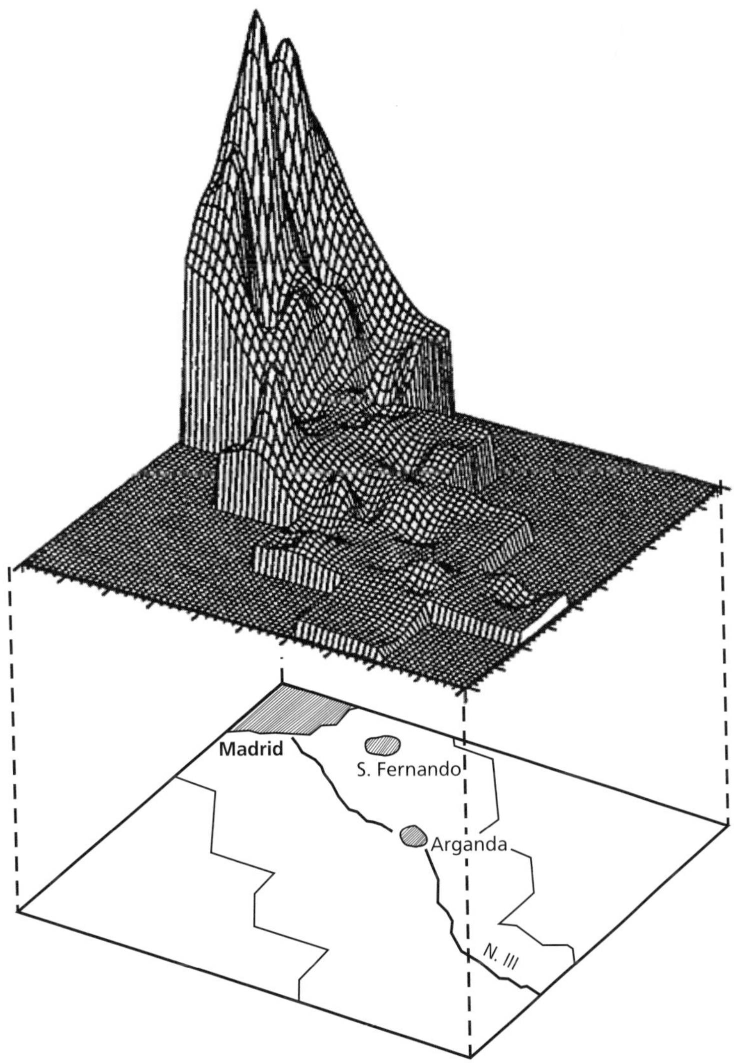

Fig. 4.1 Influence of the city on soil lead concentrations in and around Madrid.

pearing totally beyond that distance (Llamas *et al.* 1993). Charlesworth *et al.* (2007) plotted the distribution of Zn, Ni, Cu, Cd, and Pb across the city of Coventry, UK, and found "hot spots" associated with the heavily-trafficked main roads and industrial areas, but these elevated concentrations were similar to trends found in Madrid (as explained above) in that they had reduced considerably at the city limits.

Unlike the influence of atmospheric deposition, the disposal of urban and commercial wastes, and the addition of fertilizers and composted sewage sludge, have a very localized effect on the trace element content of urban soils. If present, however, these sources can contribute a larger amount of several trace elements to the urban soil than atmospheric deposition. De Miguel *et al.* (1998) found that urban soils amended with composted sewage sludge presented levels of Cu, Ni, Pb, and Zn that were two to three times higher than those in urban soils that did not receive compost additions. Consequently, the highest levels of metals in soil were detected in some of the best-kept parks and gardens in the city, where fertilizing takes place on a regular basis. Concern over the potential implications of sewage sludge application, in terms of increased trace element load in soil, has fuelled research and legislative actions in this field (Giusquiani *et al.* 1992; Tiller 1992; Chaney & Ryan 1994; Gies 1997; Berti & Jacobs 1998).

Urban soil not only acts as a net accumulator of trace elements but also provides a significant amount of them to the atmospheric aerosol and, particularly, to street dust. An example of this role of the urban soil is provided by De Miguel *et al.* (1997), who found that some of the highest concentrations of lead in the street dust of Oslo, Norway, were not associated with dense traffic but with nearby soils where lead had accumulated over long periods of time from a lead smelter that was shut down several years before the street dust sampling campaign took place.

Some of the sources of street dust have been established in this section, and the fact that they can become entrained and transported in the atmosphere. Previously, it has been established that there are hazardous contaminants stored in various urban environmental compartments. Section 4.3 considers whether their presence constitutes a risk to the environment as a whole, or arguably more importantly, whether they constitute a risk to human health. This involves a consideration of how urban geochemistry can be modified to take account of whether pollutants are bioavailable or bioaccessible, and the ways in which potential risk can be estimated.

4.3 Risk and health implications

Urban geochemistry has an obvious focus on the environmental aspects of life in the city. It is not surprising, therefore, that one of the major research interests in this field concerns the potential adverse health effects of exposure to urban pollutants. Until recently, with few exceptions, most studies had either established an inferred link between elevated concentrations of toxic elements in street dust and soil and the observed incidence of a given effect in a population, or had directly equated risk with predominance of mobile chemical species, as determined in sequential or selective extraction protocols (Banerjee 2003; Robertson *et al.* 2003). The ecotoxicological significance of trace elements in street dust has also been directly evaluated by means of bioassays (Wang *et al.* 1998), instead of being indirectly inferred from the results of a sequential extraction procedure as was introduced by Tessier *et al.* (1979). In the past few years, risk assessment strategies – extensively employed by regulatory authorities to define soil screening levels or soil guideline values – have increasingly been adopted and, when necessary, adapted to the peculiarities of urban environments to appraise the relevance of toxic elements and compounds in urban matrices (Boyd *et al.* 1999; Granero & Domingo 2002; Korre *et al.* 2002; Wcislo *et al.* 2002; Hemond & Solo-Gabriele 2004; Nadal *et al.* 2004; Ferreira-Baptista & De Miguel 2005; Kim *et al.* 2005; Lee *et al.* 2005; De Miguel *et al.* 2007).

Strategies such as those outlined above are based on the separate assessment of (a) the toxicity of the chemicals included in the analysis by exposure route (i.e., inhalation, ingestion, and dermal contact), and (b) the levels of exposure to those chemicals for the potential receptors. For non-carcinogenic toxicants, a range of exposures from zero to some finite value (reference dose or acceptable/tolerable daily intake) are assumed to be tolerated by the organism with essentially no expression of the toxic effect. If the daily dose to which a receptor is exposed exceeds the corresponding reference dose, the receptor is considered to be potentially at risk. On the other hand, there is no level of exposure to a genotoxic carcinogen that does not pose a small but finite probability of generating a carcinogenic response. Risk to the exposed individual is measured as the product of the lifetime-average daily dose times a "slope factor", defined as the incremental probability of developing cancer during a lifetime owing to chronic exposure to a unit dose of contaminant. This probability must not exceed a subjective level of risk (in the range 10^{-4}–10^{-6}) deemed acceptable by the corresponding regulatory authorities. By jointly considering toxicity and level of exposure, risk assessment allows the identification of the elements and pathways of most concern during exposure in the environment.

However, it is a common practice to estimate the concentration of trace elements in urban matrices as the fraction extracted with aqua regia or similar "strong" digestion protocols. Because only a portion of this pseudo-total content will be released in the stomach and absorbed in the intestine, this approach may lead to an overestimate of risk, particularly when these elements in urban particulate materials are strongly bound to their mineral matrix. As a result, many digestion protocols have been developed that attempt to mimic conditions whereby the metals can be assimilated into living organisms. The following sections introduce these techniques.

4.3.1 Speciation

The different digestion procedures used in speciation studies can be broadly divided in two groups: "selective" and "sequential" extractions. "Selective" extractions attempt to digest and analyze only that fraction of the sample of environmental relevance, i.e. the fraction of the total that would become dissolved and therefore mobile and available under realistic environmental conditions, or that could be incorporated by the human organism upon exposure to it. A common approach involves extraction with hydrochloric acid at different concentrations, in some cases attempting to simulate the conditions existing in the human stomach (Day *et al.* 1979; Harrison 1979; Serrano-Belles & Leharne 1997). A more sophisticated approach is mentioned by Evans *et al.* (1992), according to which the amount of bioavailable trace elements should be evaluated as for foods by digesting the sample with synthetic gastric juice, of pH 3.5, for 4 hours (Analytical Methods Committee 1985). Under these conditions the exchangeable and carbonate fractions described below should be released (Evans *et al.* 1992). Other "selective" extractions use "mild" extracting agents, such as EDTA, citric acid, acetic acid, ammonium, or sodium acetate, etc.

The most comprehensive approach to the problem of speciation involves a "sequential" extraction, in which trace elements associated with the different chemical phases present in the particulate material are extracted separately. Almost all "sequential" procedures are based on the analytical protocol proposed by Tessier *et al.* (1979) or modifications thereof (Gibson & Farmer 1984). In it, the total amount of a trace element is separated in five fractions: soluble and exchangeable; carbonate bound; bound to Fe–Mn hydrous oxides; bound to organic matter; and residual fraction. The first two fractions (exchangeable and carbonate-bound) are regarded as readily bioavailable, whereas the fraction associated with Fe–Mn hydrous oxides and organic matter should only be available under severe environmental conditions. The residual fraction is essentially unavailable. The many analytical approaches used to assess the bioavailability and transport mechanisms of particulate-associated contaminants has led to problems of comparability between studies and also the assignment of environmental relevance to the speciation found. Some techniques appear to be more efficient than others (Agemian & Chau 1976) whereas others encourage the redistribution of some elements during the fractionation process (Ajayi & Vanloon 1989). In fact, it has been suggested (Breward *et al.* 1996) that two or more schemes be used on the same samples to elucidate metal binding sites better.

The speciation of urban deposits has elicited much study. In summary, street and house dust shows a fairly good agreement with the general trends in trace-element partitioning among the different phases. Lead is preferentially associated with the carbonate and Fe–Mn oxide fractions, and to a lesser extent with the exchangeable fraction; Cu is predominantly bound to the organic fraction; Zn follows the behavior of lead and seems to be bound to the carbonate and Fe–Mn oxide fractions; and Cd is associated with the first two fractions and shows the highest affinity of all these elements for the exchangeable fraction (Harrison *et al.* 1981; Gibson & Farmer 1984; Evans *et al.* 1992; Wang *et al.* 1998; Charlesworth & Lees 1999). According to these results, the mobility of trace elements in street dust follows the sequence: Cd > Pb/Zn > Cu, and the concern about their environmental implications should, perhaps, observe the same order. The carbonate fraction becomes less important in urban soils and the relevance of the Fe–Mn oxides, organic matter, and residual fraction increases (Zimdahl & Skogerboe 1977; Harrison *et al.* 1981; Gibson & Farmer 1984; Evans *et al.* 1992; Serrano-Belles & Leharne 1997). This fact probably arises from several causes, among them the lower abundance of calcite, the lower pH, and higher concentration of organic matter in urban soil relative to street dust. Trace elements are consequently more tightly bound to soil than to street dust particles, a fact that corroborates the role of urban soil as a sink for pollutants. However, changes in the environmental conditions of the soil (pH, redox potential) might result in the release of part of the load of trace elements that have accumulated over time. However, there are many sequential digestion protocols and little consensus about which method is the most appropriate to use (Perez-Santana *et al.* 2007). It is also felt that elucidating the binding sites on particulates does not give the kind of information required when assessing impacts on living tissue.

4.3.2 Bioaccessibility and bioavailability

The toxicity values used in risk assessments for the route of ingestion are expressed in terms of absorbed doses and are often derived from assays that employ soluble salts or other easily available chemical forms of trace elements. Consequently, human health risk assessments implicitly assume that the concentration term used in the standard equations to quantify exposure represents the amount of trace elements in the sample(s) that are available for absorption (i.e. bioavailable) in the gastrointestinal tract.

Bioaccessibility is normally defined as the fraction of the trace element content that is available in the gastrointestinal tract for transport across the intestinal lumen, whereas the term bioavailability usually denotes the ingested contaminant fraction that actually reaches the systemic circulation (not all the mass of metal released during its transit in the gastrointestinal tract will be absorbed). *In vivo* assays have evaluated bioavailability but they are expensive and complicated. They can be substituted – without great loss of accuracy – by *in vitro* tests (RIVN 2006). These *in vitro* assays simulate the biochemical environment, temperature, and duration of the different stages in the process of ingestion: grinding and first dissolution in saliva in the mouth; release of metals in the gastric juice of the stomach; and further release (or removal from solution) and absorption with duodenal juice and bile in the intestine (Oomen *et al.* 2002). Although undoubtedly easier to control than *in vivo* assays, *in vitro* experiments commonly produce uncertain and little reproducible results owing, among other difficulties, to the very large number and instability of reactants and solutions, and the fact that concentrations in chyme can be near or below quantification limits.

Probably as a consequence, most *in vitro* studies of urban particulate materials have focused on the bioaccessibility of trace elements in soil and dust, operationally defined as the maximum amount of metal that is soluble in a synthetic gastric fluid (Hamel *et al.* 1998). Most of these studies have used hydrochloric acid (adjusted to pH 1.5) as a surrogate for gastric juice, as in the European Standard Toy Safety Protocol EN-71 (European Committee for Standardization 1995; Rasmussen *et al.* 2008), and some have extracted the trace elements in the sample with glycine, again adjusted to pH 1.5 with concentrated HCl (Ruby *et al.* 1999; Madrid *et al.* 2008). The variability in the – sometimes contradictory – results arrived at by different researchers reflects the complex and numerous factors that influence how much of the total trace element load in a sample is bioavailable (element investigated, granulometry and mineralogy of the sample, organic carbon content, mode of retention, anthropogenic or natural origin, acid-to-sample ratio, etc.). Madrid *et al.* (2008) report bioaccessibility values (relative to an aqua regia extract) of up to 86% for Ni and 83% for Zn, and as low as 1% for Cr in the less than 2 μm fraction of soils from two different urban environments, and an order of bioaccessibility Ni = Zn > Pb > Cu > Cr for Seville and Pb = Cu = Zn > Ni > Cr in Turin. Rasmussen *et al.* (2008) analyzed the HNO_3 + H_2O_2 "total" and the HCl bioaccessible contents of the less than 150-μm fraction of samples from urban gardens and indoor dust in Ottawa. Their results suggest an order of bioaccessibility of Cu = Zn > Ni for soil, and Zn > Cu = Ni for dust. Moreover, not only were concentrations of metals in indoor dust higher than in outdoor soil, but also their bioaccessibility was determined to be 1.5–2.5 times higher (i.e. 44% for Cu and Ni, and 65% for Zn). Most studies of metal bioavailability in soils have been conducted in locations strongly affected by industrial or mining operations (Ruby *et al.* 1996; Williams *et al.* 1998; Rodriguez *et al.* 1999; Schroder *et al.* 2003). One of the few studies on uncontaminated urban soils was performed by Ljung *et al.* (2007) in Uppsala, Sweden, where they collected samples from the upper 10 cm of playground soil and analyzed it for aqua regia and bioavailable contents following the extraction protocol of Oomen *et al.* (2003). Bioavailability for the different elements included in the study followed the order: Cd (26%) > As (16%) > Pb = Cr = Ni (4%) for the less than 50-μm fraction, considered to represent the particulate material that children ingest accidentally in their games.

Despite the variability in extraction protocols, nature of particulate material, size fraction and estimated percentages of bioaccessibility and bioavailability, several conclusions can be drawn from the those studies listed above. Firstly, concentrations of metals in the neutral pH intestinal phase have been found to be lower than in the acidic stomach juice, particularly for those elements which are more easily re-adsorbed or precipitated under near-neutral conditions (i.e. Pb). The use in a risk assessment of bioaccessible concentrations of those elements, as determined in extractions with HCl, would overestimate the risk although not to the same extent as aqua regia or similar pseudo-total digestions, in which less than 5% of the resulting concentrations of Pb, Ni, and Cr may be available for absorption in the intestine. Metals in urban particles of a natural origin are generally more strongly bound and consequently exhibit a lower bioavailability than those associated with anthropogenic sources. Lastly, indoor dust presents higher concentrations and higher bioaccessibility than urban soil. As a consequence, risk assessments – which normally integrate dust and soil as one single exposure medium – may gain in accuracy if both sources of exposure were decoupled.

4.3.3 Risk in playgrounds

A large body of knowledge has been developed over the past decades on the exposure of children to urban particulate materials (i.e. soil, street dust, indoor dust), driven by the realization that children are the most sensitive segment of the population to anthropogenic contamination and by the strong indication that toxic trace elements may reach levels of potential concern for human health in urban environments (Evans *et al.* 1992; Mielke *et al.* 1999). Specifically, some researchers have concentrated their efforts on the chemical composition of playground soil and dust (Anagnostopoulos *et al.* 1985; Duggan *et al.* 1985; Wong & Mak 1997; Ng *et al.* 2003; Ljung *et al.* 2006a,b,c), because the exposure of children to trace elements in this material is particularly high – relative to other activities and other locations – during games at school breaks and in public playgrounds after school (Roscher *et al.* 1996). Wong & Mak (1997) performed a simplified risk assessment, comparing the heavy metal concentrations found in dusts and soils in Hong Kong playgrounds with the Dutch Soil Investigation Levels and concluded that Pb and Zn might pose a health hazard for children. De Miguel *et al.* (2007) collected samples from the top 2 cm of the sandy substrate in Madrid municipal playgrounds. A detailed risk assessment revealed that the highest risk experienced by children playing in these playgrounds arose from the ingestion of soil particles during games (Dudka & Miller 1999) followed by dermal absorption, and that the element of most concern among those present in the sandy substrate of Madrid playgrounds was arsenic in terms of both carcinogenic and non-carcinogenic risk (De Miguel *et al.* 2007).

In playgrounds with chromated copper arsenate (CCA)-treated equipment, children may experience a much higher exposure to As than that found in Madrid, as a result of not only soil ingestion and dermal absorption but more importantly of direct oral ingestion of dislodgable arsenic from wood (Stilwell & Gorny 1997; Hemond & Solo-Gabriele 2004). Whether this elevated exposure might result in an unacceptable level of risk is debatable: On the one hand, the default assumptions in the standard model of risk assessment are quite conservative and elevated estimates of hazard index or carcinogenic risk do not imply that exposure to playground substrate alone should result in the advent of adverse health effects. However, it should be noted that the children's background exposure to trace elements (i.e. dietary intake, inhalation of urban aerosol, indoor exposure to dust particles at home and school, etc.) would add to the exposure to playground soil and that the overall risk to children in urban environments has not yet been reliably calculated. The quantitative estimates of risk from exposure to urban particulate materials are also affected by a high degree of uncertainty arising from the estimates of exposure rates and from the toxicity data used in the risk assessment. Despite the numerous studies attempting to quantify exposure factors relevant to a risk assessment for children during playing activities, there is a significant variability in their numerical results (Evans *et al.* 1992; Buchardt-Boyd *et al.* 1999; USEPA 1997, 2002 and references therein; Hemond & Solo-Gabriele 2004), which reflects the difficulties involved. Besides, some of these factors, like exposure frequency, cannot be directly extrapolated from one survey to another because playing habits and time spent outdoors may differ substantially from one region to another. Additionally, quantitative estimates of the toxic potency of elements and compounds found in urban matrices are being reviewed permanently with considerable changes in their values and sometimes even in the threshold or non-threshold behavior of the toxicant.

Although these considerations suggest that the numerical results of risk assessments in urban environments should be interpreted with caution, they do not invalidate the potential of risk assessment to identify the contaminants of most concern and the most relevant routes of exposure.

4.4 Urban geochemical cycles

Having established the importance of studying the geochemistry of urban environments in terms of the risk imposed by potentially increasing concentrations of hazardous elements in concert with population increase (cf Charlesworth *et al.* 2003) the question of managing the risk arises. To apply management strategies, it is important to be able to predict where "hot spots" of contamination are likely to occur. Hence there have been many attempts to model urban geochemical cycles, with varying degrees of success.

Trace elements circulate between different urban media (i.e. atmospheric aerosol, street dust, urban soil, urban sediment) in the gas phase, in aqueous solution, and as particulate solids. Figure 4.2 shows a simplified model of the sources, pathways, and sinks that constitute the urban geochemical cycle.

The most important mode of transport for trace elements within the urban environment is probably as particulate materials. The fine fraction of these solid particles is especially relevant for two reasons. Firstly because, as previously discussed, particles with a diameter less than 100 μm can be resuspended and are easily transferred between soil, street dust, and atmospheric aerosol. Secondly, it is generally agreed that particles in the silt-clay size range have the highest capacity to bind, and therefore transport, trace elements.

Along with trace elements supplied by urban and industrial sources, urban particulate materials always include an underlying component of natural material, which is associated with particles of natural soil or with airborne particles whose origin is to be found outside the city limits. Although the exact chemical makeup of this component is strongly related to the type of geological material in and around a particular city, it is probably the major source of the Ce, Ga, La, Th, and Y found in urban environments. This association of "natural" trace elements is surprisingly stable in that it is found in cities of different urban characteristics and has been found preserved all along the urban cycle: in the atmospheric aerosol, in street dust, in the urban soil, and in urban sediments. Furthermore, the same combination of elements has been discovered to mark the natural component of urban particulate materials in cities of such different characteristics as Madrid, Oslo, and Ostrava (De Miguel *et al.* 1999).

Incomplete descriptions of the urban cycle of some elements have already been reported, as in De Miguel *et al.* (1998), who followed the fate of silver in the city of Madrid. Silver can be introduced as a component of medical (X-ray plates, dental alloys), commercial (photographic film) or industrial materials (high capacity Ag–Zn and Ag–Cd batteries). Disposal of these materials ultimately results in the release and transport of Ag in urban waters. As these wastewaters are treated in urban wastewater treatment plants, Ag becomes concentrated in the sludge produced during the treatment process, where it reaches values close to $45\,\mu g\,g^{-1}$. This sludge is in turn processed into a compost that is widely used by municipalities as soil amendment in parks and gardens. Silver is

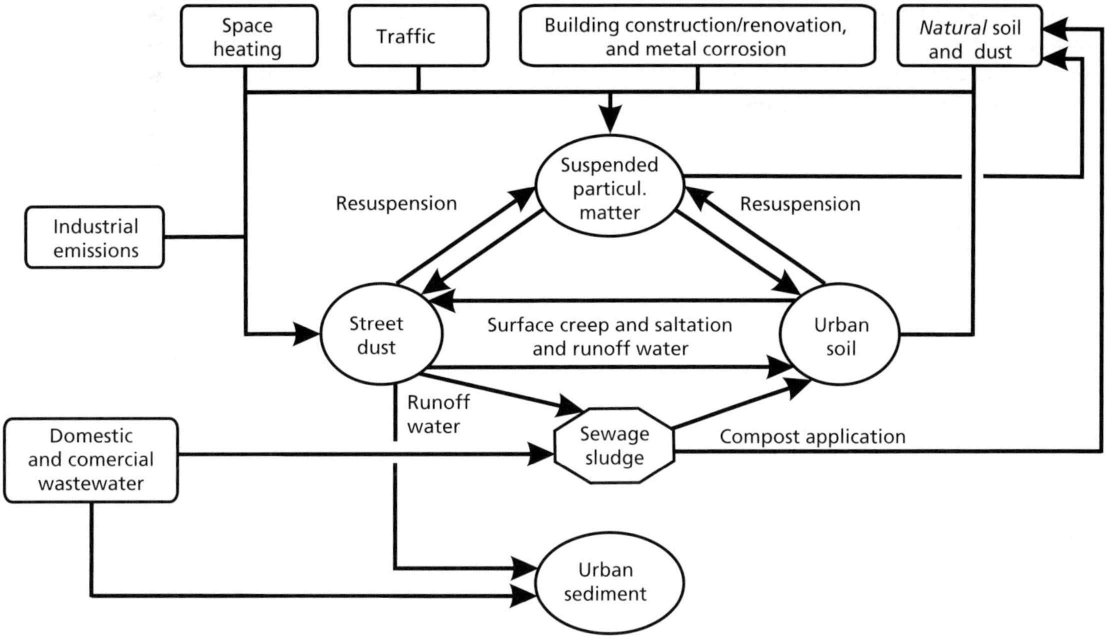

Fig. 4.2 Simplified representation of the urban geochemical cycle.

further concentrated in the compost (up to 50–70 μg g^{-1}), most likely due to the loss of mass during fermentation in the piles of maturing sewage sludge. The application of this compost on urban soil re-introduces silver, resulting in median concentrations of this element nearly five times higher in compost amended soils than in non-amended urban soils. Data published by De Miguel *et al.* (2005) strongly suggest that not all the silver that enters the urban water system is confined in sewage sludge to re-enter the urban cycle. Concentrations of up to 16 μg g^{-1} and a strong association with typically anthropogenic elements like Cu, Pb, and Zn in the sediments of the River Manzanares that runs through Madrid implies that a fraction of this silver is stored in the river sediments.

Charlesworth *et al.* (2000) envisaged the urban particulate environment as a "cascade" whereby sources included point sources, fluvial bed sediments and polluted dusts (see also section 4.2). These were then transported in water (see Poleto *et al.*, this volume) by suspended sediment in storm sewers, rivers, and streams, or in the atmosphere and were then eventually deposited in gully pots or urban lakes. However, the urban environment is complex (or "frustrating" (Charlesworth *et al.* 2000, p. 356)), with a wide variety of processes impacting on the physico-chemical characteristics of the particles as they move around the cascade. Very few relations were found between the geochemical and geophysical parameters making up the fingerprint of sediment taken from the individual compartments of the cascade.

Urban geochemistry has established that the urban environment is complex, subject to a variety of characteristic processes and impacts that lead to potentially polluted material being deposited onto soils, roads, and streets and subsequently transported in the aquatic and atmospheric spheres.

4.5 Future trends and concluding remarks

Urban geochemistry has rapidly grown in depth and complexity over the past three decades. The improvement of already existing analytical techniques and the advent of new ones have greatly widened the scope of urban geochemical research. The current routine use of ICP–AES, ICP–MS, and GC–MS has facilitated previously laborious multi-elemental determinations, and has brought detection limits down to the levels required to quantify the low concentrations of trace elements and organic compounds collected in cascade-impactor filters, dissolved in street runoff, or recovered in the individual fractions of sequential extraction procedures.

Urban geochemistry has proved to be a clearly applied scientific discipline with an obvious focus on the environmental aspects of life in the city. In the near future, therefore, it is likely that one of the major research interests will concern the adaptation and development of risk assessment tools for urban environment and has already led to the development of the subject of medical geology (Bowman *et al.* 2003). Geochemistry already plays a relevant role in risk analysis, as it helps to evaluate how a contaminant can partition between different phases and migrates from its source to the potential receptors. In turn, risk assessment provides a means to quantify the severity of the adverse health effects associated with the toxic elements and compounds that urban geochemistry investigates. Urban geochemistry and risk assessment are currently used together to characterize "brownfields", i.e. "abandoned, idled, or underused industrial and commercial facilities where expansion or redevelopment is complicated by real or perceived "environmental contamination" as defined by the USEPA. The next development will probably involve evaluation of completely urban areas from an environmental risk perspective (Beer & Ricci 1999).

Urban geochemistry will have to keep growing in pace with the increase in urban population around the world, soon becoming "the most dominant human habitat in history" (Wong *et al.* 2006, p. 12). It has already proved to have the capacity to bring about fundamental changes in urban life, as demonstrated by the gradual phasing out of leaded petrol after decades of geochemical research on urban lead. However, many questions about the urban environment have not been adequately or completely addressed yet. The modeling of the urban environment in terms of geochemical cycles, for example, is difficult owing to the complex mixtures of materials constantly undergoing change. The challenges of this "multicomponent, multiphase" environment (Turner 1992) are likely to keep geochemists busy for some time to come. Urban geochemistry has been characterized from the beginning by the high social impact of its research, an ingenious ability to modify, combine, and improve tools borrowed from other disciplines, and a wide spectrum of interests and challenges. Taking all these considerations into account, it is fair to assume that urban geochemistry will continue to generate exciting scientific results, and that it will become one of the most stimulating and dynamic branches of geochemistry.

References

Abrahams, P. W. (2002) Soils: their implications to human health. *Science of the Total Environment*, **291**, 1–32.

Adgate, J. L., Willis, R. D., Buckley, T. J., *et al.* (1998) Chemical Mass Balance source apportionment of lead in house dust. *Environmental Science and Technology*, **32**, 108–14.

Agemian, H. & Chau, A. S. Y. (1976) Evaluation of extraction techniques for the determination of metals in aquatic sediments. *Analyst*, **101**, 761–67.

Ajayi, S. O. & Vanloon, G. W. (1989) Studies on redistribution during the analytical fractionation of metals in sediments. *Science of the Total Environment*, **87/88**, 171–87.

Allott, R. W., Hewitt, C. N. & Kelly, M. R. (1990) The environmental half-lives and mean residence times of contaminants in dust for an urban Environment: Barrow-in-Furness. *Science of the Total Environment*, **93**, 403–10.

Anagnostopoulos, A., Lekkas, T., Joannas, K. & Kiouplidis, K. (1985) Lead in street dust in Thessaloniki, Part II. Lead and cadmium in school playground dust. In *International Conference on Heavy Metals in the Environment, Athens 1985*, T. D. Lekkas (ed.). CEP Consultants.

Analytical Methods Committee. (1985) The determination of trace elements in food by atomic absorption spectrometry. *Analytical Proceedings*, **22**, 48–50.

Archer, A. & Barrat, R. S. (1976) Lead levels in Birmingham dust. *Science of the Total Environment*, **6**, 275–86.

Banerjee, A. D. K. (2003) Heavy metal levels and solid phase speciation in street dust of Delhi, India. *Environmental Pollution*, **123**, 95–105.

Barbante, C., Veysseyre, A. & Ferrari, C. *et al.* (2001) Greenland snow evidence of large scale atmospheric contamination for platinum, palladium and rhodium. *Environmental Science and Technology*, **35**, 835–39.

Batterman, S. A., Dzubay, T. G. & Baumgardner, R. E. (1988) Development of crustal profiles for receptor modelling. *Atmospheric Environment*, **22**, 1821–28.

Beer, T. & Ricci, P. F. (1999) A quantitative risk assessment method based on population and exposure distributions using Australian air quality data. *Environment International*, **25**, 887–98.

Berti, W. R. & Jacobs, L. W. (1998) Distribution of trace elements in soil from repeated sewage sludge alications. *Journal Environmental Quality*, **27**, 1280–86.

Biasioli, M., Grčman, H., Kralj, T. Díaz-Barrientos, E. & Ajmone-Marsan, F. (2007) Potentially toxic elements contamination in urban soils: a comparison of three European cities. *Journal Environmental Quality*, **36**, 70—79.

Biggins, D. E. & Harrison, R. M. (1980) Chemical speciation of lead compounds in street dusts. *Environmental Science and Technology*, **14**, 336–39.

Boni, C., Caruso, E., Cereda, E., Lombardo, G., Braga Marcazzan, G. M. & Redaelli, P. (1988) Particulate matter elemental characterization in urban areas: pollution and source Identification. *Journal of Aerosol Science*, **19**, 1271–74.

Bowman, C. A., Bobrowsky, P. T. & Selinus, O. (2003) Medical geology: new relevance in the earth sciences. *Episodes, Journal of International Geosciences* **26**, 270–78.

Boyd, H. B., Pedersen, F., Cohr, K. H. *et al.* (1999) Exposure scenarios and guidance values for urban soil pollutants. *Regulatory Toxicology and Pharmacology*, **30**, 197–208.

Breward, N., Williams, M. & Bradley, D. (1996) Comparison of extraction methods for determining particulate metal fractionation in carbonate-rich Mediterranean soils. *Applied Geochemistry*, **11**, 101–04.

Buchardt-Boyd, H., Pedersen, F., Cohr, K. H. *et al.* (1999) Exposure scenarios and guidance values for urban soil pollutants. *Regulatory Toxicology and Pharmacology*, **30**, 197–208.

Carey, A. E., Gowen, J. A., Forehand, T. J., Tai, H. & Wiersma, G. B. (1980) Heavy metal concentrations in soils of five United States cities 1972 urban soils monitoring program. *Pesticides Monitoring Journal*, **13**, 150–54.

CDC (Centers for Disease Control and Prevention). (2007) *Advisory committee on childhood lead poisoning prevention*. MMWR Recomm. 56(RR-80), 1–6.

Chaney, R. L. & Ryan, J. A. (1994) *Risk Based Standards for Arsenic, Lead and Cadmium in Urban Soils*. Deutsche Gesselschaft für Chemisches Aaratewesen, Chemische Technik und Biotechnologie e.V. (DECHEMA), 136pp.

Charlesworth, S. M. & Foster, I. D. L. (2005) Gamma-emitting radionuclide and metallic elements in urban dusts and sediments, Coventry, UK: implications of dosages for dispersal and disposal. *Mineralogical Magazine*, **69**(5), 759–67.

Charlesworth, S. M. & Lees, J. A. (1999) Particulate-associated heavy metals in the urban environment: their transport from source to deposit, Coventry, UK. *Chemosphere*, **39**, 833–48.

Charlesworth, S. M., Booty, C. & Beasant, J. (2007) Monitoring the atmospheric deposition of particulate-associated urban contaminants, Coventry, UK. In *Proceedings of the 8th Highway and Urban Environment Symposium Series: Alliance for Global Sustainability Bookseries*, G. M. Morrison & S. Rauch (eds.), vol. **12**, 155–65.

Charlesworth, S. M., Everett, M., McCarthy, R., Ordóñez, A. & De Miguel, E. (2003) A comparative study of heavy metal concentration and distribution in deposited street dusts in a large and a small urban area: Birmingham and Coventry, West Midlands, UK. *Environment International*, **29**(5), 563–73.

Charlesworth, S. M., Ormerod, L. M. & Lees, J. A. (2000) Tracing sediments within urban catchments using heavy metal, mineral magnetic and radionuclide signatures. In: *Tracers in Geomorphology*, I. D. L. Foster (ed.). Wiley and Sons.

Chattopadhyay, G., Lin, K. C. P. & Feitz, A. J. (2003) Household dust metal levels in the Sydney metropolitan area. *Environmental Research*, **93**, 301–07.

Chen, J., Tan, M., Li, Y., *et al.* (2005) A lead isotope record of Shanghai atmospheric lead emissions in total suspended particles during the period of phasing out of leaded gasoline. *Atmospheric Environment*, **39**, 1245–53.

Cornille, P., Maenhaut, W. & Pacyna, J. M. (1990) Sources and characteristics of the atmospheric aerosol near Damascus, Syria. *Atmospheric Environment*, **24A**, 1083–93.

Crosby, D. G. (1998) *Environmental Toxicology and Chemistry*. Oxford University Press, 336pp.

Davies, D. J. A., Watt, J. M. & Thornton, I. (1987) Lead levels in Birmingham dusts and soils. *Science of the Total Environment*, **67**, 177–85.

Day, J. P., Fergusson, J. E. & Chee, T M (1979) Solubility and potential toxicity of lead in urban street dust. *Bulletin of Environmental Contamination and Toxicology*, **23**, 497–502.

De Miguel, E., Charlesworth, S., Ordóñez, A. & Seijas, E. (2005) Geochemical fingerprints and controls in the sediments of an urban river: River Manzanares, Madrid (Spain). *Science of the Total Environment*, **340**, 137–48.

De Miguel, E., Iribarren, I., Chacón, E., Ordoñez, A. & Charlesworth, S. (2007) Risk-based evaluation of the exposure of children to trace elements in playgrounds in Madrid (Spain). *Chemosphere*, **66**, 505–13.

De Miguel, E., Jiménez de Grado, M., Llamas, J. F., Martín-Dorado, A. & Mazadiego, L. F. (1998) The overlooked contribution of compost alication to the trace element load in the urban soil of Madrid (Spain). *Science of the Total Environment*, **215**, 113–22.

De Miguel, E., Llamas, J. F., Chacón, E. *et al.* (1997) Origin and patterns of distribution of trace elements in street dust. Unleaded petrol and urban lead. *Atmospheric Environment*, **31**, 2733–40.

De Miguel, E., Llamas, J. F., Chacón, E. & Mazadiego, L. F. (1999) Sources and pathways of trace elements in urban environments: a multi-elemental qualitative aroach. *Science of the Total Environment*, **235**, 355–57.

Doadrio, A. (1984) *Química Bioinorgánica*. Madrid, Spain: Facultad de Farmacia, Universidad Complutense de Madrid, 437pp.

Dongarrà, G., Sabatino, G., Triscari, M. & Varrica, D. (2003) The effects of anthropogenic particulate emissions on roadway dust and *Nerium oleander* leaves in Messina (Sicily, Italy). *Journal of Environmental Monitoring*, **5**, 766–73.

Drew, H. M. (1975) *Metal-Based Lubricant Composition*. Noyes Data Corporation, 332pp.

Dudka, S. & Miller, W. P. (1999) Permissible concentrations of arsenic and lead in soils based on risk assessment. *Water Air and Soil Pollution*, **113**, 127–32.

Duggan, M. J. (1984) Temporal and spatial variations of lead in air and surface dust – Implications for monitoring. *Science of the Total Environment*, **33**, 37–48.

Duggan, M. J., Inskip, M. J., Rundle, S. A. & Moorcroft J. S. (1985) Lead in playground dust and on the hands of schoolchildren. *Science of the Total Environment*, **44**, 65–79.

Edwards, R. D., Yurkow, E. J. & Lioy, P. J. (1998) Seasonal deposition of housedusts onto household surfaces. *Science of the Total Environment*, **224**, 69–80.

Ellis, J. B. & Revitt, D. M. (1982) Incidence of heavy metals in street surface sediments: solubility and grain size studies. *Water Air and Soil Pollution*, **17**, 87–100.

European Committee for Standardization. (1995) *EN–1: Safety of Toys – Part 3: Specification for Migration of Certain Elements*. British Standard EN 71-3.

Evans, E., Ma, M., Kingston, L., Leharne, S. & Chowdhry, B. (1992) The speciation pattern of lead in street dusts and soils in the vicinity of two London schools. *Environment International*, **18**, 153–62.

Fergusson, J. E. (1990) *The Heavy Elements: Chemistry, Environmental Impact and Health Effects*. Pergamon Press, 614pp.

Fergusson, J. E. & Kim, N. D. (1991) Trace elements in street and house dusts: sources and speciation. *Science of the Total Environment*, **100**, 125–50.

Ferreira-Baptista, L. & De Miguel, E. (2005) Geochemistry and risk assessment of street dust in Luanda, Angola: a tropical urban environment. *Atmospheric Environment*, **39**, 4501–12.

Foster, I. D. L. & Charlesworth, S. M. (1996) Heavy metals in the hydrological cycle: trends and explanation. *Hydrological Processes*, **10**, 227–61.

Gatz, D. (1975) Relative contribution of different sources of urban aerosols: alication of a new estimation method to multiple sites in Chicago. *Atmospheric Environment*, **9**, 1–8.

Gibson, M. J. & Farmer, J. G. (1984) Chemical partitioning of trace metal contaminants in urban street dirt. *Science of the Total Environment*, **33**, 49–57.

Gies, G. (1997) Developing compost standards in Europe. *Biocycle*, **38**, 82–83.

Giusquiani, P. L., Gigliotti, G. & Businelli, D. (1992) Mobility of heavy metals in urban waste-amended soils. *Journal of Environmental Quality*, **21**, 330–35.

Granero, S. & Domingo, J. L. (2002) Levels of metals in soils of Alcalá de Henares, Spain: human health risks. *Environment International*, **28**, 159–64.

Haines, R. C. (1984) Environmental contamination surveys of heavy metals in urban soils and hazard assessment. In *Trace Substances in Environmental Health, 18th Conference*, 450–60.

Hamel, S. C., Buckley, B. & Lioy, P. J. (1998) Bioaccessibility of metals in soils for different liquid to solid ratios in synthetic gastric fluid. *Environmental Science and Technology*, **32**, 358–62.

Hamilton, R. S., Revitt, D. M. & Warren, R. S. (1984) Levels and physico-chemical associations of Cd, Cu, Pb and Zn in road sediments. *Science of the Total Environment*, **33**, 59–74.

Harrison, R. M. (1979) Toxic metals in street and household dusts. *Science of the Total Environment*, **11**, 89–7.

Harrison, P. R., Draftz, R. G. & Murphy, W. H. (1974) Identification and impact of Chicago's ambient suspended dust. In *Proceedings of "Atmospheric-Surface Exchange of Particulate and Gaseous Pollutants (1974)", Richland, Washington, September 4–6, 1974*. Energy Research and Development Administration symposium series, CONF-740921. National Technical Information Service, U.S. Department of Commerce, 557–70.

Harrison, R. M., Laxen, D. P. H. & Wilson, S. J. (1981) Chemical associations of lead, cadmium, coer, and zinc in street dusts and roadside soils. *Environmental Science and Technology*, **15**, 1378–83.

Harrison, R. M., Smith, D. J. T., Pio, C. A. & Castro, L. M. (1997) Comparative receptor modelling study of airborne particulate pollutants in Birmingham (United Kingdom), Coimbra (Portugal) and Lahore (Pakistan). *Atmospheric Environment*, **31**, 3309–21.

Harvey, P. G., Spurgeon, A., Morgan, J., Chance, J. & Moss, E. (1985) A method for assessing hand-to-mouth activity in children as a possible transport route for toxic substances. In: *Heavy Metals in the Environment*, International Conference, Athens, T. D. Lekkas (ed.), vol. **1**. CEP Consultants, 436–37.

Hemond, H. F. & Solo-Gabriele, H. M. (2004) Children's exposure to arsenic from CCA-treated wooden decks and playground structures. *Risk Analysis*, **24**, 51–64.

Henry, R. C. (1987) Current factor analysis receptor models are ill-posed. *Atmospheric Environment*, **21**, 1815–27.

Hopke, P. K., Lamb, R. E. & Natusch, D. F. S. (1980) Multielemental characterisation of urban roadway dust. *Environmental Science and Technology*, **14**, 164–72.

Horowitz, A. (1991) *A primer on trace metal-sediment chemistry*. 2nd edition. Lewis, 136pp.

ICRP. (1991) Recommendations of the International Commission on Radiological Protection. ICRP publication 60. *Annals of the ICRP*, **21**, 1–3.

Imperato, M., Adamo, P., Naimo, D., Arienzo, M., Stanzione, D. & Violante, P. (2003) Spatial distribution of metals in urban soils of Naples city (Italy). *Environmental Pollution*, **124**, 247–56.

Jaffé, R., Carrero, H., Cabrera, A. & Alvarado, J. (1993) Organic Compounds and Heavy Metals in the Atmosphere of the city of Caracas, Venezuela – II: Atmospheric Deposition. *Water, Air and Soil Pollution*, **71**, 315–29.

Kabata-Pendias, A. & Pendias, H. (1992) *Trace Elements in Soils and Plants*. 2nd edition. CRC Press, 365pp.

Kappos, A. D., Bruckmann P., Eikmann, T. *et al.* (2004) Health effects of particles in ambient air. *International Journal of Hygiene and Environmental Health*, **207**, 399–07.

Kasparian, J., Frejafon, E., Rambaldi, P. *et al.* (1998) Characterization of urban aerosols using SEM-microscopy, X-ray analysis and lidar measurements. *Atmospheric Environment*, **32**, 2957–67.

Kim, J. Y., Kim, K. W., Ahn, J. S., Ko, I. & Lee, C. H. (2005) Investigation and risk assessment modelling of As and other heavy metals contamination around five abandoned metal mines in Korea. *Environmental Geochemistry and Health*, **27**, 193–203.

Kim, J. Y., Myung, J. H., Ahn, J. S. & Chon, H. T. (1998) Heavy metal speciation in dusts and stream sediments in the Taejon area, Korea. *Journal of Geochemical Exploration*, **64**, 409–19.

Korre, A., Durucan, S. & Koutroumani, A. (2002) Quantitative-spatial assessment of the risks associated with high Pb loads in soils around Lavrio, Greece. *Applied Geochemistry*, **17**, 1029–45.

Kowalczyk, G. S., Choquette, C. E. & Gordon, G. E. (1978) Chemical element balances and identification of air pollution sources in Washington, D.C. *Atmospheric Environment*, **12**, 1143–53.

Kowalczyk, G. S., Gordon, G. E. & Rheingrover, S. W. (1982) Identification of atmospheric particulate sources in Washington, D.C., using chemical element balances. *Environmental Science and Technology*, **16**, 79–90.

Kupiainen, K. (2007) *Road Dust from Pavement Wear and Traction Sanding*. Monograph of Boreal Environment Research, 26, Finnish Environment Institute, Helsinki.

Laidlaw, M. A. S. & Filippelli, G. M. (2008) Resuspension of urban soils as a persistent source of lead poisoning in children: a review and new directions. *Applied Geochemistry*, **23**, 2021–39.

Larson, S. M., Cass, G. R. & Gray, H. A. (1989) Atmospheric carbon particles and the Los Angeles visibility problem. *Aerosol Science and Technology*, **10**, 118–30.

Lee, J. S., Chon, H. T. & Kim, K. W. (2005) Human risk assessment of As, Cd, Cu and Zn in the abandoned metal mine site. *Environmental Geochemistry and Health*, **27**, 185–91.

Leharne, S., Charlesworth, D. & Chowdry, B. (1992) A survey of metal levels in street dusts in an inner London neighbourhood. *Environ International*, **18**, 263–70.

Lin, J. J. & Tai, H. S. (2001) Concentrations and distributions of carbonaceous species in ambient particles in Kaohsiung City, Taiwan. *Atmospheric Environment*, **35**, 2627–36.

Ljung, K., Oomen, A., Duits, M., Selinus, O. & Berglund, M. (2007) Bioaccessibility of metals in urban playground soils. *Journal of Environmental Science and Health*, **42**, 1241–50.

Ljung, K., Otabbong, E. & Selinus, O. (2006b) Natural and anthropogenic metal inputs to soils in urban Usala, Sweden. *Environmental Geochemistry and Health*, **28**, 353–64.

Ljung, K., Selinus, O. & Otabbong, E. (2006c) Metals in soils of children's urban environments in the small northern European city of Usala. *Science of the Total Environment*, **366**, 749–59.

Ljung, K., Selinus, O., Otabbong, E. & Berglund, M. (2006a) Metal and arsenic distribution in soil particle sizes relevant to soil ingestion by children. *Applied Geochemistry*, **21**, 1613–24.

Llamas, J. F., Chacón, E. & De Miguel, E. (1993) Maing of trace elements in soils in the province of Madrid. Sampling strategies and analysis of results. In *Integrated Soil and Sediment Research: A Basis for Proper Protection*, H. J. P. Eijsackers & T. Hamers (eds.). Kluwer Academic Publishers, 763pp.

Madrid, F., Biasioli, M. & Ajmone-Marsan, F. (2008) Availability and bioaccessibility of metals in fine particles of some urban soils. *Archives of Environmental Contamination and Toxicology*, **55**, 21–32.

Maxwell, C. M. & Nelson, D. W. (1978) *Development of lead emission factor for reentrained dust from paved roadways*. Presented at the 71st Annual Meeting of the Air Pollution Control Association, Houston, Texas, June 25–30, 1978.

Mazzei, F., D'Alessandro, A., Lucarelli, F. *et al.* (2008) Characterisation of particulate matter sources in an urban environment. *Science of the Total Environment*, **401**, 81–89.

Mielke, H. W., Gonzales, C. R., Smith, M. K. & Mielke, P. W. (1999) The urban environment and children's health: soils as an integrator of lead, zinc and cadmium in New Orleans, Louisiana, USA. *Environmental Research A*, **81**, 117–29.

Mock, G. (2000) *Domesticating the World: Conversion of Natural Ecosystems*. Earth Trends 2000, World Resources Institute.

Nadal, M., Schuhmacher, M. & Domingo, J. L. (2004) Metal pollution of soils and vegetation in an area with petrochemical industry. *Science of the Total Environment*, **321**, 59–69.

Ng, S. L., Chan, L. S., Lam, K. C. & Chan, W. K. (2003) Heavy metals and magnetic properties of playground dust in Hong Kong. *Environmental Monitoring and Assessment*, **89**, 221–32.

Nicholson, K. W. (1988) A review of particle resuspension. *Atmospheric Environment*, **22**, 2639–51.

Oberdörster, G., Gelein, R., Ferin, J. & Weiss, B. (1995) Association of particulate air pollution and acute mortality: involvement of ultrafine particles? *Inhalation Toxicology*, **7**, 111–24.

Oomen, A. G., Hack, A., Minekus, M., *et al.* (2002) Comparison of five in vitro digestion models to study the bioaccessibility of soil contaminants. *Environmental Science and Technology*, **36**, 3326–34.

Oomen, A. G., Rompelberg, C. J. M., Bruil, M. A., Dobbe, C. J. G., Pereboom, D. P. K. H. & Sips, A. J. A. M. (2003) Development of an in vitro digestion model for estimation of bioaccessibility of soil contaminants. *Archives of Environmental Contamination and Toxicology*, **44**, 281–87.

Ordóñez, A., Loredo, J., De Miguel, E. & Charlesworth, S. M. (2003) Distribution of heavy metals in the street dusts and soils of an industrial city in Northern Spain. *Archives of Environmental Contamination and Toxicology*, **44**, 160–70.

Pacyna, J. M. (1991) Chemical tracers of the origin of Arctic air pollution. In *Pollution of the Arctic Atmosphere*, W. T. Sturges (ed.) *Environmental Management Series* (J. Cairns Jnr & R. M. Harrison, eds). Elsevier.

Pacyna, J. M. (1983) *Trace Element Emission from Anthropogenic Sources in Europe*. Technical Report 10/82) Norsk Institutt for Luftforskning. Ref. 24781, 107pp.

Palacios, M. A., Gómez, M., Moldovan, M. & Gómez, B. (2000) Assessment of environmental contamination risk by Pt, Rh and Pd from automobile catalyst. *Microchemical Journal*, **67**, 105–13.

Paterson, E., Sanka, M. & Clark, L. (1996) Urban soils as pollutant sinks – a case study from Aberdeen, Scotland. *Applied Geochemistry*, **11**, 129–31.

Patra, A., Colvile, R., Arnold, S. *et al.* (2008) On street observations of particulate matter movement and dispersion due to traffic on an urban road. *Atmospheric Environment*, **42**, 3911–26.

Perez-Santana, S., Pomares Alfonso, M., Villanueva Tagle, M., Peña Icart, M., Brunori, C. & Morabito, R. (2007) Total and partial digestion of sediments for the evaluation of trace element environmental pollution. *Chemosphere*, **66**, 1545–53.

Pierson, W. R. & Brachaczek, W. W. (1976) *Particulate matter associated with vehicles on the road*. Automotive Engineering Congress and Exposition, Detroit, Michigan. Paper 760039.

Pryor, S. C., Barthelmie, R. J., Schoof, J. T., Binkowski, F. S., Delle Monache, L. & Stull, R. (2008) Modeling the impact of sea-spray on particle concentrations in a coastal city. *Science of the Total Environment*, **391**, 132–42.

Rasmussen, P. E., Beauchemin, S., Nugent, M., Dugandzic, R., Lanouette, M. & Chénier, M. (2008) Influence of matrix composition on the bioaccessibility of coer, zinc, and nickel in urban residential dust and soil. *Human and Ecological Risk Assessment*, **14**, 351–71.

Raunemaa, T., Riihiluoma, V. & Kulmala, M. (1986) Particle emission from gasoline powered vehicles: emission, deposition and re-emission under different traffic density situations. *Journal of Aerosol Science*, **17**, 973–83.

Ravindra, K., Bencs, L. & Van Grieten, R. (2004) Platinum group elements in the environment and their health risk. *Science of the Total Environment*, **318**, 1–43.

RIVN (Dutch National Institute for Public Health and the Environment). (2006) *How Can Information on Oral Bioavailability Improve Human Health Risk Assessment for Lead Contaminated Soils?* RIVM report 711701042.

Robertson, D. J., Taylor, K. G. & Hoon, S. R. (2003) Geochemical and mineral magnetic characterisation of urban sediment particulates, Manchester, UK. *Applied Geochemistry*, **18**, 269–82.

Rodriguez, R. R., Basta, N. T., Casteel, S. W. & Pace, L. W. (1999) An in vitro gastrointestinal method to estimate bioavailable arsenic in contaminated soils and solid media. *Environmental Science and Technology*, **33**, 642–49.

Roscher, E., Liebl, B., Schwegler, U., Schmied, R. & Kerscher, G. (1996) Richtwerte für polyzyklische aromatische Kohlenwasserstoffe im Boden von Kinderspielplätzen – Ableitungskriterien und Empfehlungen. *Gesundheitswesen*, **58**, 470–76.

Rose, N., Juggins, S., Watt, J. & Battarbee, R. (1994) Fuel type characterization of spheroidal carbonaceous particles using surface chemistry. *Ambio*, **23**, 296–99.

Ruby, M. V., Davis, A., Schoof, R., Eberle, S. & Sellstone, C. M. (1996) Estimation of lead and arsenic bioavailability using a physiologically based extraction test. *Environmental Science and Technology*, **29**, 422–30.

Ruby, M. V., Schoof, R., Brattin, W. *et al.* (1999) Advances in evaluating the oral bioavailability of inorganics in soil for use in human health risk assessment. *Environmental Science and Technology*, **33**, 3697–705.

Rundle, S. A. & Duggan, M. J. (1986) Lead pollution from the external redecoration of old buildings. *Science of the Total Environment*, **57**, 181–90.

Rundle, S. A., Inskip, M. J., Duggan, M. J. & Moorcroft, J. S. (1985) Lead-in-dust on children's hands. In: *Heavy Metals in the Environment*, vol. **1**, T. D. Lekkas (ed.), 457–59, International Conference, Athens, CEP Consultants.

Sadiq, M. & Mian, A. A. (1994) Nickel and vanadium in air particulates at Dharahn (Saudi Arabia) during and after the Kuwait oil fires. *Atmospheric Environment*, **28**, 2249–53.

Sánchez-Camazano, M., Sánchez-Martín, M. J. & Lorenzo, L. F. (1994) Lead and cadmium in soils and vegetables from urban gardens of Salamanca (Spain). *Science of the Total Environment*, **146/147**, 163–68.

Sansalone, J., Koran, J., Smithson, J. & Buchberger, G. (1998) Physical Characteristics of urban roadway solids transported during rain events. *ASCE Journal of Environmental Engineering* **124**, 427–40.

Schäfer, J., Eckhardt, J. D., Berner, Z. A. & Stüben, D. (1999) Time-dependent increase of traffic-emitted platinum group elements (PGE) in different environmental compartments. *Environmental Science and Technology*, 33, 3166–70.

Schroder, J. L., Basta, N. T., Si, J., Casteel, S. W., Evand, T. & Payton, M. (2003) In vitro gastrointestinal method to estimate relative bioavailable cadmium in contaminated soil. *Environmental Science and Technology*, **37**, 1365–70.

Schwar, M. J. R., Moorcroft, J. S., Laxen, D. P. H., Thompson, M. & Armorgie, C. (1988) Baseline metal-in-dust concentrations in Greater London. *Science of the Total Environment*, **68**, 25–43.

Sehmel, G. A. (1980) Particle resuspension: a review. *Environment International*, **4**, 107–27.

Serrano-Belles, C. & Leharne, S. (1997) Assessing the potential for road release from road dust and soils. *Environmental Geochemistry and Health*, **19**, 89–100.

Sharaf, N. E., Abdel-Shakour, A., Amer, N. M., Abou-Donia, M. A. & Khatab, N. (2008) Evaluation of children's blood lead level in Cairo, Egypt. *American–Eurasian Journal of Agricultural and Environmental Science*, 3(3), 414–19.

Shi, G., Chen, Z., Xu, S., *et al.* (2008) Potentially toxic metal contamination of urban soils and roadside dust in Shanghai, China. *Environmental Pollution*, **156**(2), 251–60.

Stigliani, W. M. & Anderberg, S. (1992) *Industrial metabolism at the regional level: the Rhine Basin. Laxenburg*, Austria: International Institute for Applied Systems Analysis, WP-92-10. 40 pp.

Stilwell, D. E. & Gorny, K. D. (1997) Contamination of soil with coer, chromium, and arsenic under decks built from pressure treated wood. *Environ. Contam. Toxicol. B*, 58, 22–29.

Strnad, M., Sáňka, M., Boháček, Z., Borkovcová, I. & Vondra, J. (1993) Soil pollution by heavy metals and polycyclic aromatic hydrocarbons in the town of Brno (Czech Republic). *International Journal of Environmental Analytical Chemistry*, **54**, 233–48.

Sturges, W. T. & Harrison, R. M. (1986) Bromine: lead ratios in airborne particles from urban and rural sites. *Atmospheric Environment*, **20**, 577–88.

Tessier, A., Campbell, P. G. C. & Bisson, M. (1979) Sequential extraction procedure for the speciation of particulate trace metals. *Analytical Chemistry*, **51**, 844–51.

Tiller, K. G. (1992) Urban soil contamination in Australia. *Australian Journal of Soil Research*, **30**, 937–57.

Tomza, U. (1984) *Trace Elements in the Atmospheric Aerosol at Katowice, Poland*. Technical Report. Instituut voor Nucleaire Wetenschappen, Rijksuniversiteit Gent, Gent, Belgium.

Tong, S. T. Y. (1998) Indoor and outdoor household dust contamination in Cincinnati, Ohio, USA. *Environmental Geochemistry and Health*, **20**(3), 123–33.

Tong, S. T. Y. & Lam, K. C. (2000) Home sweet home? A case study of household dust contamination in Hong Kong. *Science of the total environment*, **256**(2–3), 115–23.

Turner, D. R. (1992) The chemistry of metal pollutants in water. In: *Pollution: Causes, Effects and Controls*, R. A. Harrison (ed.). Royal Society of Chemistry.

Turner, A. & Simmonds, L. (2006) Elemental concentrations and metal bioaccessibility in UK household dust. *Science of the Total Environment*, **371**, 74–81.

UN Department of Economic and Social Affairs. (2008) *World Urbanization Prospects: The 2007 Revision*. United Nations.

USEPA (US Environmental Protection Agency). (1997) *Exposure Factors Handbook – General Factors*. EPA/600/P–5/002Fa, vol. I. Office of Research and Development. National Center for Environmental Assessment. US Environmental Protection Agency. Washington, DC: http://www.epa.gov/ncea/pdfs/efh/front.pdf.

USEPA (US Environmental Protection Agency). (2002) *Child-Specific Exposure Factors Handbook*. EPA–00-P-00-002B. National Center for Environmental Assessment. US Environmental Protection Agency. Washington, DC. http://cfpub.epa.gov/ncea/cfm/recordisplay.cfm?PrintVersion=True&deid=52047.

Van Dingenen, R., Raes, F., Putaud, J. P. *et al.* (2004) A European aerosol phenomenology – 1: physical characteristics of particulate matter at kerbside, urban, rural and background sites in Europe. *Atmospheric Environment*, **38**, 2561–77.

Vermette, S. J., Irvine, K. N. & Drake J. J. (1991) Temporal variability of the elemental composition in urban street dust. *Environmental Monitoring and Assessment*, **18**, 69–77.

Viana, M., Querol, X., Alastuey, A., Gil, J. I. & Menéndez, M. (2006) Identification of PM sources by principal component analysis (PCA) coupled with wind direction data. *Chemosphere*, **65**, 2411–18.

Wadge, A., Hutton, M. & Peterson, P. J. (1986) The concentrations and particle size relationships of selected trace elements in fly ashes from U.K. coal-fired power plants and a refuse incinerator. *Science of the Total Environment*, **54**, 13–27.

Wang, W. H., Wong, M. H., Leharne, S. & Fisher, B. (1998) Fractionation and biotoxicity of heavy metals in urban dusts from Hong Kong and London. *Environmental Geochemistry and Health*, **20**, 185–98.

Warren, R. S. & Birch, P. (1987) Heavy metal levels in atmospheric particulates, roadside dust and soil along a major urban highway. *Science of the Total Environment*, **59**, 253–56.

Watt, J., Thornton, I. & Cotter-Howells, J. (1993) Physical evidence suggesting the transfer of soil Pb into young children via hand-to-mouth activity. *Applied Geochemistry*, **Suppl. Issue**, no. **2**, 269–72.

Wcislo, E., Ioven, D., Kucharski, R. & Szdzuj, J. (2002) Human health risk assessment case study: an abandoned metal smelter site in Poland. *Chemosphere*, **47**, 507–15.

Widory, D., Roy, S., Le Moullec, Y. & Goupil, G., Cocherie, A. & Guerrot, C. (2004) The origin of atmospheric particles in Paris: a view through carbon and lead isotopes. *Atmospheric Environment*, **38**, 953–61.

Williams, M. (2004) Air pollution and policy – 1952–2002. *Science of the Total Environment*, **334–335**, 15–20.

Williams, T. M., Rawlins, B. G., Smith, B. & Breward, N. (1998) In-vitro determination of arsenic bioavailability in contaminated soil and mineral beneficiation waste from Ron Phibun, southern Thailand: a basis for improved human risk assessment. *Environmental Geochemistry and Health*, **20**, 169–77.

Wong, C. S. C., Li, X. & Thornton, I. (2006) Urban environmental geochemistry of trace metals. *Environmental Pollution*, **142**, 1–16.

Wong, J. W. C. & Mak, N. K. (1997) Heavy metal pollution in children's playgrounds in Hong Kong and its health implications. *Environmental Technology*, **18**, 109–15.

Yesilonis, I. D., Pouyat, R. V. & Neerchal, N. K. (2008) Spatial distribution of metals in soils in Baltimore, Maryland: role of native parent material, proximity to major roads, housing age and screening guidelines. *Environmental Pollution*, **156**, 723–31.

Yuan, C. S., Cheng, S. W., Hung, C. H. & Yu, T. Y. (2003) Influence of operating parameters on the collection efficiency and size distribution of street dust during street scrubbing. *Aerosol and Air Quality Research*, **3**(1), 75–86.

Zimdahl, R. L. & Skogerboe, R. K. (1977) Behavior of lead in soil. *Environmental Science and Technology*, **11**, 1202–07.

5 Urban aquatic sediments

Cristiano Poleto[1], Susanne Charlesworth[2] & Ariane Laurenti[3]

[1]*Hydraulic Research Institute, Federal University of Rio Grande do Sul, Brazil*
[2]*Department of Geography, Environment and Disaster Management, Coventry University, UK*
[3]*Department of Pathology, Federal University of Santa Catarina, Brazil*

5.1 Introduction

The process of urbanization begins with the removal of vegetation and exposure of bare soil. In the classic study by Wolman & Schick (1967), in Maryland, USA, the increase in sediment delivery after land clearance was over 1000 times that of natural erosion rates and illustrates the ongoing problems experienced by urban areas the world over. In association with changes to the terrestrial environment, the city's aquatic systems are not only impacted by the greater delivery of sediment, but also physical changes to channel morphology such as straightening, channelization and canalization (Fig. 5.1). Such changes lead to disconnection of the river continuum, increasing stream discharge, flooding, and increasing sediment erosion, which adds further layers of complexity to what was already a complex natural ecosystem and only a holistic approach to its study will enable better understanding to be gained. Thus the urban aquatic environment can in some ways be considered unique in that water is transported in artificial conduits, enabling it to exit the city as quickly as possible. Owing this speed, it can carry relatively large loads of particulate-associated pollutants (PAPs), which eventually reach receiving watercourses. Urban runoff is now one of the major sources of pollutants to the aquatic environment (Jefferies *et al.* 2007), and Deletic *et al.* (2000) cite sediments as "the most important potential pollutant" (p. 3386) carried in association with that runoff. If this is not acknowledged and is subsequently untreated it has the potential to downgrade receiving water quality.

As point source emissions of pollutants are increasingly brought under control, diffuse sources are emerging as a serious and continuing threat to the aquatic environment, and this is reflected in an increase in legislation and initiatives designed to reduce and control them.

5.2 The urban aquatic environment

The urban hydrological cycle consists of modified natural features normally involved in processes governing the transport and deposition of fine sediment, such as channelized river reaches, lakes, and ponds with concrete banks and culverted inflows, as well as uniquely urban landforms such as storm drains and gully pots. These features enable the rapid removal of water from the urban area (Pearson 1990) and as such are designed with transport in mind rather than deposition. The smooth profiles of storm sewers for instance, coupled with high water discharges, change the magnitude and frequency of flooding (Douglas 1999) and therefore do not encourage sediment deposition. However, it has been found (Butler & Davies 2000) that up to "80% of urban drainage systems in the UK have at least some permanent sediment deposits" (p. 315). Sediments are themselves considered to be contaminants (Horowitz 1995), but their importance in terms of urban geochemistry lies in the contaminants that can become adsorbed and transported with them (Lick 1987; Striegel 1987). It is difficult to divorce water and sediment quality because the first is the transport medium of the second.

The overall environmental quality of water bodies located in urban areas is generally quite poor owing to pollutants of domestic, industrial, hospital, or agricultural origin, leading to anoxic processes,

Sedimentology of Aqueous Systems, 1st edition.
Edited by Cristiano Poleto and Susanne Charlesworth.

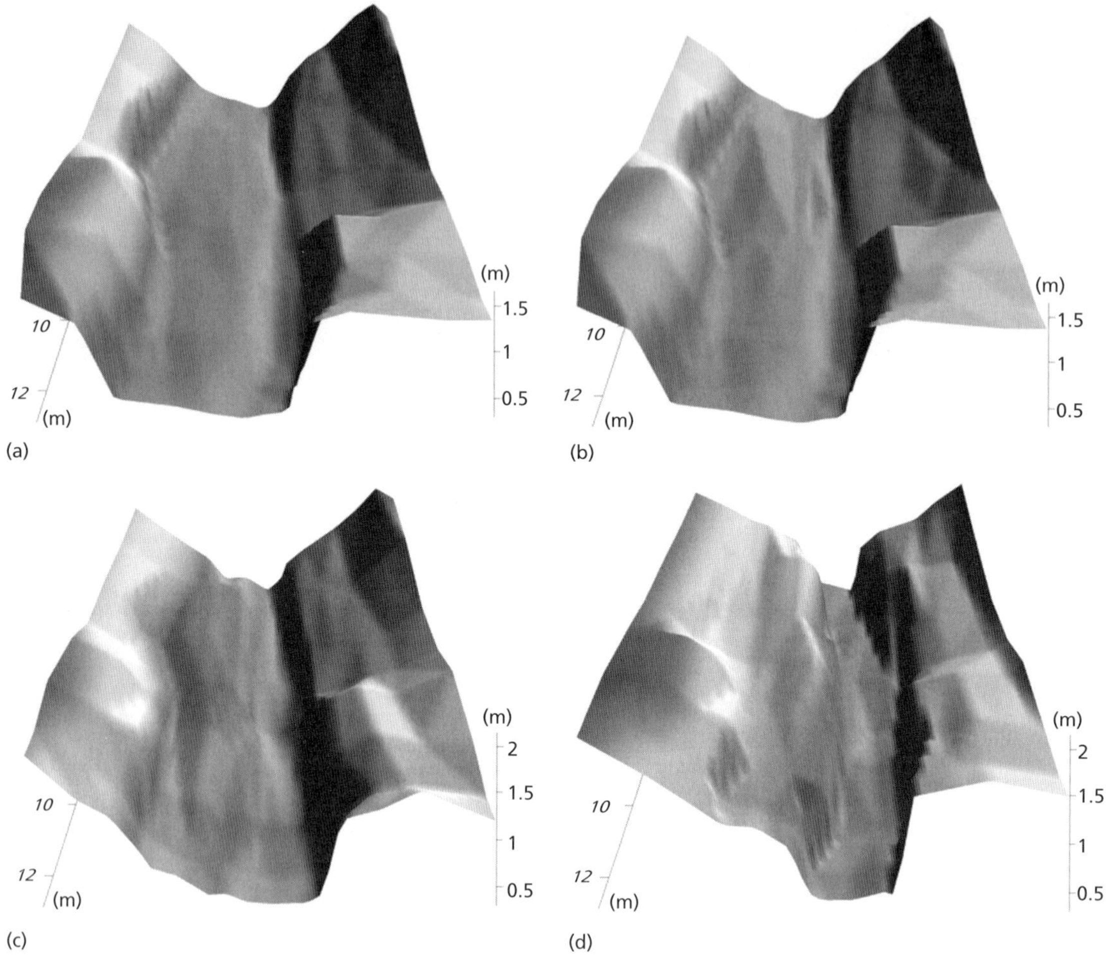

Fig. 5.1 Morphological changes in an urban channel verified by bathymetries: (a) November, 2003; (b) September, 2004; (c) January, 2005; (d) September, 2005.
Adapted from Poleto & Merten (2007).

which can generate toxic gases, offensive odors, and the presence of toxic organic and inorganic substances in the water column. The disequilibria caused by the interaction of human beings with terrestrial ecosystems have implications for the ecological cycle and therefore water quality (Jorge 2007). The severity of the problem is exacerbated by the presence of contaminated sediments with which the water column is permanently in contact. The alterations that occur in the aquatic ecosystem reflect, in part, the impacts suffered by the terrestrial ecosystem, sediment being an important link between the two. In this context, studies characterizing sediments can be held up as good indicators of these alterations in urban areas. The impacts of the erosion and accumulation of such sediment on receiving watercourses results in a feedback mechanism whereby the river adjusts to changes caused by urbanization (see, for example, Poleto & Merten 2007). These changes are worse, for instance, with the quantity of solid residues present in Brazilian rivers whereby the quantities involved can actually alter river morphology (see Poleto *et al.* 2005).

Humans have simultaneously increased the sediment transport by global rivers through soil erosion (by 2.3 ± 0.6 billion tonnes per year), yet reduced the

(a) (b)

Fig. 5.2 (a) Solid residues and sediments causing changes in Brazilian rivers; (b) dredging in Brazilian urban river.

flux of sediment reaching the world's coasts (by 1.4 ± 0.3 billion tonnes per year) because of retention for instance within reservoirs (Syvitski *et al.* 2007), which creates the necessity for dredging of urban watercourses (Fig. 5.2).

In association with the movement of sediment in urban aquatic environments are sediment (or particulate)-associated pollutants (or PAPs), principally linked to the finer-particle-sized sediments (silts and clays). Once in the river channel, PAPs can be transported for long distances and, when present in high concentrations, can cause serious environmental problems. Among the diverse pollutants transported in this way particularly in urban environments, heavy metals deserve special attention because they are not biodegradable, but are bioaccumulative, an example of which is metallic mercury, and can cause detrimental health effects in biota. Charlesworth *et al.*, this volume, detail the evaluation of risk that such substances are capable of causing to those living in urban environments.

To mitigate the effects of pollution and prevent its distribution, and to estimate rates of transfer and the final fate of the contaminants, it is necessary to identify the sources of pollution. The flow of sediments in suspension transported by a river is a mixture of particles originating in different locations and sources, which influences fine sediment quality and permits an understanding of the dynamic process of sediment transfer through the river channel. The following sections explore the sources of such sediment and the mechanisms for transport of the pollutants in association with them.

5.3 The characteristics of urban sediment

Urban sediments comprise fragments of rocks and degraded soil produced by the processes of weathering and erosion. The mineralogy of such urban sediments depends on the underlying lithology of the urban area, and its soil type. Hence particles of quartz, clay, and carbonates may be present, which can form aggregates with organic matter and/or Fe–Mn oxides, as well as anthropogenic material including glass, that produced from industrial processes and construction waste. Hence sediments include both minerogenic and organic particulates, air and water, and can include biofilms and biota which mature to become microecosystems in their own right (Droppo 2002). These microecosystems are an aggregate of water, inorganic and organic matter with functions or physical behaviors, chemical and biological autonomies, and the ability to interact with the environment. However, the urban environment creates a varied and complex mixture of pollutants both intrinsic and extrinsic, which can be both anthropogenic or natural in origin (Foster & Charlesworth 1996; Dawson & Macklin 1998; Singh *et al.* 2005).

Fluvial sediments therefore have varying characteristics due to their particle size, mineralogy, and organic matter content. These variations are the result of factors such as geology, relief, land use, climate, and anthropogenic impacts in the river basin itself. Sediments in rivers draining urban watersheds tend to be more organic because of, for example, domestic effluents (Rocha & Martin 2005), which are responsible for deleterious effects on water quality (Gromaire *et al.* 2001; Chebbo & Gromaire 2004).

5.4 Urban sediment quality

Environmental quality is normally expressed in terms of water chemistry, but some of the most toxic elements, particularly metals, are not transported in solution; rather, they form associations with particulates by adsorption and precipitation (Sigg 1998). Table 5.1a,b shows some of the limited data available on heavy metal concentrations in suspended sediment loads of urban rivers and lakes. In fact, of the 128 priority pollutants listed by the United States Environment Protection Agency (Bartram & Ballance 1996), 65% are either only, or mostly, found in association with biota and particulates.

Many studies have found that the concentration of these contaminants increases with decreasing particle size owing to the large surface area of all the particles combined (Wilber & Hunter 1979). However, coarser particles can also be associated with significant contaminant loads and hence their larger mass compared with fine particles makes coarser particles also an important factor (Horowitz 1991). The usual emphasis is on the finer fraction such as silts and clays, which have negative surface charges (Striegel 1987), because these are perceived to be more significant in terms of their contaminant-carrying capacity. Transportation attached to river sediments is considered the most important way metals are circulated in diverse river catchments (Castilhos 1999; Guerra 2000).

Three groups of major contaminants are frequently identified in aquatic urban sediments and because of this they are used in sediment guidelines (see MacDonald and Ingersoll, this volume), which inform management strategies for protection of the environment and biota:

- trace elements;
- hydrocarbons derived from petroleum; and
- synthetic organochlorides.

Trace elements, especially those called "heavy metals," are among the most frequently found pollutants in the environment; geographically they are spread globally, with above background values found in polar ice cores. Hence, this Chapter will concentrate on the distribution of these elements in the urban aquatic environment. Chemically, these elements are characterized by their high densities (Baird 2002), and in contrast with organic pollutants, are not biologically or chemically degraded and hence are conservative in nature. Heavy metals such as cadmium (Cd), copper (Cu), chromium (Cr), lead (Pb), mercury (Hg), nickel (Ni), zinc (An), arsenic (As), cobalt (Co), and selenium (Se), make up the group of chemical elements that appear frequently in urbanized areas (Porto 1995; Gromaire *et al.* 2001; Banerjee 2003; Poleto & Laurenti 2008). These elements can accumulate locally and be transported long distances (Marchand *et al.* 2006). For example, studies by Nichols *et al.* (1991) of the sediments deposited in the upper connecting channels of the Great Lakes found that although pollution was at its heaviest closest to the industrial source areas, there were still significant concentrations up to 60 km downstream. Decreasing concentrations may be either associated with dilution effects as the sediments are mixed with less contaminated material as they are transported further from the urban centre, or may be due to the smaller volumes of finer, relatively contaminated, particulates being preferentially transported in the water body, whereas the relatively less contaminated coarser material settles out earlier in the journey. Patchy areas of higher concentration may be associated with specific urban sources such as storm sewer outfalls (see, for example, Foster *et al.* 1996; Rhoads & Cahill 1999). Table 5.1c summarizes the concentrations of contaminants found in some studies of urban rivers and streams. It includes data from De Miguel *et al.* (2005) from rivers passing through both Coventry, UK, and Madrid, Spain, and highlights the variability in concentrations that have been found in urban streams.

There is evidence, however, that most metals specifically generated in urban areas tend not to travel far from their source areas (Foster & Charlesworth 1996). Foster & Charlesworth (1996) compared sediments deposited in paired lake catchments in the city of Coventry, rural Warwickshire, and the Scilly Isles, all in the UK. It was found that there was an order of magnitude difference in the concentration of Pb and Zn in bottom sediments from city-centre

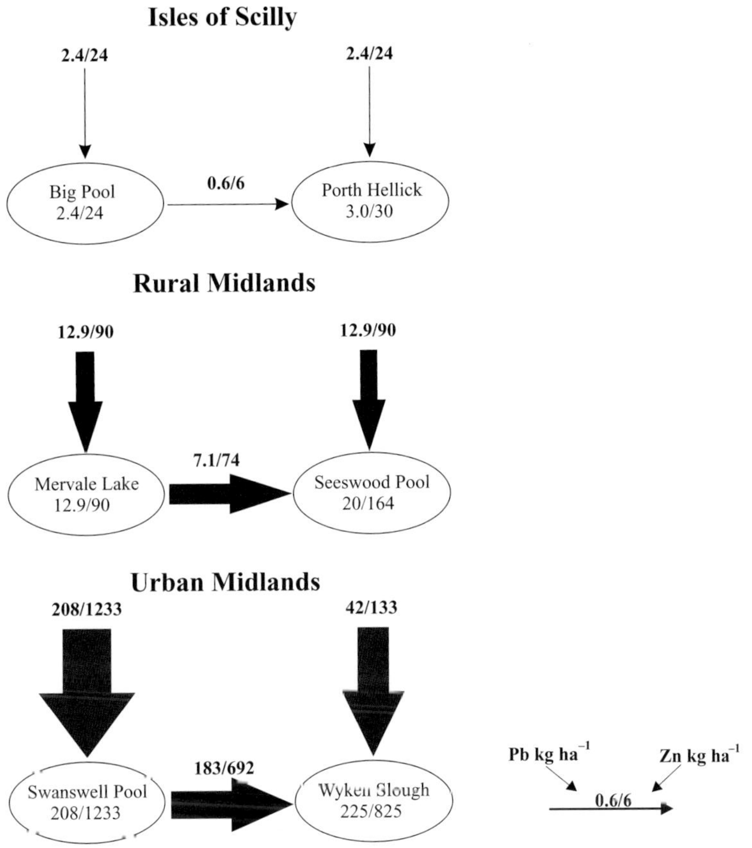

Fig. 5.3 Estimated anthropogenic excess loading from atmospheric and catchment sources for six lowland England lakes and reservoirs: (1) Pb; (2) Zn.
From Foster & Charlesworth (1996).

sites to urban peripheral lakes, then rural and finally to more isolated sites (Fig. 5.3). Urban lakes can therefore provide a significant sink for contaminated sediments, and although there are few studies of such features, Table 5.1b shows the concentrations that can be found in their bed sediment.

The sediments that do accumulate in urban lake basins reflect processes occurring in their catchments (Schueler & Simpson 2001) and as such can provide information on the pollution history of their catchments (Charlesworth & Foster 1993; Christopher *et al.* 1993; Graney & Eriksen 2004). In a study of two urban lakes in Coventry, UK, Charlesworth & Foster (1993) were able to show the decline in metals from approximately 1970 to the early 1990s. Such studies can be applied to health concerns (see, for example, Christopher *et al.* 1993; Graney & Eriksen 2004;) and management strategies (Charlesworth & Foster 1991; Rao *et al.* 2004). Although it has been established that urban sediments are significantly contaminated, if the trace element concentrations in the bed sediments of lakes receiving urban runoff are compared with those impacted by non-ferrous metal extraction processes (Foster & Charlesworth 1996), then it is found that urban lake concentrations are a magnitude lower. There have been very few studies of natural wetlands that have been subsequently impacted by urbanization (see Bentivegna *et al.* 2004), although there are many of constructed wetlands designed for runoff contaminant mitigation (see, for example, Lai and Lam 2009). Table 5.1d includes the concentrations in wetland substrate

Table 5.1 Maximum concentrations (μg g^{-1}) of trace elements found in association with particulate matter in the urban aquatic environment

	Elements				
	Pb	Cu	Ni	Zn	Cd
A: Suspended sediment					
Lake Ellyn[1]: inlets and outlets	1600	210	nd	950	6
River Sowe[2]	719	852	141	1482	33.7
B: Lakes					
Wyken Pool[3]	476	490	163	1000	29
Swanswell Pool[3]	312	292	140	3600	17
Holmer[4]	407	51	129	2045	21
Sudbury[3]	7700	nd	6200	1400	Nd
Lake Ellyn[1]	1750	275	nd	228	Nd
Summer Palace Lake[5]	27	27	10	68	Nd
Palace Moat Tokyo[5]	320	300	45	1400	Nd
Lake Michigan[5]	130	55	30	350	Nd
St John's Lakes[6]	600	nd	30	850	2.3
C: Fluvial substrate					
River Sowe[2]	957	270	843	1586	24.3
River Thame (urban)[7]	71	184	117	659	2.5
River Thame (below urban)[7]	480	1214	640	3420	24.5
Saddle River[8]	200	104.8	22.3	275.1	2.9
R. Rhine[9]	369	286	175	1240	13
Manoa Stream[10]	1078	300	439	510	1.04
Wyken Brook[3]	210	340	800	7000	15
Manzares[11]	371	347	47	591	Nd
River Sherbourne[11]	134	183	183	817	6.9
D: Wetland					
London marsh[12]	1180	320	nd	990	20
Wyken[3]	1952	300	420	2800	55
Kearney Marsh, New Jersey, USA[13]	23.7	11.3	42.7	315	5.25

[1] Striegl (1987); [2] Foster & Charlesworth (1996); [3] Charlesworth & Foster (1993); [4] Gaskell (1992); [5] Zhou *et al.* (1989); [6] Christopher *et al.* (1993); [7] Thoms (1987); [8] Wilber & Hunter (1979); [9] Förstner & Müller (1976); [10] Sutherland (2000); [11] De Miguel *et al.* (2005); [12] Zhang *et al.* (1990); [13] Bentivegna *et al.* (2004).

found by Charlesworth & Foster (1993), from Wyken Slough Marsh, Coventry, UK, which indicates that wetlands can act to mitigate against contamination by acting as a temporary sink. This study calculated the total stores of Zn and Pb in marsh, lake, and river sediments, and found that although a significant amount of contaminants were stored in the marsh substrate, 12 times as much Pb and nearly 7 times as much Zn was stored in the lake sediments. Mungur *et al.* (1995) studied the ability of a natural wetland in northwest London to treat highway runoff from a main road and found that a combination of the reduction of water velocity through the wetland, and interaction with plant stems encouraged PAP settlement. Wetland plants also systemically took up pollutants, but to a limited extent.

Particulates associated with urban activities can thus be significantly contaminated with toxic metals. If these should subsequently be deposited on hard urban surfaces such as pavements and roads, the likelihood is that they will be washed off during storms to enter rivers and lakes. The following section explores the PAP transfer mechanisms unique to urban environments and how sources of these contaminants can be traced to mitigate contamination of the urban aquatic environment at source.

5.4.1 Sources of particulate-associated pollutants in urban areas

Many studies around the world have identified the principal sources of contaminants in urban centres in association with heavily trafficked areas (Jansson 2002; Charlesworth *et al.* 2003a; Adachi & Tainosho 2005), municipal wastewater systems (Gromaire *et al.* 2001; Pardos *et al.* 2004; Brown & Peake 2006), construction activities using materials such as bricks, gravel, sand, and concrete (OMEE 1993; Tucci 2003), and industry. Construction sites have been considered one of the urban land uses with high pollution potential, especially because of erosion of unprotected soil surfaces (Wolman & Schick, 1967; Sonzogny *et al.* 1980; Harbor 1999; Burton & Pitt 2002).

However, non-point sources such as emissions from vehicles, by definition are difficult to investigate. For instance, the toxic effects of Pb are well known, and where there is one identifiable source of Pb, for example Pb-based paints or leaded petrol, it can be legislated for. However, low-level environmental exposure to Pb can be associated with multiple sources (petrol, industrial processes, paint, solder in canned foods, water pipes) and pathways (air, household dust, street dirt, soil, water, food). Evaluation of the relative contributions of sources is therefore complex and likely to differ between areas and population groups (Von Schirnding 1999; Tong *et al.* 2000). In a study by Poleto (2007), the principal sources of sediment to a small urban catchment in Brazil included the river channel itself, paved and unpaved streets. Three suspended sediment samples were collected at hourly intervals during the storm event. Figure 5.4 shows how the relative importance of these sources changes with time with that from paved street declining while both unpaved streets and the river channel increase in importance.

Studies such as that by Stigliani *et al.* (1993) found that contaminated sediments deposited in the River Rhine were dominated by diffuse sources. And in a study of an urban stream in Oahu, Hawaii, Sutherland (2000) found that the sediments were significantly polluted with Pb and less so with Ba, Cd, and Zn, sources of which were directly attributable to the

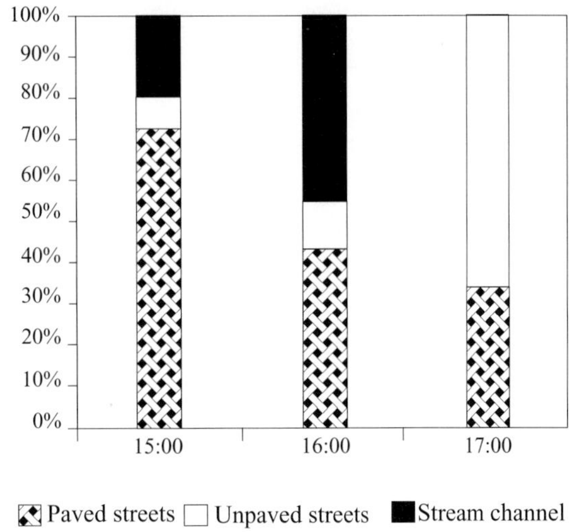

Fig. 5.4 Sediment sources in a small urban watershed. From Poleto *et al.* (2009).

wear of vehicles, wear of tires, spillages of fluids, and exhaust emissions. Further, owing to the predominance of impervious surfaces in the urban environment, when there is a storm or sudden snowmelt, surface runoff flows rapidly across the urban catchment because infiltration and storage are greatly reduced (Robinson *et al.* 2000). This was shown in Poleto & Merten's (2007) study of Pb, Zn, and Cr removal from paved streets in Brazil, which showed the transport downstream of these pollutants in association with stormwater flow and their distribution in the basin (Fig. 5.5).

This rapid runoff will carry with it what Robinson *et al.* (2000) term a "cocktail" of associated contaminants whose concentrations increase with increasing imperviousness. These contaminants are then carried into storm sewers and thence to receiving waters where they have been identified as the cause of degradation in urban watercourses (Pitt *et al.* 1993). In Eastern England, Perdikaki & Mason (1999) found that most of the problems associated with contaminant-rich run-off from urban areas were associated with particulate material. The importance of water as the transport medium for particulate-associated material is further emphasized during storms when pollutants are flushed out into the stream, while overall discharge remains unaffected (Brinkmann 1985). This "first flush" effect is virtually unique to

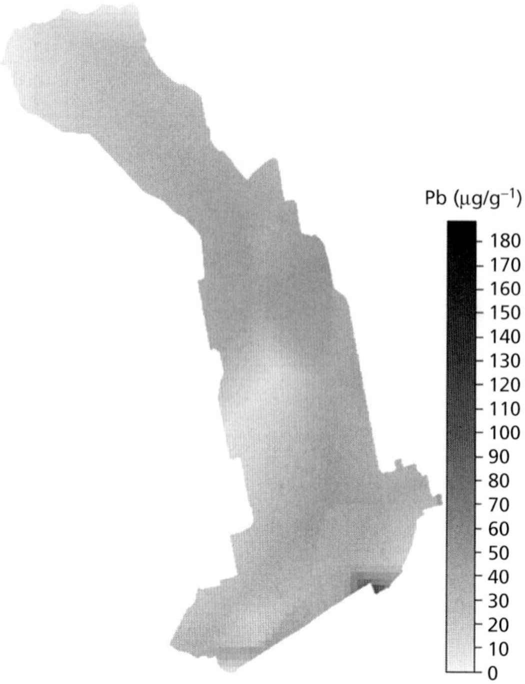

Fig. 5.5 Distribution of lead in a Brazilian urban watershed. From Poleto & Merten (2007).

the urban environment (Ellis *et al.* 1986; Morrison *et al.* 1989), and can result in a peak in contaminant loading before the peak in discharge on the urban storm hydrograph. It is due to the structure of traditionally hard-engineered drainage systems in urban areas, which encourages PAPs adhering to the road or pavement surface or associated with road gutters and gully pots (Charlesworth & Foster 2005) to be removed in the first 10% of rainfall during a storm. Other factors involved in the production of a first flush and cited by Ashley *et al.* (1992) include:

- rainfall intensity;
- antecedent dry period;
- cleaning protocol;
- localization and type of drainage system;
- drainage system gradient.

Although there is some argument about whether the first flush is common to all urban environments (e.g. Deletic 1998), or may be a site and storm specific phenomenon (Lawler *et al.* 2006), there is sufficient evidence for allowance to be made for it in management strategies targeting water quality in urban environments (Kayhanian & Stenstrom 2008). Ellis *et al.* (1986) further suggest that these urban flushing effects can be cyclical owing to local climate leading to periods of dry and then wet conditions where material tends to accumulate on surfaces and is then washed off after rain. Pollutant concentrations can therefore be dependent on storm duration and frequency (Marsalek 1990). However, Ellis *et al.* (1986) found that first flush was not necessarily present during all storm events, but when it was, secondary peaks may accompany it as material was flushed from contributing surfaces further from the site of discharge. Brinkmann (1985) suggests that both urban sediment and water quality are dependent on site-specific characteristics, leading to results obtained being applicable to the specific stormwater catchment where the study was performed. The following section further explores the issue of site specificity, which has important ramifications for subsequent stormwater management strategies in urban areas.

5.5 Transport of particulate-associated pollutants in urban aquatic environments; partitioning and speciation

Although section 4 stated that most pollutants are transported in the environment attached to particulates, the environmental conditions prevalent in urban areas are subject to rapid and constant change. With changing conditions, such as lowering of pH, or changing redox (Brikker 1999), formally particulate-associated contaminants can be released into solution. There have been few studies of changing partitioning of contaminants between sediments and water, but Morrison *et al.* (1984) were able to use a storm hydrograph to show that the partition coefficient between dissolved and particulate-associated heavy metals changed little as the storm progressed, and that it remained similar for a storm following 2 days after the initial one. Many recent studies (e.g. Glenn *et al.* 2001, 2004; Fan *et al.* 2004; Hallberg *et al.* 2007; Sansalone & Ying 2008) have applied the partition coefficient of these contaminants in the design of urban water treatment processes. However, pollutants are not just adsorbed to particulate surfaces, but they can be chemically linked by adsorption, co-precipitation, formation of organometallic co-ordination complexes, and incorporation into the

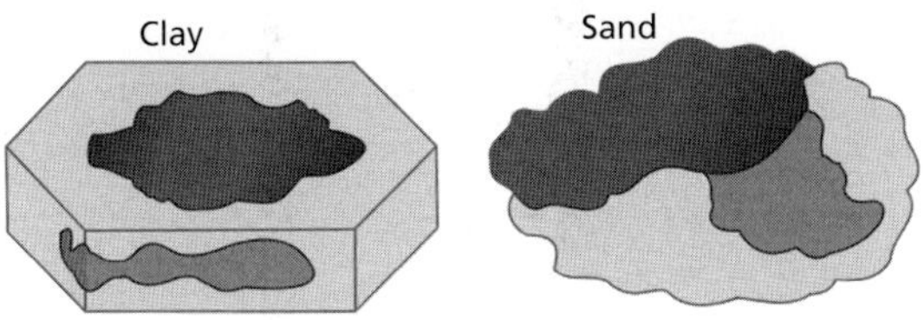

Fig. 5.6 Diagram of typical sediment (mineral + organic matter + oxides).
Adapted from Federal Interagency Stream – Restoration Working Group (1998).

crystalline structure of minerogenic materials, as shown in Fig. 5.6, making sediments all the more important in the treatment process.

There have been many studies of such speciation of trace elements in urban aquatic sediments, which have sought to identify the trace elements that were potentially the most likely to be released into solution, should environmental conditions change. Many such studies have found (e.g. Gibbs 1973) that PAPs are transported in discrete phases, reflected by Tessier *et al.* (1979) in their sequential extraction protocol (see also Charlesworth *et al.*, and Bortoluzzi *et al.*, both this volume).

Sutherland (2000) and Charlesworth & Foster's (1993) studies in urban lakes have shown that the dominant binding site can vary both with the individual element and with time. The former found Pb to be potentially in the most available form, whereas the latter found Cu and Zn were mainly associated with exchangeable sites. Other studies identified other elements that were of concern; for example, Revitt & Morrison (1987) found that 59% of Cd in association with particulates in stormwater could be considered bioavailable. These data show the range of inorganic contaminants that have the potential, should environmental conditions change, to cause problems upon release (Bird *et al. 2003*).

However, many of these phases are operationally defined and thus it is difficult to apply them back to the environment on the one hand, but also it is difficult to compare different studies of sequentially extracted heavy metals because different analytical methods tend to be used (Sutherland 2000). When attempting to apply partitioning to urban sediments, however, another problem is that the environment is rarely stable. Pulses of sediment are constantly being released by phases of construction (Hollis 1988). This sediment may be contaminated topsoil, or may simply supply silt and clay particles, which provide binding sites for contaminants already present in the environment. These may remain on impermeable urban surfaces until the first storm transports them by storm sewers to receiving streams. Morrison *et al.* (1988, 1995) studied processes in gullypots, coining the term "biochemical reactors" and found that speciation is constant until the onset of a storm event, when acidic rainwater flushing through the gully pot leads to increased solubility of trace elements and their consequent removal in solution, adding significantly to the first flush effect. Adding the possibility of first flush to the mobilization of sediments by street cleansing activities and gully pot emptying (Ellis & Revitt 1982; Morrison *et al.* 1995), a great deal of contaminated material can eventually reach receiving waters by storm sewers (Harrison *et al.* 1985; Anderberg & Stigliani 1994; Charlesworth & Foster 1999).

5.5.1 Site specificity

Table 5.1 shows the wide range of concentrations of trace elements found in association with urban aquatic sediment. This reflects differences in the layout of the individual city: trafficked areas, pedestrianization, distribution of green space, treatment of watercourses, etc. As was discussed by Charlesworth & Lees (1999), and further improved by Wong *et al.* (2006), processes unique to the urban environment impact significantly on the transport, deposition, and storage of urban sediment. These processes include, but are not exclusive to the frequency of street cleansing, climate which determines the intensity and frequency of rainfall, the type of industry present, and the structure of the hard-engineered sewer system. As a result, Charlesworth & Lees (1999) have shown that the frequency distribution of urban sediment samples changes as they pass from source to deposit. The source groups have a highly positively skewed distribution, the transported group less so, and the distribution curve of the deposited group becomes near normal (Fig. 5.7). This reflects, firstly, the great variability in sources of these elements within urban

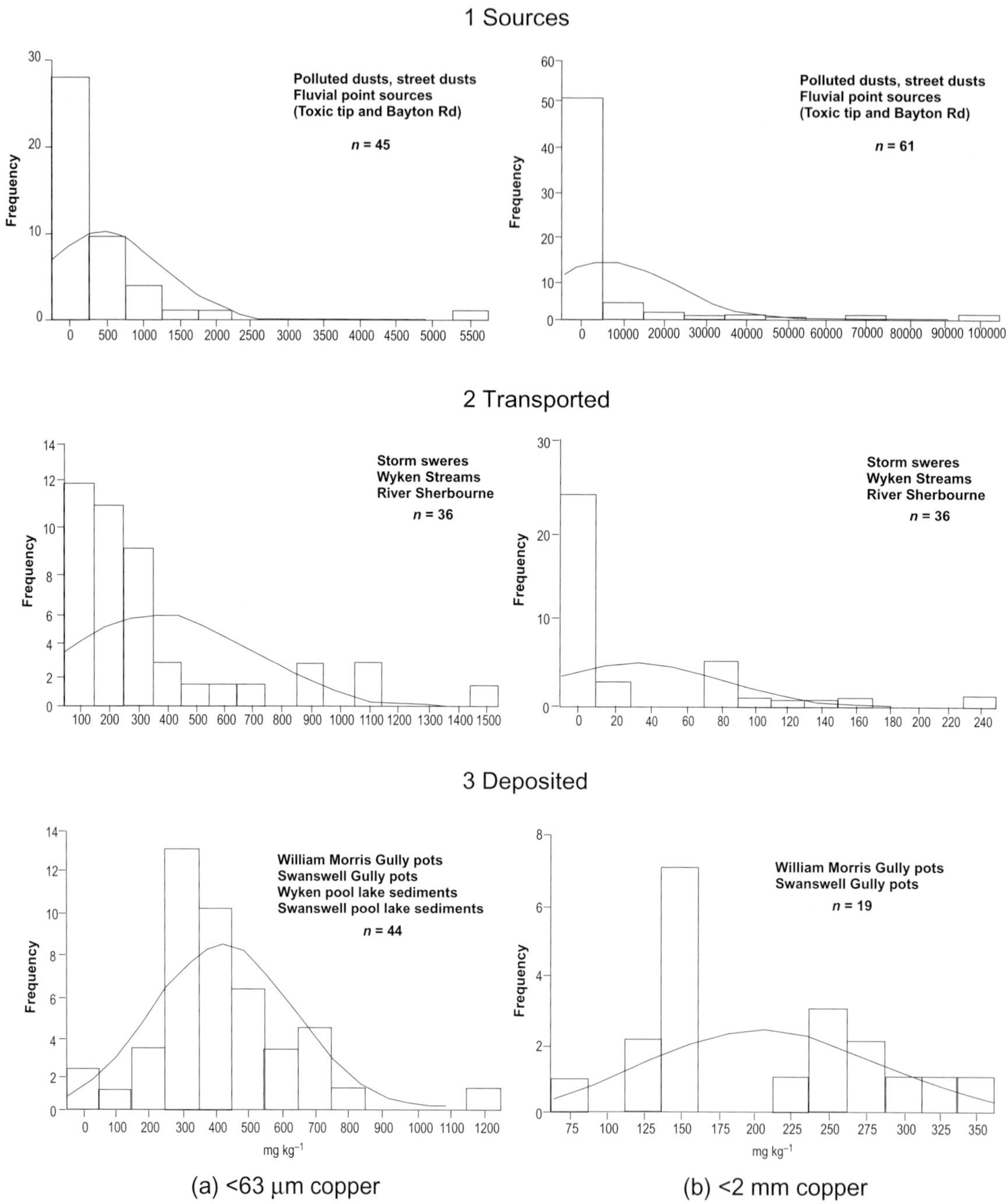

Fig. 5.7 Frequency diagram for Cu in (1) sources, (2) transported, and (3) deposited sediments according to particle size (a) <63 μm (b) <2 mm in samples from Coventry city centre.
From Charlesworth & Lees (1999).

areas that can be related back to the historical pattern of urbanization and industrialization leading to the development of industrial estates and traffic-free precinct areas. Secondly, the trend in frequency distribution also reflects the many geochemical and physical changes that occur during transport of urban sediments. These are a function of a great many processes not only unique to urban areas in general, but possibly unique to particular urban centers. The data collected from urban areas may therefore be site- or ecosystem specific (Jennett *et al.* 1980), and even event specific, where runoff produced from separate storms from the same outfall varies according to prevailing conditions related to phase of construction, traffic movements, industrial discharges, etc. (Morrison *et al.* 1984). Charlesworth & Lees' (1999) attempt to simplify sediment transport in the urban environment into a source–transport–deposit cascade highlighted the difficulty of such an approach; the urban environment is a dynamic system in which sediments accumulating for example in gully pots, or the bed of a river, can become remobilized during storms and become sources of contamination themselves (Deletic *et al.* 2000) and be transported to the next environmental compartment dependant on prevailing environmental conditions.

It is not only the sediment characteristics that can be unique to the specific urban centre, but this sediment will support an ecosystem that may well reflect site specificity. After traffic as a major source of sediment, construction activity in the catchment will provide regular pulses of particulate material. Studies have shown that the high concentration of calcium found in some lakes may be due to construction activities. This kind of impact can cause an ecological imbalance, such as, for example, the generation of a large population of mollusks due to the increased availability of calcium, which is used in constructing a shell (Beasley & Kneale 2002). Thus sediments can influence the development of macroinvertebrates at the bottom of the food chain, which can lead to the modification of ecosystems. Such environmental changes initiate qualitative modifications in the biodiversity of local species (Pompeu *et al.* 2005), which, according to Wolman & Schick (1967), are reflected in all aquatic organisms. These impacts are reflected in changes at the cellular level. For example, Ono *et al.* (2000) found positive correlations between sediments generated by construction activity and genotoxicity (damage caused to DNA) in urban areas.

Thus, to mitigate the adverse impacts of stormwater pollution, it is essential to have appropriate management strategies and efficient treatment designs (Egodawatta & Goonetilleke 2007). The identification of the origins of urban sediments allows an understanding of the processes of their transference to the river channel (Walling *et al.* 2002; Walling 2005) which is fundamental to developing these strategies (Taylor 2007; Owens *et al.* 2001) and a holistic approach to these problems is therefore needed. However, urban catchments are rarely "joined up", with many having been artificially cut off from their catchments historically (Charlesworth & Foster 1993), and many studies have historically broken down the urban aquatic environment into "road reaches" (Hamilton *et al.* 1984; Harrison *et al.* 1985), individual roofs (Quek & Forster 1993; Thomas & Greene 1993), and gullypot catchments (Morrison *et al.* 1989). The following sections therefore concentrate on methods by which sediment source tracing and apportionment may be investigated, the results of which can be used to inform strategies to manage their impacts in urban environments. This is not an easy task in urban areas (Charlesworth *et al.* 2000), and some of the techniques developed for pristine or simpler catchments do not transfer well to such multi-impacted catchments.

5.5.2 Sediment source tracing in urban areas

Source identification would enable accurate evaluation of the potential for pollution, which in turn could be used to deduce the impacts and finally make possible the selection of an appropriate means of control for sources that are actively producing sediments (Porto 1995). However, there are few studies of the urban environment in which it is possible to trace the movement of sediment and its associated contaminants from source to deposit in a complete catchment (Charlesworth & Lees 1999, 2001; Charlesworth *et al.* 2000; Carter *et al.* 2003). Add to these difficulties the multiplicity of sources and the many biogeochemical reactions that change the chemistry of the sediment, and few studies have found a chemical or physical characteristic that would provide a means of tracing specifically in the urban environment (Charlesworth *et al.* 2000).

Methods of source identification have included the use of natural tracers (Russell *et al.* 2001), mineral magnetic studies (Charlesworth & Lees 1999), and the generation of a sediment fingerprint (Peart & Walling 1986; Carter *et al.* 2003). This last method was based on comparison of the properties of the unknown sediment with material from potential sources; it was developed by identifying a small group of geochemical variables that could explain the variability of the sources. Following this, samples of suspended sediment were classified using multivariate statistics. The focus of the work was on distinguishing sources originating from rivers beds and surface soils and although results were reasonably good, there were limitations. Yu & Oldfield (1989) then used artificial mixtures of sediments of diverse sources to evaluate the capacity of the method in correctly separating the different sources. The results showed that the mathematical procedure that they used was a practical and efficient method of establishing the relation between sediments in suspension and the multiple sources involved. This work shows that quantitative calculations are more useful than purely qualitative descriptions, allowing the identification of sources contributing to river sediments (Minella 2003). However, the validity of establishing a geochemical fingerprint depends on whether the average properties of sediments in suspension can be compared directly with the same properties of potential sources, using conservative properties.

The value of the study of urban environmental quality has enabled the impacts of sediment and associated pollutants to be assessed. Point sources of sediment can now be controlled, for example on construction sites (see, for example, USEPA 2005) and concern can now turn to the management of diffuse sources. With the reduction in release of Pb to the environment owing to the introduction of unleaded petrol and the removal of Pb in paints, sediment studies have shown that the concentration of Pb in the environment has reduced. This has led to the focus on other metals such as Zn and Cu in urban environments. However, mercury is generally considered the most toxic of metals (Jitaru & Adams 2004; Boszke & Kowalski 2006), and although this is associated with release during specific industrial processes and to a lesser extent waste disposal, in some cities of the world it is one of the most concerning of metal pollutants. It is one of the most likely of metals to impact on human health because of its propensity to bioaccumulate up the food chain and to methylate in water or sediment into its most toxic form (Hortellani *et al.* 2005). The following section therefore provides a case study of mercury in urban environments.

5.5.3 Mercury in urban areas

Many studies have shown that industrial and agricultural activities, waste disposal, gold mining, and the use of fossil fuels are sources of mercury to the environment (Sanders *et al.* 2006). Annually, in Brazil, more than 85 million light bulbs are thrown away in sanitary landfill, totaling about 3.5 tonnes of mercury. As well as this source of mercury, it is estimated that, from 1983 to 1993 more than 900 tonnes of mercury contaminated the Amazon because of prospecting activities. For each kilogram of gold, 1.3 kg of mercury is lost to the environment and of that, between 55% and 65% is released to the atmosphere with the rest finding its way into aquatic ecosystems. Hence the sources of mercury can be both point and diffuse. According to Boszke and Kowalski (2006), coal and lignite combustion in Poland are responsible for the release of 44% and 18.5% of atmospheric mercury respectively, with cement production and the disposal of fluorescent light tubes emitting 16.6% and 6.4% each. Mercury is thus mainly emitted into the atmosphere and water where it has high mobility, organic matter affinity, and a biomagnification capacity that makes this element one of the most harmful metals to biota Gorski *et al.* (2003). However, the main property of concern is that the metal is capable of conversion to methylmercury, which is highly toxic. This can accumulate in the tissue of fish and mollusks in much greater quantities to those found in the environment. The World Health Organization has therefore established a maximum limit of $0.5\,\mu g\,g^{-1}$ total mercury in fish, and a recommended maximum consumption of 400 g per week of fish and/or fish products (USEPA 2000). These values are alarming compared with the amounts of fish eaten by riverside populations of some Amazonian regions, where their daily consumption of fish is about 250 g per individual (Poleto & Castilhos 2008).

In a study of mercury cycling in eight streams in three states in the USA, Marvin-DiPasquale *et al.* (2009) found that sediments in the three urban streams contained an average of three times as much total mercury as that found in non-urban stream sediments. However, it appeared that the capacity to convert to methylmercury in the urban streams was less than that of non-urban ones. Because the production of methylmercury is controlled by methylating bacteria present in the sediments, it would seem in this case that the sedimentary environment in urban areas was not conducive to their presence and hence less methylmercury was produced. Mercury is the same as other metals in its sorption to particulates and hence its concentration increases with decreasing particle size (Hunerlach *et al.* 2004). It preferentially binds to organic matter (Mason & Sullivan 1998; Machado *et al.* 2008) and sulfur (Marins *et al.* 1998) and hence its concentration increases with rising amounts of these elements in the sediment.

This chapter has given evidence of the pollution of the urban aquatic environment caused by anthropogenic activities. In particular, this section has highlighted mercury, and lead and zinc, as elements of particular concern. It has been shown that traditional drainage techniques, which are designed to transport water and its associated contaminants out of the urban area as quickly as possible, do a disservice to the receiving environment, providing as they do a significant transport mechanism for PAPs. The following section presents a means of application of some of the physico-chemical studies discussed in this chapter to provide one of the most promising techniques for efficient and sustainable remediation of pollution in urban areas.

5.6 Sustainable drainage systems

Although a detailed consideration of alternative drainage techniques is beyond the scope of this chapter, an introduction to sustainable drainage systems (SUDS) will be given here. Also sometimes called low-impact development in the USA (Dietz 2007), the functions of SUDS are threefold, as exemplified by the SUDS triangle (Woods-Ballard *et al.* 2007; Charlesworth *et al.* 2003b) in which there is an equal balance between water quality, quantity, and biodiversity or amenity. This approach mimics nature, whereby water is encouraged to infiltrate through a permeable surface that provides a means of "cleaning" the water of its contaminants. These surfaces will be incorporated in such individual devices (or best management practices (BMPs) in the USA) as porous paving (PPS), constructed wetlands (see section 5.3), ponds, or swales, or the devices can be deployed as a SUDS "train" in which several are used together (for further details see Charlesworth *et al.* 2003b).

Hence, the contaminated sediments generated in an urban area can become treated such that, in some cases, a significant amount of the pollutants finding their way into the SUDS devices can be removed (Pratt 2004) by such processes as physical entrapment in vegetated devices or the structure of a PPS, systemic take-up by vegetation, or incorporation in the micro-ecosystem that develops on the geotextile associated with some PPSs (Newman *et al.* 2004).

Urban impacts are unsustainable should the status quo prevail. Approaches such as SUDS provide a means of sustainable urban living in the long term by managing the behavior of human beings to take account of water, rather than trying to modify the behavior of water to suit the activities of society.

5.7 Conclusions

Weathering and erosion of material in aquatic environments is a natural process leading to deposition of sediment in rivers, lakes, and wetlands. However, anthropogenic activities in urbanized catchments pollute these environments, resulting in deterioration of water and sediment quality in urban rivers and lakes. This environmental degradation has become a serious problem around the world owing to accelerated urbanization and industrialization. Water in cities is perceived as a nuisance at best, but at times it can serve as a main water resource for surrounding areas and thus water quality improvement becomes important. It can also provide a means of enhancement of urban areas, providing amenity and aesthetics as well as being part of a sustainable drainage approach, mitigating quality degradation as well as flooding hazard.

This chapter has presented some of the characteristics and mechanisms whereby sediment and its associated pollutants are produced, transported, and deposited in the city, and the way in which these are traced and monitored. To make best use of the water present in many urban areas, it is necessary to gather this kind of knowledge, which can provide the means to formulate management strategies and quality standards (Förstner 2009) so as to protect the urban environment and revitalize degraded sediments.

References

Adachi, K. & Tainosho, Y. (2005) Single particle characterization of size-fractionated road sediments. *Applied Geochemistry*, **20**(5), 849–59.

Anderberg, S. & Stigliani, W. M. (1994) An integrated approach for identifying sources of pollution: the example of cadmium in the Rhine River basin. *Water Science and Technology*, **29**(3), 61–67.

Ashley, R. M., Wotherspoon, D. J. J. & Coghlan, B. P. (1992) The erosion and movement of sediments and associated pollutants in combined sewers. *Water Science Technology*, **25**(8), 101–14.

Baird, C. (2002) *Química Ambiental.* Maria Angeles Lobo e Luiz Carlos Marques (Trads.). 2.ed. Porto Alegre: Bookman, 622p.

Banerjee, A. D. K. (2003) Heavy metal levels and solid phase speciation in street dusts of Delhi, India. *Environmental Pollution*, **123**, 95–105.

Bartram, J. & Ballance, R. (1996) *Water Quality Monitoring*. E & FN Spon, 383pp.

Beasley, G. & Kneale, P. (2002) Reviewing the impact of metals and PAHs on macroinvertebrates in urban watercourses. *Progress in Physical Geography*, **26**(2), 236–70.

Bentivegna, C. S., Alfano, J. E., Bugel, S. M. & Czechowicz, K. (2004) Influence of sediment characteristics on heavy metal toxicity in an urban marsh. *Urban Habitats*, **2**(1), 91–111.

Bird, G., Brewer, P. A., Macklin, M. G. *et al.* (2003) The solid state partitioning of contaminant metals and As in river channel sediments of the mining affected Tisa Drainage Basin, northwestern Romania and eastern Hungary. *Applied Geochemistry*, **18**, 1583–95.

Boszke, L. & Kowalski, A. (2006) Spatial distribution of mercury in bottom sediments and soils from Poznan, Poland. *Polish Journal of Environmental Studies* **15**(2), 211–18.

Brikker, O. P. (1999) An overview of the factors involved in evaluating the geochemical effects of highway runoff on the environment. US Geological Survey Open-File Report 98-630.

Brinkmann, W. L. F. (1985) Urban stormwater pollutants: sources and loadings. *GeoJournal*, **11**, 277–83.

Brown, J. N. & Peake, B. M. (2006) Sources of heavy metals and polycyclic aromatic hydrocarbons in urban stormwater runoff. *Science of the Total Environment*, **359**(1–3), 145–55.

Burton, G. & Pitt, R. (2002) *Stormwater Effects Handbook: A Toolbox for Watershed Managers, Scientists, and Engineers*. Boca Raton, FL, USA: CRC Press.

Butler, D. & Davies, J. W. (2000) *Urban Drainage*. E & FN Spon, 489pp.

Carter, J., Owens, P. N., Walling, D. E. & Leeks, G. J. L. (2003) Fingerprinting suspended sediment sources in a large urban river system. *Science of the Total Environment*, **314–316**, 513–534.

Castilhos, Z. C. (1999) *Gestão em Poluição Ambiental: Análise da Contribuição dos Garimpos de Ouro na Contaminação por Mercúrio da Ictiofauna e das águas fluviais na Região do Rio Tapajós, Estado do Pará, Brasil.* Tese de Doutorado. Curso de Pós-Graduação em Geociências-Geoquímica Ambiental. Universidade Federal Fluminense, Niterói-RJ, 194pp.

Charlesworth, S. M. & Foster, I. D. L. (1991) Pollution, rescue and management: the problem of two shallow urban lakes, Coventry, UK. IAHS Conference Report, Vienna 1991, 115–24.

Charlesworth, S. M. & Foster, I. D. L. (1993) The effect of urbanisation on sedimentation: the history of two lakes in Coventry, UK. In: *Geomorphology and Sedimentology of Lakes and Reservoirs*, J. McManus & R. W. Duck (eds). Chichester, UK: John Wiley.

Charlesworth, S. M. & Foster, I. D. L. (1999) Sediment budgets and metal fluxes in two contrasting urban lake catchments in Coventry, UK. *Applied Geography*, **19**, 199–210.

Charlesworth S. M. & Foster, I. D. L. (2005) Gamma emitting radionuclides and metallic elements in urban dusts and sediments, Coventry, UK: implications of dosages for dispersal and disposal. *Mineralogical Magazine*, **69**, 759–67.

Charlesworth, S. M. & Lees, J. A. (1999) Particulate-associated heavy metals in the urban environment: their transport from source to deposit, Coventry, UK. *Chemosphere*, **39**, 833–48.

Charlesworth S. M. & Lees J. A. (2001) The application of some mineral magnetic measurements and heavy metal analysis for characterising fine sediments in an urban catchment, Coventry, UK. *Journal of Applied Geophysics*, **48**(2), 113–125.

Charlesworth, S. M., Everett, M., McCarthy, R., Ordóñez, A. & Miguel, E. (2003a) A comparative study of heavy metal concentration and distribution in deposited street dusts in a large and a small urban area: Birmingham and Coventry, West Midlands, UK. *Environment International*, **29**, 563–73.

Charlesworth, S. M., Harker, E. & Rickard, S. (2003b) Sustainable drainage systems (SuDS): a soft option for hard drainage questions? *Geography*, **88**(2), 99–107

Charlesworth, S. M., Ormerod, L. M. & Lees, J. A. (2000) Tracing sediment within urban catchments using heavy metal, mineral magnetic and radionuclide signatures.

In *Tracers in Geomorphology*, I. D. L. Foster (ed.). Chichester, UK: John Wiley. 560pp.

Chebbo, G. & Gromaire, M. C. (2004) The experimental urban catchment "Le Marais" in Paris: what lessons can be learned from it? *Journal of Hydrology*, **299**, 312–23.

Christopher, T. K., Davenport, P. H. & Burden, E. T. (1993) The effect of urban and industrial development on the geochemistry of the watersheds in the St. John's area: preliminary results. In *Current Research: Newfoundland Department of Mines and Energy, Geological Survey Branch*. Report 93-1; 419–33.

Dawson, E. J. & Macklin, M. G. (1998) Speciation of heavy metals in floodplain and flood sediments: a reconnaissance survey of the Aire Valley, West Yorkshire, Great Britain. *Environment Geochemistry*, **20**, 67–76.

Deletic, A. (1998) The first flush load of urban surface runoff. *Water Research*, **32**(8), 2462–70.

Deletic, A., Ashley, A. & Rest, D. (2000) Modelling input of fine granular sediment into drainage systems via gully-pots. *Water Research* **34**, 3386–44.

De Miguel, E., Charlesworth, S., Ordóñez, A. & Seijas, E. (2005) Geochemical fingerprints and controls in the sediments of an urban river: River Manzanares, Madrid (Spain). *Science of the Total Environment*, **340**, 137–148.

Dietz, M. E. (2007) Low impact development practices: a review of current research and recommendations for future directions. *Water, Air and Soil Pollution*, **186**, 351–363.

Douglas, I. (1999) Physical problems of the urban environment. In *Applied Geography: Principles and Practice*, A. Pacione (ed.), 124–34. Routledge.

Droppo, I. G. (2002) *A New Definition of Suspended Sediment: Implications for the Measurement and Prediction of Sediment Transport*. Oslo, Norway: International Association of Hydrological Sciences, publication 283, 3–12.

Egodawatta, P. & Goonetilleke, A. (2007) Understanding road surface pollutant wash-off and underlying physical processes using simulated rainfall. In *11th International Conference on Diffuse Pollution and 1st Joint Meeting of the IWA Diffuse Pollution and Urban Drainage Specialist Groups, Belo Horizonte-MG, Brazil, August 26th–31st, 2007.*

Ellis, J. B. & Revitt, D. M. (1982) Incidence of heavy metals in street surface sediments: solubility and grain size studies. *Water, Air and Soil Pollution*, **17**, 87–100.

Ellis, J. B., Revitt, D. M., Shutes, R. B. E. & Hamilton, R. S. (1986) The effects of urbanisation on receiving water quality: heavy metal toxicity. In *Effects of Land Use on Fresh Waters: Agriculture, Forestry, Mineral Exploitation, Urbanisation*, J. F. L. G. Solbe, (ed.), 473–77. Ellis Horwood.

Fan, C-Y., Field, R., Sullivan, D. & Lai, F-h. (2004) Toxic pollutants in urban wet-weather flows: an overview of the multi-media transport, impacts and control measures. In *World Water and Environmental Resources Congress, Orlando, Florida, USA.*

Federal Interagency Stream – Restoration Working Group. (1998) Stream Corridor Restoration: Principles, Process and Practices. http://www.abe.msstate.edu/Tools/csd/strm-cor-res/index.php. Accessed March 2007.

Förstner, U. (2009) Sediments and priority substances in river basins. *Journal of Soils and Sediments*, **9**, 89–93.

Förstner, U. & Müller, G. (1976) Heavy metal pollution monitoring by river sediments. *Fortschritte der Mineralogie*, **53**, 271–88.

Foster, I. D. L. & Charlesworth, S. M. (1996) Heavy metals in the hydrological cycle: trends and explanation. *Hydrological Processes*, **10**, 227–61.

Foster, I. D. L., Charlesworth, S. M. & Proffitt, S. B. (1996) Sediment-associated heavy metal distribution in urban fluvial and limnic systems; a case study of the River Sowe, UK. *Archiv für Hydrobiologie Special Issues in Advanced Limnology*, **47**, 537–45.

Gaskell, R. (1992) Lake sediment properties from an urban catchment: South Telford, Shropshire. B.Sc. Dissertation, Coventry University.

Gibbs, R. J. (1973) Mechanisms of trace metal transport in rivers. *Science*, **10**, 71–73.

Glenn, D.W., Liu, D. & Sansalone, J. J. (2001) Influence of highway runoff chemistry, hydrology and residence time on nonequilibrium partitioning of heavy metals: implications for treatment at the highway shoulder. *Transportation Research Record: Journal of the Transportation Research Board*, **1755**, 129–40.

Glenn, D. W, Liu, D. & Sansalone, J. J. (2004) Influence of chemistry, hydrology and suspended solids on partitioning of heavy metals to particles – considerations for in-situ controls of urban stormwater quality. *9th International Conference on Urban Drainage. Portland, Oregon, USA.*

Gorski, P. R., Cleckner, L. B., Hurley, J. P., Sierszen, M. E. & Armstrong, D. E. (2003) Factors affecting enhanced mercury bioaccumulation in inland lakes of Isle Royale National Park, USA. *Science of the Total Environment*, **304**(1–3), 327–48.

Graney, J.R. & Eriksen, T.M. (2004) Metals in pond sediments as archives of anthropogenic activities: a study in response to health concerns, *Applied Geochemistry*, **19**, 1177–88.

Gromaire, M. C., Garnaud, S., Saad, M. & Chebbo, G. (2001) Contribution of different sources to the pollution of wet weather flows in combined sewers. *Water Research*, **35**(2), 521–33.

Guerra, T. (2000) *Estudo da Contaminação Hidrogeoquímica Fluvial e sua Inserção na Avaliação Econômico-ambiental da Mineração de Carvão na Região do Baixo Jacuí, Rio Grande do Sul, Brasil.* Tese de Doutorado em Geoquímica Ambiental, UFF, Niterói.

Hallberg, H., Renman, G. & Lundbom, T. (2007) Seasonal variations of ten metals in highway runoff and their partition between disspolved and particulate matter. *Water, Air and Soil Pollution* **181**, 183–91.

Hamilton, R. S., Revitt, D. M. & Warren, R. S. (1984) Levels and physico-chemical associations of Cd, Cu, Pb and Zn in road sediments. *Science of the Total Environment*, **33**, 59–74.

Harbor, J. (1999) Engineering geomorphology at the cutting edge of land disturbance: erosion and sediment control on construction sites. *Geomorphology*, **31**, 247–63.

Harrison, R. M., Johnston, W. R., Ralph, J. C. & Wilson, S. J. (1985) The budget of lead, copper and cadmium for a major highway. *Science of the Total Environment*, **46**, 137–45.

Hollis, G. E. (1988) Rain, roofs and runoff: hydrology in cities. *Geography*, **73**, 9–18.

Horowitz, A. J. (1991) *A Primer on Sediment-Trace Element Chemistry*, 2nd edn. Chelsea, EUA: Lewis Publishers. 136pp.

Horowitz, A. J. (1995) The *Use of Suspended Sediment and Associated Trace Elements in Water Quality Studies*. International Association of Hydrological Sciences, Special Publication, no. 4, 58pp.

Hortellani, M. A., Sarkis, J. E. S., Bonetti, J. & Bonetti, C. (2005) Evaluation of mercury contamination in sediments from Santos-São Vicente estuarine system, São Paulo State, Brazil. *Journal of the Brazilian Chemical Society* **16**(6A), 1140–49.

Hunerlach, M. P., Alpers, C. N., Marvin-DiPasquale, M., Taylor, H. E. & De Wild, J. F. (2004) Geochemistry of mercury and other trace elements in fluvial tailings upstream of Daguerre Point Dam, Yuba River, California, August, 2001; Scientific investigations report 2004-5165. *US Geological Survey: Reston VA*.

Jansson, M. B. (2002) Determining Sediment Source Areas in a Tropical River Basin, Costa Rica. *Catena*, **47**, 63–84.

Jefferies, C., Napier, F., Heal, K. V., Fogg, P. & D'Arcy, B. (2007) The fate of traffic related pollutants in soft engineering SUDS: an experimental and field approach. In *11th International Conference on Diffuse Pollution and 1st Joint Meeting of the IWA Diffuse Pollution and Urban Drainage Specialist Groups, Belo Horizonte-MG, Brazil, August 26th–31st, 2007*.

Jennett, J. C., Effler, S. W. & Wixson, B. G. (1980) Mobilization and toxicological aspects of sedimentary contaminants. In: *Contaminants and sediments*, volume **1**, *Fate and Transport Case Studies, Modelling, Toxicity*, R. A. Baker (ed.). Ann Arbor Science.

Jitaru, P. & Adams, F. (2004) Toxicity, sources and biogeochemical cycle of mercury. *Journal de Physique*, **121**, 185–93

Jorge, M. T. E. (2007) Monitoring Of Diffuse Pollution Supported By Geographic Information Systems. In *11th International Conference on Diffuse Pollution and 1st Joint Meeting of the IWA Diffuse Pollution and Urban Drainage Specialist Groups, Belo Horizonte-MG, Brazil, August 26th–31st, 2007*.

Kayhanian, M. & Stenstrom, M. K. (2008) First flush characterisation for stormwater treatment. Stormwater. http://www.stormh20.com/march-april-2008/pollutants-run-off.aspx. Accessed October 29, 2008.

Lai, D. Y. F. & Lam, K. C. (2009) Phosphorus sorption by sediments in a subtropical constructed wetland receiving stormwater runoff. *Ecological Engineering*, **35**, 737–43.

Lawler, D. M., Petts, G. E., Foster, I. D. L. & Harper, S. (2006) Turbidity dynamics during spring storm events in an urban headwater river system: the Upper Tame, West Midlands, UK. *Science of the total environment*, **360**(1–3), 109–26.

Lick, W. (1987) The transport of sediments in aquatic systems. In *Fate and Effects of Sediment-Bound Chemicals in Aquatic Systems*, K. L. Dickson, A. W. Maki, & W. A. Brungs (eds), 61–74. Oxford: Pergamon Press.

Machado, W., Santelli, R. E., Loureiro, D. D., *et al* (2008) Mercury accumulation in sediments along a eutrophication gradient in Guanabara Bay, Southeast Brazil. *Journal of the Brazilian Chemical Society* **19**(3), 569–75.

Marchand, C., Lallier-Verger, E., Baltzer, F., Albéric, P., Cossa, D. & Baillif, P. (2006) Heavy metal distribution in mangrove sediments along the mobile coastline of French Guiana. *Marine Chemistry*, **98**, 1–17.

Marins, R. V., Lacerda, L. D., Paraquetti, H. H. M., Paiva, E. C. & Villas-Boas, R. C. (1998) Geochemistry of mercury in sediments of a sub-tropical coastal lagoon, Sepetiba Bay, Southeastern Brazil. *Bulletin of Environmental Contamination and Toxicology*, **61**, 57–64.

Mason, R. P. & Sullivan, K. A. (1998) Mercury and methylmercury transport through an urban watershed. *Water Research* **32**(2), 321–30.

Marsalek, J. (1990) Evaluation of pollutant loads from urban non-point sources. *Water Science and Technology*, **22**(10–11), 23–30.

Marvin-Dipasquale, M., Lutz, M. A. Brigham, M. E., *et al.* (2009) Mercury cycling in stream ecosystems: 2. Benthic methylmercury production and bed sediment-pore water partitioning. *Environmental Science and Technology* **43**(8), 2726–32.

Minella, J. P (2003) Identificação de fontes de produção de sedimentos em uma pequena bacia rural. Dissertação de Mestrado apresentada ao PPG em Recursos Hídricos e Saneamento Ambiental da UFRGS. Porto Alegre. 80p. il.

Morrison, G. M., Revitt, D. M., Ellis, J. B., Svensson, G. & Balmer, P. (1984) Variations in dissolved and suspended solid heavy metals through an urban storm hydrograph. *Environmental Technology Letters*, **7**, 313–18.

Morrison, G. M., Revitt, D. M., Ellis, J. B., Svensson, G. & Balmer, P. (1988) Transport mechanisms and processes for metal species in a gully pot system. *Water Research*, **22**, 1417–27.

Morrison, G. M., Revitt, D. M. & Ellis, J. B. (1989) Sources and storm loading variations of metal species in a gullypot catchment. *Science of the Total Environment*, **80**, 267–78.

Morrison, G. M., Revitt, D. M. & Ellis, J. B. (1995) The gully pot as a biochemical reactor. *Water Science and Technology*, **31**, 229–36.

Mungur, A. S., Shutes, R. B. E., Revitt, D. M. & House, M. A. (1995) An assessment of metal removal from highway runoff by a natural wetland. *Water Science and Technology* **32**(3), 169–75.

Newman, A. P., Puehmeier, T., Kwok, V., *et al.* (2004) Protecting groundwater with oil-retaining pervious pavements: historical perspectives, limitations and recent developments. *Quarterly Journal of Engineering Geology and Hydrogeology*, **37**(4), 283–91.

Nichols, S. J. & Manny, B. A., Schloesser, D. W. & Edsall, T. A. (1991) Heavy metal contamination of sediments in the upper connecting channels of the Great Lakes. *Hydrobiologia*, **219**, 307–15.

OMEE (Ontario Ministry of the Environment and Energy). (1993) *Guidelines for the Protection and Management of Aquatic Sediment Quality in Ontario*. Toronto: Ontario Ministry of the Environment and Energy.

Ono, Y., Uemura, H., Kanjo, Y., Kawara, O. & Ayano, T. (2000) Genotoxicity of substances extracted from construction materials. *Journal of Material Cycles and Waste Management*, **2**, 1, 38–42.

Owens, P. N., Walling, D. E., Carton, J., Meharg, A. A., Wright, J. & Leeks, G. J. L. (2001) Downstream changes in the transport and storage of sediment-associated contaminants (P, Cr and PCBs) in agricultural and industrialized drainage basins. *Science of the Total Environment*, **266**, 177–86.

Pardos, M., Benninghoff, C., Alencastro, L. F. & Wildi, W. (2004) The impact of a sewage treatment plant's effluent on sediment quality in a small bay in Lake Geneva (Switzerland–France). Part 1: Spatial distribution of contaminants and the potential for biological impacts. *Lakes & Reservoirs: Research and Management*, **9**, 41–52.

Pearson, D. (1990) *Highway drainage and its problems. Conference on Developments in Storm Drainage*. Paper 4.1. IWEM & CIRIA, 152–65.

Peart, M. R. & Walling, D. E. (1986) Fingerprinting sediment source: the example of a drainage basin in Devon, UK. In *Drainage Basin Sediment Delivery*, R. F. Hadley (ed.), 41–55. Wallingford: IAHS publications, no. 159.

Perdikaki, K. & Mason, C. F. (1999) Impact of road runoff on receiving streams in eastern England. *Water Research*, **33**, 1627–33.

Pitt, R., Lalor, M., Field, R. & Brown, M. (1993) The investigation of source area controls for the treatment of urban storm water toxicants. *Water Science and Technology*, **28**(3–5), 271–82.

Pratt, C. J. (2004) *Sustainable Urban Drainage: A Review of Published Material on the Performance of Various SUDS components*. Environment Agency.

Poleto, C. (2007) *Fontes Potenciais e Qualidade dos Sedimentos Fluviais em Suspensão em Ambiente Urbano*. Tese de Doutorado. Porto Alegre: UFRGS, 137pp.

Poleto, C. & Castilhos, Z. C. (2008) *Sedimentos e Poluição Difusa em Bacias Urbanas*. In *Ambiente e Sedimentos*, C. Poleto (ed.). Porto Alegre: ABRH, 342pp.

Poleto, C. & Merten, G. H. (2007) Urban watershed studies in southern *Brazil. Journal of Urban and Environmental Engineering*, **1**(2), 70–78.

Poleto, C. & Laurenti, A. (2008) *Sedimentos Urbanos e Corpos D'água*. In *Ambiente e Sedimentos*, C. Poleto (ed.). Porto Alegre: ABRH, 342pp.

Poleto, C., Merten, G. H. & Minella, J. P. (2009) The identification of sediment sources in a small urban watershed in southern Brazil: an application of sediment fingerprinting. *Environmental Technology*, **30**(11), 1145–53.

Poleto, C., Merten, G. H. & Silveira, A. L. L. (2005) socio-economic impacts on fluvial system an urban watershed in southern Brazil. In *International Conference on Urban Drainage – ICUD, X, Copenhagen, Dinamarca*.

Pompeu, P. S., Alves, C. B. M. & Callisto, M. (2005) The Effects of Urbanization on Biodiversity and Water Quality in the Rio das Velhas Basin, Brazil. In *American Fisheries Society Symposium*. **47**, 11–22.

Porto, M. F. A. (1995) Aspectos Qualitativos do Escoamento Superficial em Áreas Urbanas. In *Drenagem Urbana*, C. E. M. Tucci, R. L. Porto & M. T. Barros (eds), 387–428. Porto Alegre: ABRH/Editora da Universidade/UFRGS, **5**, 387–428.

Quek, U. & Forster, J. (1993) Trace metals in roof runoff. *Water, Air and Soil Pollution*, **68**, 373–89.

Rao V. V. S. G., Yoshida, M., Prakash, B. A., Chandrasekhar, S. V. N. & Kumar, K. M (2004) Environmental impact of human activities to urban lake sediments: potentially toxic elements (PTEs) contamination in Hussainsagar Lake, Hyderabad. *The 11th National Symposium on Hydrology, National Institute of Hydrology, Roorkee, India.*

Revitt, D. M. & Morrison, G. M. (1987) Metal speciation variations within separate stormwater systems. *Environmental Technology Letters*, **8**, 373–80.

Rhoads, B. L. & Cahill, R. A. (1999) Geomorphological assessment of sediment contamination in an urban stream system. *Applied Geochemistry*, **14**, 450–83.

Robinson, M., Boardman, J., Evans, R., Heppell, K., Packman, J. & Leeks, G. (2000) Land use change. In: *The Hydrology of the UK*, M. Acreman (ed.). Routledge Publishers.

Rocha, R. R. A. & Martin, E. S. (2005) Análise Preliminar do Estado Ambiental do Córrego Água da Lavadeira,

Rancharia-SP: Análise Física e Química da Água. *Revista Eletrônica da Associação dos Geógrafos Brasileiros*, 2(2), 116–30.

Russell, M. A., Walling, D. E. & Hodgkinson, R. A. (2001) Suspended sediment sources in two small lowland agricultural catchments in the UK. *Journal of Hydrology*, **252**, 1–24.

Sanders, C. J., Santos, I. R., Silva-Filho, E. V. & Patchineelam, S. R. (2006) Mercury flux to estuarine sediments, derived from Pb-210 and Cs-137 geochronologies (Guaratuba Bay, Brazil). *Marine Pollution Bulletin*, **52**, 1085–89.

Sansalone, J. J. & Ying, G. (2008) Partitioning and granulometric distribution of metal leachate from urban traffic dry deposition particulate matter subject to acidic rainfall and runoff retention. *Water Research*, **42**, 4146–62.

Schueler, T. & Simpson, J. (2001) Why urban lakes are different. *Urban Lake Management, Watershed Protection Techniques*. 3(4), 747–50.

Sigg, L. (1998) Partitioning of metals to suspended particles. In: *Metals in surface waters*, H. E. Allen, A. W. Garrison, & G. W. Luther III (eds.). Ann Arbor Press.

Singh, K. P., Mohan, D., Singh, V. K. & Malik, A. (2005) Studies on distribution and fractionation of heavy metals in Gomti river sediments – a tributary of the Ganges, India. *Journal of Hydrology*, **312**, 14–27.

Sonzogny, W., Chesters, G., Coote, D., *et al.* (1980) Pollution from land runoff. *Environmental Science & Technology*, **14**(2), 148–53.

Stigliani, W. A., Jaffe, P. R. & Anderberg, S. (1993) Heavy metal pollution in the Rhine basin. *Environ Sci Technol*, **27**, 787–793.

Striegel, R. G. (1987) Suspended sediment and metals removal from urban runoff by a small lake. *Water Resources Bulletin*, **23**, 985–96.

Sutherland, R. A. (2000) Bed sediment-associated trace metals in an urban stream, Oahu, Hawaii. *Environmental Geology*, **39**, 611–27.

Syvitski, J. P. M., Vörösmarty, C. J., Kettner, A. J. & Green, P. (2007) Impact of humans on the flux of terrestrial sediment to the global coastal ocean. *Science*, **308**, 376–80.

Taylor, K. (2007) Urban environments. In: *Environmental Sedimentology*, C. Perry & K. Taylor (eds). UK: Blackwell, 441pp.

Tessier, A., Campbell, P. G. C. & Bisson, M. (1979) Sequential extraction procedure for the speciation of particulate trace metals. *Analytical Chemistry*, **51**, 844–51.

Thomas, P. R. & Greene, G. R. (1993) Rainwater quality from different roof catchments. *Water Sci Technol*, **28**, 291–299.

Thoms, M. C. (1987) Channel sedimentation within the urbanized River Thame, UK. *Regulated Rivers: Research and Management*, **1**, 229–46.

Tong, S., von Schirnding, Y. E. & Prapamontol, T. (2000) Environmental lead exposure: a public health problem of global dimensions. *Bulletin of the World Health Organization*, **78**(9), 1068–77.

Tucci, C. E. M. (2003) Inundações e Drenagem Urbana. In: *Inundações Urbanas na América do Sul*. Tucci, C. E. M., Bertoni, J. C. (eds.). Porto Alegre: Associação Brasileira de Recursos Hídricos – ABRH.

USEPA. (2000) *Mercury Research Strategy*. Cincinatti: EPA 600/R – 00/073. 57pp.

USEPA. (2005). *Stormwater Phase II Final Rule: Construction Site Runoff Control. Minimum Control Measure*. Fact Sheet 2.6. EPA 833-F-00-008. 4pp.

Von Schirnding, Y. E. (1999) The impact of lead poisoning on the workforce and society. In: *Proceedings of the International Conference on Lead Poisoning, Bangalore, India, 8–10 February 1999. Bangalore, 41–47.* The George Foundation.

Walling, D. E. (2005) Tracing suspended sediment sources in catchments and river systems. *Science of the Total Environment*, **344**, 159–84.

Walling, D. E., Russell, M. A., Hodgkinson, R. A. & Zhang, Y. (2002) Establishing sediment budgets for two small lowland agricultural catchments in the UK. *Catena*, **47**, 323–53.

Wilber, W. G. & Hunter, J. V. (1979) The impact of urbanization on the distribution of heavy metals in the bottom sediments of the Saddle River. *American Water Resources Association*, **15**, 790–800.

Wolman, M. G. & Schick, A. P. (1967) Effects of construction on fluvial sediment, urban and suburban areas of Maryland. *Water Resources Research*, 3(2), 451–64.

Wong, C. S. C., Li, X. & Thornton, I. (2006) Urban environmental geochemistry of trace metals. *Environmental Pollution*, **142**, 1–16.

Woods-Ballard, B., Kellagher, R., Martin, P., Jefferies, C., Bray, R. & Shaffer, P. (2007) *The SUDS Manual*. CIRIA C697.

Yu, L. & Oldfield, F. (1989) A multivariate mixing model for identifying sediment source from magnetic measurements. *Quaternary Research*, **32**, 168–81.

Zhang, T., Ellis, J. B., Revitt, D. M. & Shutes, R. B. E. (1990) Metal uptake and associated pollution control by *Typha latifolia* in urban wetlands. In: *Constructed Wetlands in Water Pollution Control*, P. F. Cooper & B. C. Findlater (eds), 451–59. Oxford: Pergamon Press.

Zhou, M. M., Li, L. J., Bertine, K. K., Koide, M. & Goldberg, E. D. (1989) Atmospheric pollution in Beijing, China, as recorded in sediments of the Summer Palace Lake. *Environmental Conservation*, **8**, 233–36.

6 Biomarkers in integrated ecotoxicological sediment assessment

Mark G.J. Hartl[1]

[1]*Centre for Marine Biodiversity and Biotechnology, School of Life Sciences, Heriot-Watt University, Edinburgh, UK*

6.1 Introduction

After the publication of *Silent Spring* by Rachel Carson (1962), which saw the first attempt to separate environmental toxicology from classical toxicology, Truhaut (1977) introduced the term "ecological toxicology" or "ecotoxicology" to describe the toxicological impact of environmental contaminants in the ecosystem, beyond the level of the individual organism. This was defined further by Chapman (2002) to include toxicity on all levels of biological organization and the environmental fate of contaminants.

Environmental contaminants have been linked to population-level variations of sensitive keystone species, whose reduced vitality or removal can erode community structure and potentially reduce diversity and stability (Paine 1966). Such impacts on higher levels of biological organization only become apparent when a significant part of the population is already affected. Therefore, much effort has been devoted to establishing protocols for monitoring more subtle sublethal effects on various levels of biological organization. These are referred to as biomarkers for flagging up potential adverse anthropogenic impacts at an early and manageable stage. A biomarker, as defined by Depledge *et al.* (1993), is "... a biochemical, [genetic] cellular, physiological or behavioural variation that can be measured in tissue or body fluid samples or at the level of the whole organism (either individuals or populations), that provides evidence of exposure and/or effects of one or more chemical pollutants (and/or radiation)".

Sedimentology of Aqueous Systems, 1st edition.
Edited by Cristiano Poleto and Susanne Charlesworth.

Biomarkers are powerful tools for detecting sublethal exposure to a given substance or a complex chemical mixture, enabling the evaluation of more subtle effects on organisms and can be applied as an early warning system. Biomarkers can be loosely categorized as those of exposure, effect, and susceptibility. A biomarker of exposure indicates that an organism has come into contact with a contaminant or contaminant mixture, and can give qualitative and quantitative estimates of bioavailability (Schlenk 1999; Chambers *et al.* 2002), but provides little information about the cause of the observed interaction. Causality can be established by applying biomarkers of effect that relate to a specific contaminant or contaminant class through a well-described mode of action. As the response to exposure may depend on various environmental and physiological conditions, it will not be identical in all individuals of the same species reducing the dose–response resolution. Therefore, in addition to biomarkers of exposure and effect, biomarkers of susceptibility are required to help identify areas of uncertainty that may occur between the exposure to a contaminant and the emergence of clinical symptoms (Schlenk 1999).

The use of biomarkers in aquatic ecotoxicology has traditionally been limited to the exposure of sentinel organisms or *in vitro* test systems to pollutants in aqueous solutions or suspensions. These approaches have been instrumental in providing guidelines for legislative measures aimed at reducing the impact of anthropogenic pressure on marine and freshwater environments. In recent years, however, the relative improvement of water quality in many areas and the recognition that sediments may serve not just as sinks but also as secondary sources for many persistent chemicals (Harris *et al.* 1996), has shifted the focus of ecotoxicological studies toward sediments and the potential deleterious effects persistent pollutants have on benthic ecosystems (Anderson *et al.* 1996; Martin-Diaz *et al.* 2004).

This chapter discusses various approaches to sediment ecotoxicology, the advantages and limitations, and its role in environmental monitoring and impact assessment.

6.1.1 Sediments

Aquatic sediments represent an open, dynamic, and heterogeneous biogeochemical system (Sundby 1991), formed by an accumulation of particulate matter introduced to the aquatic environment from a variety of sources, such as continental run-off, coastal erosion or atmospheric fall-out, which is then deposited on the bottom of a water body. Typically, sediments are a structured accumulation of particulate mineral matter, inorganic matter of biogenic origin, organic matter in various stages of decomposition or synthesis, and water (Knezovich *et al.* 1987; ASTM 1994b). Sediments usually consist of an inorganic matrix (silica, alumina, and carbonates) which is coated with organic matter, manganese, and iron oxides, but can be anything from pure inorganic to pure organic in composition (Rand *et al.* 1997), giving rise to a wide variety of physical, chemical, and biological characteristics. For experimental purposes, sediments can be formulated from particulate matter of known origin and characteristics to create specific controlled conditions (ASTM 1994b; Suedel & Rodgers 1994; Hartl *et al.* 2000; Quevauviller & Ariese 2001).

6.1.2 Assessing the characteristics of aquatic sediments

Sediments are largely site specific, depending on a multitude of physico-chemical parameters, most notably, salinity, grain size, sedimentation rates, and the organic carbon fraction.

Water, as the universal solvent, dissolves more substances than any other liquid. Five anions (Cl^-, SO_4^{2-}, HCO_3^-, Br^-, H_3BO_3) and five cations (Na^+, Mg^{2+}, Ca^{2+}, K^+, Sr^{2+}) constitute 99.9% of ions dissolved in seawater. These "conservative" constituents always maintain their relative proportions, regardless of salinity. The proportions of "non-conservative" constituents, such as nutrients (phosphor, nitrate, ammoniac) and dissolved gasses (O_2, CO_2), are greatly influenced by geochemical and biological processes in the surrounding catchments. CO_2 has a high solubility in water. Chemical reactions with water molecules enable 100–200 times more CO_2 to dissolve in water than would be possible through physical interactions alone. CO_2 dissolves in water to form carbonic acid (H_2CO_3), which in turn dissociates to hydrogen carbonate (HCO_3^-) and carbonate ions (CO_3^-), depending on pH. Whereas the pH of freshwater does vary, depending on the underlying geology and organic input, the pH of normal seawater is around 8, caused by the surplus of acid cations. Under these slightly alkaline conditions, all three forms of carbonate ions can be found. Initially, addition of protons (acid) or hydroxyl ions (base) causes the carbonate species to shift in the relative amounts, but does not release free hydrogen ion. Therefore, seawater is generally very well buffered, whereas the buffering capacity of freshwater depends on the geological setting (Ott 1988; Brown *et al.* 1992; Libes 1992).

The salinity in estuaries can vary greatly, owing to tidal action, and the magnitude of river discharge, which facilitates fluctuations in pH and shifts in the sorption behavior and bioavailability of many chemical compounds. Granulometric characteristics are largely governed by hydrodynamic conditions. In fast-flowing water, sediments tend to consist of coarse sand, gravel, and pebbles, whereas slow-moving water allows for the deposition of fine-grained sand, silt, and clay (Riedl & Ott 1982). In fact, the grain size–frequency distribution is fundamental to the biogeochemistry of sediments and is used in their classification. Like sedimentologists, benthic ecologists use arbitrarily graded scales, either logarithmic or geometric, to classify sediment grain size spectra for habitat characterization (Buchanan 1971; Ott 1988; Libes 1992).

Grain size, shape, and packing density determine the porosity and in turn the volume of pore or interstitial water of sediments (Ott 1988). Coarse-grained, loosely packed sediments are characterized by a small surface area and high porosity, whereas fine-grained, densely packed sediments exhibit a very large surface area with low porosity. Low porosity increases the pore water residence time, reduces oxygen supply and favors the establishment of distinct biogeochemical gradients, giving rise to oxidation–reduction conditions (expressed as redox potential, E_h), which can differ greatly from those in the overlying water (Brassard 1991).

There are a variety of dedicated sources that address methods for the physical and chemical characterization of sediments (Buchanan 1971; Horowitz & Elrick 1988; ASTM 1990; US Environmental Protection Agency 1992; Reynoldson & Rodriguez 1999; Rodriguez & Reynoldson 1999). These are also discussed in detail in this volume (see Chapter 3).

6.1.3 Significance of sediments in ecotoxicology

Many contaminants, especially less polar organic compounds and the most toxic of the trace elements, show a strong affinity to suspended particulate matter and are thereby sequestered from the water column and incorporated into the sediment (Harris & Cleary 1987; Ragnarsdottir 2000). Redox conditions in sediments drive shifts in ion ratios that can change the chemical speciation, sorption behavior and partition coefficients of incorporated compounds and trace elements, resulting in the pore water displaying a very different natural chemical composition than the overlying water. Undisturbed sediments accumulate chemical compounds and so can become sinks and eventually reservoirs for contaminants potentially toxic to aquatic organisms. The retention capacity of sediments for many contaminants is, however, reversible, owing to changes in salinity, pH, E_h or mechanical disturbance. Sediments therefore act not only as sinks, but also as secondary sources of accumulated contaminants, directing often highly concentrated pulses of toxic substances at benthic organisms; that is, organisms that during part or their life cycles are intimately associated with sediments, as a source of food or refuge. Fine-grained, organically rich sediments, therefore, play a major role in the biogeochemical fate of chemicals, both of natural and anthropogenic origin (Eggleton & Thomas 2004; Atkinson *et al.* 2007).

Unlike water-quality criteria (WQC) that have been implemented for some time (US Environmental Protection Agency 1972, 1986), sediments have traditionally been regarded as a final sink for many, especially, non-polar pollutants. Guidelines for sediment quality criteria (SQC), mostly based on the equilibrium partitioning approach (Di Toro *et al.* 1991), have been drawn up by several organizations (US Environmental Protection Agency 1993; CCME 1995; OSPAR 1997). Detailed accounts of methods for establishing sediment quality criteria have been extensively reviewed elsewhere (Ingersoll 1995; Simpson *et al.* 2005).

6.2 Approaches to assessing sediment toxicity

6.2.1 General considerations

The purpose of sediment toxicity tests is to determine whether sediments contain substances harmful to benthic organisms. They can cover a range of issues, including determining bioavailability of contaminants, the potential interaction among contaminants, spatial and temporal distribution, and establishing causality of observed effects. Furthermore, sediment toxicity tests are used in tiered decision trees for the assessment of contaminated sediments to earmark areas for cleanup and to monitor the effectiveness of remediation and management initiatives.

Clearly, the methods applied will depend on the aims of a given study and can range from acute tests that measure the effects of an individual contaminant on a single species to complex chronic tests with an increased level of ecological relevance that determine effects of chemical mixtures on the structure and function of communities and ecosystems. Accordingly, the sediment phase examined can range from whole sediment to pore water and elutriates of varying volumina (Ni Shuilleabhain *et al.* 2003) and use test organisms covering all trophic levels, including algae and macrophytes, benthic invertebrates, pelagic invertebrates with benthic life stages, and fish (Burton 1992).

Toxicity tests should ideally be simple, inexpensive, have a rapid turnaround time, and a high level of ecological relevance. As can be seen from Fig. 6.1, this is very rarely realized, as the complexity and cost of sediment toxicity tests increases with ecological relevance, and thus large-scale integrated field tests, the most ecological relevant, are very rarely performed. Consequently, there has been a rapid development of laboratory-based tests using field-collected sediment samples, that present far fewer logistical difficulties and at the same time allow for the control and correction of confounding

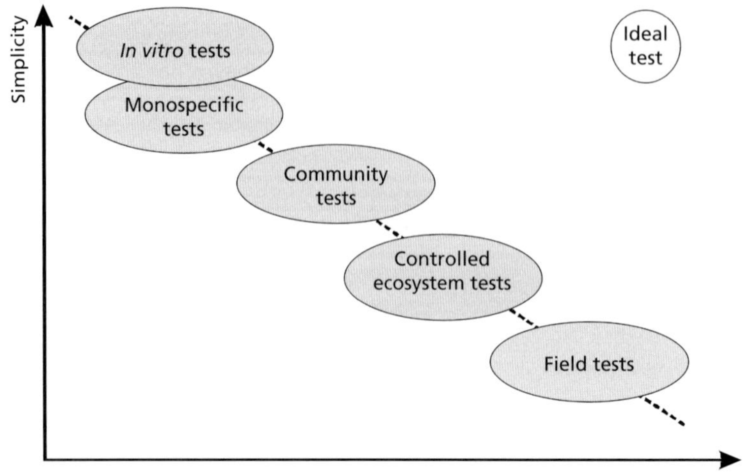

Fig. 6.1 Relation between ecological relevance and procedural simplicity for various types of aquatic ecotoxicological test.

environmental variables (Hartl *et al.* 2005); these may provide data for contaminant fate and impact assessment models.

However, laboratory-based approaches to ecotoxicology require procedures for collection, storage, and preparation of sediments (ASTM 1990), which all induce unavoidable geochemical changes, in particular to the pH and redox status (E_h) after handling (Luoma & Ho 1998; Hutchins *et al.* 2007). Natural sediment deposits are structured systems of oxic sediments on top of anoxic ones (Fenchel 1969). The depth of the oxic layer is a function of grain size, sedimentation rate, and the biological oxygen demand of the system (Aller 1978), the latter driven in turn by the organic content of the sediment and ambient temperature. The redox potential discontinuity is the depth where oxygen demand begins to exceed supply and separates the oxygenated from the reduced sediment layers beneath (Elskens *et al.* 1991). Usually horizontally orientated, the redox potential discontinuity can be complicated by the burrows of invertebrate infauna, such as lugworms and mudshrimps. The two zones display very different chemical conditions (Machan & Ott 1972; Aller 1978): in oxidized sediments, iron and manganese oxides occur mainly as reactive species. Under oxic conditions, contaminants, such as other metals, will bind to these and other available surfaces. In anoxic sediments oxidized iron and manganese are rare. The soluble reduced forms, together with other metal sulfides, are the dominant species and accumulate in the pore water.

Therefore, before any sampling, several details of the sampling procedure will need careful consideration, as proper handling of samples during the collection process is essential for maintaining quality standards and avoiding distortion of analytical results (Chapman 1989; ASTM 1990; Power & Chapman 1992; US Environmental Protection Agency 1992; Chapman & Wang 2001). Suitable sampling methods will depend on site accessibility and the type and amount of sediment required, which are representative of the conditions at the site and, where possible, the integrity of the sample is maintained. The advantages and limitations of commonly used sampling methods have been extensively discussed elsewhere: dredges (ASTM 1990), grabs and box corers (Carlton & Wetzel 1985; Papucci *et al.* 1986; Webb 1989; Weaver & Schultheiss 1990; Flower *et al.* 1995; Santschi *et al.* 2001), mega corers (Black *et al.* 2002), and hand-held corers or scoops (Byrne & O'Halloran 1999; Coughlan *et al.* 2002).

A complete integrated ecotoxicology assessment approach should include chemical analysis. Therefore it is imperative that suitable containers are used to collect environmental and biological samples, such as high-density polyethylene or polytetrafluroethylene

(PTFE), pre-cleaned with a strong detergent and rinsed with 10% HNO_3, to minimize adsorption of contaminants to the collection vessel and to maintain the chemical integrity of samples containing complex mixtures. This is not always feasible. Therefore, depending on the aims of the survey, either appropriately cleaned brown borosilicate glass with Teflon lid liners (organics, inorganic metals) or plastic or polycarbonate (inorganic metals) containers are most commonly used. Generally, these should be filled to capacity with a little 'head' space for expansion in case of frozen storage. For anoxic sediment collection containers should be purged with an inert gas such as nitrogen to allow anoxic conditions to be maintained (Ankley & Schubauerberigan 1994; Bufflap & Allen 1995; Carr & Chapman 1995; US Environmental Protection Agency 2001b). A vital aspect is sample labeling, which should be clear, indelible, and reliable even in field conditions, to prevent confusion in sample identity.

In most cases, sediments will have to be stored. For chemical analysis, the effect of storage has been studied on the stability of sediment-associated contaminants (ASTM 1990; US Environmental Protection Agency 1992; Gomez-Ariza *et al.* 1999), their extractability (Thomson *et al.* 1984), or general sediment characteristics (Watson *et al.* 1985). The effect of storage on the toxicity of compounds is unclear. For example, the effects of freezing ranged from decreased toxicity in *Daphnia magna* (Malueg *et al.* 1984) to no effects at all in polychaetes (Carr *et al.* 1989). Accordingly, the recommended storage periods for sediments range from 5 (Swartz 1987) or 7 days (Anderson *et al.* 1987) to less than two weeks (Shuba *et al.* 1978; ASTM 1990). Therefore, storage of sediments for prolonged periods should be avoided. Where this is not possible, sediments should be stored at 4 °C and storage time kept to an absolute minimum (Luoma & Ho 1998).

Finally, health and safety regulations will vary from country to country. However, as field-collected sediments can contain complex mixtures of potentially toxic substances, including mutagens and carcinogens, some basic safety precautions should be considered. It is desirable for toxicity tests to be performed as soon as possible after collection, which often leaves little or no time for chemical analysis and it is therefore necessary to minimize the direct contact of workers with sediment by using gloves, protective clothing, and safety goggles. In extreme cases, the presence of volatile compounds may warrant the use of well-ventilated areas, fume hoods or respirators (Ingersoll 1995). Special precautions should be observed where potential radioactive sediment may be sampled (US Army Corps of Engineers 1997).

6.2.2 Test systems for sediment toxicity assessment

The endpoints in sediment toxicity testing will vary with the question being addressed and may include acute and long-term toxicity, endocrine, reproductive, and genotoxic effects.

A comprehensive assessment of potential sediment toxicity requires a tiered approach considering multiple exposure phases and test models representing different trophic levels, levels of biological organization, and sediment related habitats (Davoren *et al.* 2005; Hartl *et al.* 2005). This integrated approach should involve the use of short-term general tests using sediment extracts (tier 1); the application of hazard identification models, and more specific (multiple) endpoints in multi-organism experiments, representing different trophic levels, habitats associated with sediments, routes of exposure, and bioavailability by using both sediment extracts and whole sediments (tier 2) (Hartl *et al.* 2006); and the assessment of *in situ* ecosystem function through lifetime reproductive success and components of biodiversity (tier 3) (Nendza 2002).

The primary criteria for the selection of test species include the species' ecological and/or economical importance and their relative sensitivity to sediment contamination, life expectancy, predictable and consistent response of control organisms, ease of culture and maintenance, reproducibility, cost, and in the case of tiers 2 and 3 ecological relevance and exposure history (Boisson *et al.* 1998; Nendza 2002; Ownby *et al.* 2002).

6.2.2.1 Tier 1 tests

Although simulation of *in situ* exposure of aquatic organisms to contaminated sediments is most realistic using a whole sediment approach, this often involves considerable infrastructural investment, logistical considerations, animal experimentation licenses (in the case of vertebrates[1]), and does not lend itself readily to a rapid initial assessment of potential sediment toxicity. Therefore, to generate such data, tier 1 tests use various sediment phases: (1) solid phase tests (Cook & Wells 1996; Côte *et al.* 1998; Kemble *et al.* 2000); (2) pore water or interstitial water occupying the spaces between sediment particles. Contaminants present in pore water represent the water-soluble, bioavailable fraction, a major route of exposure to benthic organisms (see Bufflap & Allen 1995; Adams *et al.* 2001; US Environmental Protection Agency 2001b; Dalmacija *et al.* 2006; Lewis *et al.* 2006); (3) elutriates originally developed by Keely & Engler (1974) to determine the solubility of contaminants released during physical disturbance, such as dredging operations (Beg *et al.* 2001; Doherty 2001; Casado-Martinez *et al.* 2007; Losso *et al.* 2007); (4) organic solvent extraction procedures that liberate contaminants that would otherwise not be found in elutriates or pore water fractions (Looser *et al.* 2000; Vella & Adami 2001; Huang 2004), especially those that simulate the release of organic compounds through digestive processes in the guts of deposit feeding invertebrates (see Nakajima *et al.* 2006). The test phase and extraction processes of choice will depend on the type of sediment and the question being addressed (Tables 6.1–6.5).

As the main concern of these tests is to establish potential toxicity of sediment contaminants rather than ecological relevance, many tier 1 tests use an *in vitro* approach to bioassays. These may include various commercially available bacterial bioluminescence and invertebrate test kits as well as tissue cultures allowing for standardized procedures with highly reproducible results. Consequently, many tier 1 test systems have been accredited by governments and regulatory bodies as monitoring tools for environmental impact assessments contributing to relevant legislation (US Environmental Protection Agency 1977). Selected tier 1 tests, designed around various extraction phases, are compared in Tables 6.2–6.5. A more in-depth discussion of acute and *in vitro* approaches to sediment toxicity assessment, including details and an evaluation of fractionation techniques, can be found in Ni Shuilleabhain *et al.* (2003).

[1]Regulations will vary from country to country and should be consulted before starting any toxicology work.

6.2.2.2 Tier 2 tests

As with tier 1 tests, several standardized whole-sediment bioassays using a variety of sentinel organisms are well accepted by regulatory authorities in several countries (Keddy *et al.* 1995; Environment Canada 1997; US Environmental Protection Agency 2001a; Simpson *et al.* 2005) and may include the use of outdoor simulated field studies (Graney *et al.* 1997); the responses measured have in many cases been successfully related to effects in the field (Day *et al.* 1995; Côte *et al.* 1998).

As mentioned above, sediments are heterogeneous systems. Accordingly, the distribution of sediment-associated contaminants and their behaviour, in terms of sorption expressed as particle-water partitioning coefficient (K_d) and bioavailability, determined by zonation patterns of pH and redox potential (E_h), are usually very patchy (Luoma & Ho 1998; Simpson *et al.* 2005). This can make the reproducibility and the interpretation of data from exposure experiments difficult. A tier 2 test system will therefore typically involve one of two or a combination of both of the following approaches: whole homogenized sediment and spiked sediment formulations.

With both preparation methods, ecological relevance is to varying degrees compromised in favor of standardization, cause and effect relations, and reproducibility (Luoma & Ho 1998).

6.2.2.2.1 Whole-sediment tests. Toxicity bioassays using whole sediment collected from the environment have been developed for many relevant taxa; examples are given in Table 6.6. Whole sediment toxicity bioassays using benthic fish as sentinel organisms have been reviewed by Hartl (2002).

After collection, homogenization of the sediments should be carried out as soon and as quickly as possible; prolonged mixing can change the particle size distribution causing oxidation (Ankley *et al.* 1996). This can be avoided by restricting the sampling of sediments, where possible to the oxidized layer, as most macrofaunal species will only ingest or be exposed to oxidized sediments (Coughlan *et al.* 2002; Hartl *et al.* 2007). This is not always practical because of difficulties in determining the depth of the

Table 6.1 Advantages and limitations of selected sediment fractionation procedures.

Phase	Method	Environment		Speed	Operation	Sample volume	Sediment texture				Loss of contaminants		Potential artifact		Cost	Ecological realism	Procedure
		In situ	*Ex situ*				Fine	Medium	Coarse	Sandy	Adsorp.	Vol.	Ox.	Temp.			
Pore water	Centrifugation		+	Rapid	Easy	Large	+	+	+				+	+	high	low	stand.
	Squeezing		+	Rapid	Easy	Large	+	+	+	+	+		+	+	low	low	stand.
	Peeper	+		t.c.	Difficult	Small	+	+							high	high	n.stand.
	Suction	+		Rapid	Easy	Small	+	+	+	+					low	low	n.stand.
Elutriates			+	t.c.	Difficult		+	+	+	+						low	stand.
Solvent extracts			+	t.c.	Difficult		+	+	+	+						low	stand.

t.c., time consuming; Adsorp., adsorption; Vol., volume; Ox., oxidation; Temp., temperature; stand., standardized; n. stand., non-standardized.
Summarized from: Ni Shuilleabhain *et al.* (2003)

Table 6.2 Selected procedures using pore water extracts in tier 1 sediment toxicity assessments.

Test phase	Endpoint	Taxa	Chemicals identified	*In vivo*	*In vitro*	Reference
Porewater	Genotoxicity	Bacteria				
		Mutatox©	Metals, organochlorines		+	Lewis *et al.* (2006)
	Enzyme inhibition/ bioluminescence	Bacteria				
		Microtox©	PAHs, POPs, metals		+	Cheung *et al.* (1997)
		Microtox©	PAHs		+	Viguri *et al.* (2007)
	Fertilization	Invertebrates				
		Echinoderms	PAHs	+		Lee *et al.* (2003)
	Embryogenesis	Invertebrates				
		Echinoderms	Metals	+		Wauhob *et al.* (2007)
	Survival	Micro algae				
		Dunaliella	Organotin	+		Cheung *et al.* (2003)
		Invertebrates				
		Crustaceans	PAHs	+		Müller *et al.* (2002)
			Organotin	+		Cheung *et al.* (2003)
		Hydra	PAHs, POPs	+		Côte *et al.* (1998)
	Motility	Invertebrates				
		Crustaceans	Metals	+		Spencer *et al.* (2006)

Table 6.3 Selected procedures using elutriates extracts in tier 1 sediment toxicity assessments.

Test phase	Endpoint	Taxa	Chemicals identified	*In vivo*	*In vitro*	Reference
Elutriates	Enzyme inhibition/ bioluminescence	Bacteria				
		Microtox©	PAHs, POPs, metals		+	Mueller *et al.* (2003)
		LUMIStox(R)	PAHs, PCBs, metals		+	Dellamatrice *et al.* (2006)
		MetPAD©	Metals		+	Boularbah *et al.* (2006)
		ToxiChromotest©	PAHs, POPs		+	Cheung *et al.* (1997)
	Motility	Micro-algae				
		FW Diatom		+		Cohn & McGuire (2000)
		Various	Metals	+		Mucha *et al.* (2003)
		Invertebrates				
		Crustaceans		+		Faimali *et al.* (2006)
	Embryogenesis	Invertebrates				
		Echinoderms	PAHs, metals	+		Fernandez *et al.* (2008)
		Ascidians	Metals	+		Geffard *et al.* (2007)
		Bivalves	PAHs, metals	+		Fernandez *et al.* (2008)
	Survival	Invertebrates				
		Crustaceans	Metals	+		Antunes *et al.* (2007)
	Enzyme inhibition	Vertebrates				
		Fish	PAHs, metals	+		Davoren *et al.* (2005)

Table 6.4 Selected procedures using organic extracts in tier 1 sediment toxicity assessments.

Test phase	Endpoint	Taxa	Chemicals identified	*In vivo*	*In vitro*	Reference
Organic solvents	Enzyme inhibition/ bioluminescence	Bacteria				
		LUMIStox(R)	PAHs		+	Papadopoulou & Samara (2002)
		Thamnotoxkit FTM	PAHs, PCBs		+	Côte *et al.* (1998)
	Stress response	Vertebrates				
		Fish	PAHs, metals	+		Hallare *et al.* (2005)
	Reproductive cycle	Microalgae		+		Schwab & Brack (2007)
	Survival	Invertebrates				
		Cnidarians	PAHs, POPs	+		Côte *et al.* (1998)
	Immobilization	Invertebrates				
		Crustaceans	PAHs	+		Schwab & Brack (2007)

Table 6.5 Selected procedures using solid phase in tier 1 sediment toxicity assessments.

Test phase	Endpoint	Taxa	Chemicals identified	*In vivo*	*In vitro*	Reference
Solid phase	Genotoxicity	Invertebrates				
		Clams	PAHs, metals			Coughlan *et al.* (2002)
		Vertebrates				
		Fish	PAHs, metals			Kilemade *et al.* (2004)
	Enzyme inhibition/ bioluminescence	Bacteria				
		Microtox©	PAHs, POPs, metals		+	Kemble *et al.* (2000)
		Microtox©	PAHs		+	Mueller *et al.* (2003)
		Microtox©	PAHs		+	Stronkhorst *et al.* (2003)
		ToxiChromoPad©	PAHs ,POPs		+	Côte *et al.* (1998)
		ToxiChromoPad©	PAHs		+	Mueller *et al.* (2003)
	Enzyme inhibition	Macroalgae				
		Entomoneis cf punctulata	PAHs	+		Simpson *et al.* (2007)
	Survival	Microorganisms				
		Yeast	PCBs, HCHs, DDTs, PAHs	+		Weber *et al.* (2006)
		Invertebrates				
		Amphipods	AHCs, metals	+		Melo & Nipper (2007)
		Insects	PAHs, POPs, metals	+		Côte *et al.* (1998)

Table 6.6 Examples of chronic toxicity bioassays using field-collected whole sediment in a variety of taxa.

Sentinel organism		Measured chemicals	Biomarker	Reference
Polychaetes				
	Hediste diversicolour	Metals	Behaviour; AChE; LDH; GST; SOD	Moreira *et al.* (2006)
Molluscs				
	Manilla clam (*Tapes semidecussatus*)	Metals	Lipofuchsin accumulation Burial activity	
Insects			Condition indices	Byrne & O'Halloran (2000)
	Chironomus riparius	Metals	Bioaccumulation	Roulier *et al.* (2008)
Crustaceans				
	Copepod (*Amphiascus tenuiremis*)	PAHs, PCB, metals	Reproductive success	Chandler & Green (1996)
	Amphipod (*Ampelisca abdita*)	PAHs, organotins	HSPs	Werner *et al.* (1998)
	Various amphipods	PAHs	Survival; behaviour; community structure	Lenihan *et al.* (1995)
Fish				
	Turbot (*Scophthalmus maximus*)	PAHs, metals	DNA damage; EROD	Kilemade *et al.* (2004); Hartl *et al.* (2006)
	Senegalese sole (*Solea senegalensis*)	PAHs, metals	EROD; metallothioneins	Jimenez-Tenorio *et al.* (2007)

AChE, acetylcholine esterase; LDH, lactatedehydrogenase; GST, glutathione S-transferase; SOD, superoxide dismutase; HSPs, heat-shock proteins; PAHs, polycyclic hydrocarbon; PCBs, polycyclic biphenols; EROD, ethoxyresorufin-*O*-deethylase.

redox potential discontinuity, which can be highly variable (Luoma & Ho 1998). Where required, it is recommended that homogenized sediments should be stored at 4 °C for an absolute minimum, but for no longer than two weeks (Shuba *et al.* 1978; ASTM 1990; Luoma & Ho 1998).

Although far more time-consuming, expensive and difficult to standardize than tier 1 tests using sediment extracts (see above), whole-sediment toxicity tests are considered to be more relevant, because they provide more realistic chronic exposure pathways (Hartl *et al.* 2005). Chronic bioassays should ideally use exposures spanning multiple organism life cycles (Luoma 1995). A limitation of whole-sediment toxicity tests is the development of "bottle effects", where owing to the closed nature in the test system, redox conditions change over time that can make chronic exposure bioassays difficult to perform (Luoma & Ho 1998).

Thus, depending on the life expectancy of the test organism, chronic whole-sediment toxicity bioassays are often at best "sub-chronic", which can underestimate chronic toxicity by several orders of magnitude, a fact that must be considered when developing model systems for predicting sediment toxicity (Pesch & Stewart 1980).

In addition, most sediment toxicity tests are unable to establish causality to one contaminant or contaminant group because of contaminant interaction, exhibiting, synergistic, additive, or antagonistic effects. Among methods aimed at addressing this problem are sediment manipulation techniques that eliminate the effects of certain contaminant groups, thus allowing the empirical identification of causal agents. Promising approaches at toxicity identification evaluation (TIE) of sediments contaminated with complex mixtures include the following: (1) the use of anionic exchange resins that reduce the concentrations and toxicity of sediment-associated metals but have negligible effects on ammonia and non-polar constituents (Burgess *et al.* 2000, 2007); (2) the removal of ammonia from interstitial water through the addition of intact fronds of sea lettuce, *Ulva lactuca* (Ho *et al.* 1999) or zeolite (Burgess *et al.* 2003); (3) the addition of powdered coconut charcoal to sequester polycyclic aromatic hydrocarbons (PAHs), polychlorinated biphenyls (PCBs), and pesticides, thus reducing toxicity (Ho *et al.* 2004).

The strength of whole-sediment toxicity bioassays lies in the realistic route of exposure making the data more ecological relevant than tier 1 acute tests alone. However, inconsistencies between the results of bulk and sediment extract bioassays, probably caused by differing routes of exposure, mean that more developmental work in this area is necessary to standardize procedures making data more comparable.

6.2.2.2.2 Spiking sediments. To establish causality in toxicological bioassays, sediments can be spiked to create controlled sediment conditions, which allow the empirical unraveling of the mechanisms involved in sediment–chemical–organism interactions (Lamberson & Swartz 1992). Although ecological relevance may be compromised, owing to extensive handling of sediment, the route of exposure is still maintained and can provide valuable information on contaminant fait modeling. Furthermore, spiked sediments can be used to check the recovery of analytes for quality assurance purposes.

The American Society for Testing and Materials (ASTM) has defined spiking as "the experimental addition of a test material, such as a chemical or mixture of chemicals, sewage sludge, oil, particulate matter or highly contaminated sediment, to a clean negative control or reference sediment to determine the toxicity of the material added. After the test material is added, sometimes with a solvent carrier, the sediment is thoroughly mixed to evenly distribute the test material throughout the sediment" (ASTM 1993). Spiking can avoid possible additive, synergistic, or antagonistic effects of complex chemical mixtures commonly found in natural sediments so that the fate of specific compounds can be studied, which is also useful for generating sediment quality criteria. Furthermore, spiked sediments can be used for modeling pore water conditions for use in *in vitro* studies (Ni Shuilleabhain *et al.* 2003).

Regardless if formulated or natural sediment is used, spiking invariably involves major changes to sediment properties. It is therefore vital to monitor all relevant parameters, such as pH, E_h, total organic carbon (TOC), acid volatile sulfides (AVS), and so on, to interpret the results from sediment bioassays. In addition, further considerations for spiking sediments are the equilibration times of pore water metals, temperature, contaminant loss due to tank wall adhesion and degradation, the use of carrier solvents and carrier controls (in the case of organics), and mixing techniques (ASTM 1990; Hartl *et al.* 2000; Northcott & Jones 2000; Simpson *et al.* 2005). Table 6.7 contains a selection of procedures for spiking marine and freshwater sediments with organic compounds and various inorganic metals.

6.2.2.2.3 Control and reference sediments. Control sediments, by definition, are contaminant-free, apart from any solvents used in spiked sediments, but otherwise comparable to the test sediment. This allows for the distinction between effects of the sediment or solvent themselves and the pollutant(s) of interest (ASTM 1994a,b). However, in many cases it has become increasingly difficult to find sediments free of pollutants, in relative proximity to the test site, with comparable geochemical characteristics. To fulfill these criteria, reference sediments (ASTM 1994b), which are relatively clean sediments with similar physical properties to the test sediments, are increasingly complementing or even replacing the traditional control sediment in ecotoxicological studies (Coughlan *et al.* 2002; Hartl *et al.* 2007).

6.2.2.3 Tier 3 tests

As can be seen from the above analysis of tier 1 and 2 tests, laboratory-based bioassays can be very useful tools for rapidly generating general toxicity data, establishing cause and effect and organism–contaminant–interaction models, and for method development. As outlined above, chronic sediment toxicity tests with long-lived organisms are often unsuitable for tier 1 and 2 laboratory application. The toxicology of contaminants may require bio- or photoactivation and/or biomagnification, and their toxicity would thus be underestimated in short-term laboratory experiments. To an extent, this may be taken into account by testing only the most sensitive species of a certain ecosystem and applying appropriate safety factors (Boxall *et al.* 2002). They are, however, limited in their ecological relevance, because exposure dynamics and interactions occurring in the field

Table 6.7 Selected procedures for spiking sediments.

Contaminant	Solvent	Sediment: Natural	Sediment: Collection	Sediment: Mass	Sediment treatment: Sieved	Sediment treatment: Stored	Sediment treatment: Procedure	Sediment treatment: Mixing	Quality assessment: Equillibration	Quality assessment: Efficiency	Quality assessment: Degradation	Reference
Inorganic metals												
$CdCl_2$	–	Yes	–	400 g	4 mm	5 wk	Seawater	Water/sed.	6–9 wk	100	0	Simpson *et al.* (2000)
$Cu(NO_3)_2$	–	Yes			0.64 µm	<1 d, 4 °C	Seawater	Water/sed.	0	100	0	Phelps *et al.* (1983)
$CuCl_2/ZnCl_2$	Deox. dH_2O	Yes	–	30 kg	2 mm	2 wk, 4 °C	Wet	Shake/rolled	40d	–	–	Hutchins *et al.* (2007)
$CuSO_4$	–	Yes	–	400 g	4 mm	5 wk	Seawater	Water/sed.	6–9 wk	100	0	Simpson *et al.* (2000)
$ZnSO_4$	Methanol	Yes	Ekman dredge	–	–	> mth	Slurry	Stirring	> mth	–	–	Watanabe *et al.* (1997)
	–	Yes	–	400 g	4 mm	5 wk	Seawater	Water/sed.	6–9 wk	100	0	Simpson *et al.* (2000)
$ZnCl_2$	HCl	Yes	–	400 g	1 mm	–	Wet	Blender	60 d; 4 °C		–	Mayer *et al.* (2001)
Organotins												
TBTCl	TBA	Yes	Jeskins sampler	6–7 kg	–	47 d	Wet	Wet rolling	47 d	72	0.25	Stronkhorst *et al.* (1999)
	Methanol	Yes	Ekman dredge	–	–	> mth	Slurry	Stirring	> mth	–	–	Watanabe *et al.* (1997)
	Methanol	PACS-2	–	2 g	–	? – 20 °C	Slurry	–	–	–	–	Chiron *et al.* (2000)
	Acetic acid	no	–	20 g	–	–	Slurry	Stirring	–	–	0.4	Hartl *et al.* (2000)
TBT (paint chips)	–	Yes	–	–	–	? –20 °C	Wet	Mixing	24 h 18 °C	–	0.03	Schratzberger *et al.* (2002)
DBTCl	Methanol	Yes	Ekman dredge	–	–	> mth	Slurry	Stirring	> mth	–	–	Watanabe *et al.* (1997)
TPhTCl	Methanol	PACS-2	–	2 g	–	? –20 °C	Slurry	–	–	–	–	Chiron *et al.* (2000)
	Acetic acid	no	–	20 g	–	–	Slurry	Stirring	–	–	0.4	Hartl *et al.* (2000)
Dioxins												
TCDD	Hexane	Yes	–	200 g	–	? 4 °C	Dried	Wet rolling	14 d	93	0.675	Barber *et al.* (1998)
HCBD	Hexane	Yes	–	–	–	6 wk, 4 °C		Wet rolling	6 wk, 4 °C		1.8	Fuchsman *et al.* (2000)

PCBs												
PCB77	Acetone	Yes	–	–	1 mm	–	Wet	Mixing	6 wk	–	–	Sormunen *et al.* (2008)
PAHs												
BaP	DMSO	desal.	–	–	500 µm	–	Wet	Wet rolling	–	–	–	Looise *et al.* (1996)
	Acetone	Yes	–	–	125–500 µm	–	Wet	Stirring	–	57	–	Kolok *et al.* (1996)
	Acetone	Yes	–	400 g	1 mm	–	Wet	Blender	60 d; 4 °C		–	Mayer *et al.* (2001)
Flouranthene	Acetone	Yes	Ekman dredge	2 g	125–500 µm	Room temp	Dried	Mixed	–	–	–	Duan *et al.* (2000)
Chlorinated hydrocarbons												
dieldrin	Acetone	Desal.	–	–	500 µm	–	Wet	Wet rolling	–	–	–	Looise *et al.* (1996)
Organophosphates												
TCP	–	Yes	Kayak-type corer	–	–	4 mths, 4 °C	Wet	Stirring	5–7 d	5–20	6	Penttinen *et al.* (1996)
Lindane	Acetone	Yes	Grab		500 µm	<14d, 4 °C	Wet	Wet rolling	0	64	–	Ciarelli *et al.* (1997)
	Acetone	Desal.	–	–	500 µm	–	Wet	Wet rolling	–	–	–	Looise *et al.* (1996)
Chlorpyrifos	–	Yes	–	–	–	–	Wet	wet rolling	–	–	–	Ankley *et al.* (1994)

TBT, tributyltin; PACS, polycyclic aromatic compounds; TCDD, 2,3,7,3-treatchloro-dibenzo-*p*-dioxin; HCBD, hexachlorobutadiene; PAH, polycyclic aromatic hydrocarbons; BaP, benzo[a]pyrene; DMSO, dimethyl sulphoxide; TCP, trichlorophenol; d, days; mth, months; wk, weeks.

can often not be replicated accurately under laboratory conditions. *In situ* studies, on the other hand, allow not only the observation of the effects of chronic (often multi- generation) exposure to contaminated sediments, but will also provide more realistic information concerning contaminant fate. Relevant examples of *in situ* tests can be found in Table 6.8.

In situ effects assessments can be described as environmental measurements made in the field with limited manipulation and disturbance of the sediment physico-chemical gradients and associated microhabitats. Therefore mitigation of handling and laboratory-induced artifacts is likely to give a more accurate representation of biotic and abiotic factors that are likely to affect routes of exposure and interactions with sediment-associated organisms, either a single species, populations, or communities; the last of these will depend to a certain extent on community structure. Contaminant exposure will manifest itself in various direct and indirect ways: direct effects include changes to distribution and abundance of taxa in proportion to their sensitivity; indirect effects can arise from changes to fecundity and alterations to food-web structure. Therefore changes to benthic community structure can be used to assess sediment quality.

Although understanding the fate of contaminants in the field is of great significance in environmental impact assessments, there are several problems and constraints with *in situ* contaminant effect studies. Most notable are the sophisticated and often costly logistics involved in identifying suitable study (and control) sites, the collection and deposition (in the case of organism transfer experiments) of indicator species, and the complexity of data interpretation. Chronic studies often require the monitoring (and repeated observation) of individual animals, which

Table 6.8 Examples of field studies using a variety of taxa.

Taxa	Species	Enlosure	*In situ*	Contaminants	Biomarker	Reference
Polychaetes						
	Hediste diversicolor	Acrylic tubes	+	Metals	Behaviour; AChE; LDH; GST; SOD	Moreira *et al.* (2006)
Molluscs						
	Elliptio complanata	–	+	PAHs	DNAdamage	Humphries (2006)
	Mytilus galloprovincialis	Caged	+	Metals	DNAdamage; GST	Regoli *et al.* (2004)
	Ruditapes philippinarum	Caged	+	PAHs, PCBs, metals	EROD; MTLP; GST; GPX	Martin-Diaz *et al.* (2008)
Crustaceans						
	Hyalella azteca	Chambers	+	Metals	Survival	Robertson & Liber (2007)
	Carcinus maenas	Caged	+	PAHs, PCBs, metals	EROD; MTLP; GST; GPX	Martin-Diaz *et al.* (2008)
Insects						
	Chironomus riparius	PVC tubes	+	PM; lindane	AChE	Maycock *et al.* (2003)
Invertebrate communities						
	Invertebrate communities	–	+	Metals	Community structure	
Fish						
	Gobius niger	–	+	PAHs	EROD	Ramsak *et al.* (2007)
	Ameiurus nebulosus	–	+	PAHs, PCBs, DDT	HIS	Yang & Baumann (2006)
	Micropterus salmoides	–	+	PAHs	EROD, ALAD, GST, SULT, UGT	Schreiber *et al.* (2006)
	Parophrys vetulus	–	+	PAHs, PCBs, AHs, Dioxins	CYP1A, histopathology, DNA damage	Malinsetal *et al.* (2004)
	Gillichthys mirabilis	Caged	+	PAHs, PCBS, DDT, Metals	Growth	Forrester *et al.* (2003)

necessitates some form of enclosure, such as, micro- or mesocosms (Graney *et al.* 1997; Hartl & Ott 1999). Careful planning and experimental design is needed to avoid or minimize handling stress of sentinel organisms, and the installation and removal of enclosures are possible sources of artifacts.

Furthermore, although general guidelines have been developed, unlike tier 1 and tier 2 approaches, there are no standardized procedures available for *in situ* experiments, partly because of site-specific or temporal conditions (seasonal variations), thus making inter-experimental comparative analyses often difficult.

6.3 Integrated approaches to environmental impact assessments

Policies governing environmental impact and management in general and sediment quality assessment in particular increasingly require a shift from lower level to more ecosystem-based, holistic approaches incorporating chemical, biological, and ecological objectives, with a focus on large-scale risk evaluation and management (Hagger *et al.* 2006; Apitz *et al.* 2007).

As in any complex procedure, each level of a tiered approach has its advantages and limitations; the latter can be overcome or mitigated by the application of multiple lines of evidence (LOEs) integrated in a weight-of-evidence (WOE) approach (Hyland *et al.* 1998; Neuparth *et al.* 2005; Hagger *et al.* 2006; Pereira *et al.* 2007). Such an integrated assessment typically includes relevant bioassays within different tiers, levels of biological organization, trophic compartments, as well as chemical analysis, and physicochemical characterization of the representative sediment samples. This will address issues that arise through complex biogeochemical properties of sediments, such as bioavailability and the route of exposure of contaminants.

As risk is defined as the product of toxicity and exposure, an isolated data input describing one or the other would be insufficient for the purpose of an ecological impact or quality assessment. In addition to risk assessment, the information gained from an integrated approach aims to provide insights into sediment quality that allow conclusions to be drawn that would otherwise not be supported by the data. Several recent case studies applying various degrees of integration to environmental sediment quality assessments are presented in Table 6.9.

6.4 Conclusions

Variability in approach and resulting data significantly contribute to the uncertainty in results of sediment toxicity testing. Efforts to standardize methods for sediment sampling, handling, aqueous and organic elutriation, and pore-water extraction through inter-calibration exercises have led to improvements in this area. Nevertheless, despite these targeted initiatives and the establishment of international working groups (SETAC, SEDNET), it would appear that the science of sediment quality assessment is still somewhat behind that of water or soil, because there are relatively few recognized standard test methods for evaluating sediment toxicity compared with those established for water and soil.

The consensus within the scientific community requires a sediment monitoring strategy to incorporate both chemical characterization and ecotoxicological analysis in a balanced way without over emphasizing one tier or single test result. Furthermore, there is wide recognition that toxicity is not merely a chemical property but rather a function of the test organism and the test conditions. Consequently, although sediments might contain relatively high concentrations of contaminants, these may not necessarily lead to adverse effects on test organisms. Contaminant fate in a sediment–water system is highly dependent on sorption behavior, which in turn determines bioavailability and toxicity. Therefore, quantitative chemical analysis in environmental samples is not necessarily indicative of biological and ecological effects of identified contaminants, because the bioavailable fraction may vary greatly, making toxicity predictions based on chemical analysis alone difficult and unreliable. Moreover, varying sensitivities displayed by different species to a particular contaminant further highlights the need for a battery-style bioassay approach within a tiered ecotoxicological assessment. A battery of toxicity tests for evaluation of sediment toxicity should include representatives from different trophic levels, because utilization of only a few test species would clearly constrain contaminated sediment evaluations that rely solely on bioassay results. Test species should have a wide geographic distribution

Table 6.9 Examples of integrated sediment toxicity assessment.

Ecosystem	Tier	LBO	TL	CA	Contaminants	Species	Duration	Reference
Marine	3	B; C; P	1	Yes	PAHs, metals	*Perna perna*	1 season	Pereira *et al.* (2007)
Marine	1, 2	D; S	2	Yes	PAHs, PCBs, HCs, HCBs, dioxins; metals	*Corrophium volutator; Paracentrotus lividus*	–	Apitz *et al.* (2007)
Marine	1,	T	2	Yes	PAHs, PCBs, metals	*Scrobicularia plana; Solea senegalensis*	1 year	Riba *et al.* (2005)
Marine	1, 2	B; H	1	Yes	PAHs	*Sparus auratus*	SC	Morales-Caselles *et al.* (2006)
Estuary	1	DNA; G; S	2	Yes	PAHs, PCBs, metals	*Vibrio fisheri; Ampelisca abdita; A. verilli; Mercenaria mercenaria*	2 SS	Hyland *et al.* (1998)
Estuary		B; C; DNA; P; M	2	–	PAHs; TBT; metals	*Carcinus maenas; Cerastoderma edule; Littorina littorea*	SC	Galloway *et al.* (2004)
Estuary	3	Pop	1	Yes	PAHs, PCBs, DDT, Metals	*Palaemonetes pugio*	1 year	Fulton *et al.* (2006)
River	1, 2	DNA; B; C; D	–	–	–	RTL-W1 cell line	SC	Keiter *et al.* (2006)

B, biochemical; C, cellular; CA, chemical analysis; D, development; DNA, genetic; G, growth; H, histopathology; LBO, level of biological organisation; M, morphological; P, physiological; Pop, population; RTL-W1, rainbow trout liver cell line; S, survival; SC, single collection event; SS, summer season; T, tissue; TL, trophic level.

and possess direct ecological relevance to a range of locations. The use of test species native to the particular region of interest can improve the ecological relevance of a given test, but data may not be comparable to those obtained with more cosmopolitan species.

In conclusion, it is essential to employ an integrated battery approach for the assessment of sediments, using both chemical characterization and ecotoxicological testing that comprises economically viable multi-exposure routes and multi-trophic tests to provide an ecologically relevant perspective on the sediment quality.

6.5 Recommendations and future research

This chapter has attempted to give a general overview of sediment quality assessment procedures broken down to their individual levels or tiers. Although each individual level within a tiered decision tree provides useful information, only a fully integrated approach will yield the necessary data to evaluate the extent of the likely impact of a proposed development or remediation activity on the aquatic environment in general and its sediments in particular. Therefore, in addition to performing bioassays within different tiers, elements of an integrated approach to sediment quality assessment should also include the following:

- the monitoring of sediment chemistry through qualitative and quantitative contaminant analysis and the physico-chemical characterization through graniometry and water content;
- the assessment of contaminant fate through determining the bioavailability and bioaccumulation potential of contaminants under site-specific conditions;
- the use of indicator species from different trophic levels to determine the tendency of contaminants to travel through the food chain and biomagnify;

• the application of biomarkers on multiple levels of biological organization, from biochemical responses to benthic community structure.

Artificial dialysis samplers, such as semi-permeable membrane devices, for collecting pore-water contaminants in sediments have been known for some time (Hesslein 1976; Mayer 1976). However, because these methods generally produce relatively small volumes of pore water, are unreliable at low contaminant concentrations, and have cost implications and problems associated with logistical issues, they have not been widely used *in situ* (Ni Shuilleabhain *et al.* 2003; Boehm *et al.* 2005). An interesting and possibly cost-effective approach is the development of "artificial mussels" as monitoring devices (Leung *et al.* 2008). Thus, providing these devices can overcome the limitations of semi-permeable membrane devices, by removing variability between individual "mussels" and allowing for a more standardized assessment, they may be able to replace the use of filter-feeding bivalves in biomonitoring programs, and it is conceivable that such devices could be suitably adapted for use in sediment.

Furthermore, the discovery of increasing concentrations of minute plastic fragments in sediments and their tendency to attract persistent organic contaminants (Mato *et al.* 2001) as well as the rapid development of nanotechnology, in particular the increased use of nanoparticles, presents a new challenge for sediment quality assessment.

References

Adams, W. J., Berry, W. J., Burton, G. A. J., *et al.* (2001) Summary of SETAC technical workshop pore water toxicity testing. In *Biological, Chemical, and Ecological Considerations with a Review of Methods and Applications, and Recommendations for Future Areas of Research*, R. S. Carr, & M. Nipper (eds), 1–29. Pensacola, FL: SETAC.

Aller, R. C. (1978) Experimental studies of changes produced by deposit feeders on pore water, sediments and overlying water chemistry. *American Journal of Science*, **278**, 1185–234.

Anderson, J. W., Birge, W. J., Gentile, J., Lake, J., Rodgers, J. J. (1987) Biological effects. Bioaccumulation and ecotoxicology of sediment-associated chemicals. In: *Fate and Effects of Sediment-bound Chemicals in Aquatic Systems*, K. L. Dickson & A. W. Maki (eds), 267–96. New York: Pergamon.

Anderson, B. S., Hunt, J. W., Hester, M. & Phillips, B. M. (1996) Assessment of sediment toxicity at the sediment-water interface. In: *Techniques in Aquatic Toxicology*, ed. G. K. Ostrander (ed.), 609–24. Boca Raton, FL: CRC Lewis Publishers.

Ankley, G. T., Call, D. J., Cox, J. S., Kahl, M. D., Hoke, R. A. & Kosian, P. A. (1994) Organic carbon partitioning as a basis for predicting the toxicity of chlorpyrifos in sediments. *Environmental Toxicology and Chemistry*, **13**, 621–26.

Ankley, G. T., DiToro, D. M., Hansen, D. J. & Berry, W. J. (1996) Technical basis and proposal for deriving sediment quality criteria for metals. *Environmental Toxicology and Chemistry*, **15**, 2056–66.

Ankley, G. T. & Schubauerberigan, M. K. (1994) Comparison of techniques for the isolation of sediment pore water for toxicity testing. *Archives of Environmental Contamination and Toxicology*, **27**, 507.

Antunes, S. C., Pereira, R. & Goncalves, F. (2007) Evaluation of the potential toxicity (acute and chronic) of sediments from abandoned uranium mine ponds. *Journal of Soils and Sediments*, **7**, 368–76.

Apitz, S. E., Barbanti, A., Bernstein, A. G., Bocci, M., Delaney, E. & Montobbio, L. (2007) The assessment of sediment screening risk in Venice lagoon and other coastal areas using international sediment quality guidelines. *Journal of Soils and Sediments*, **7**, 326–41.

ASTM. (1990) *Standard Guide for Collection, Storage, Characterization and Manipulation of Sediments for Toxicity Testing*. Philadelphia: American Society for testing and Materials. E1391-90.

ASTM. (1993) *Standard Guide for Conducting 10-Day Static Sediment Toxicity Tests with Marine and Estuarine Amphipods*. Philadelphia: American Society for testing and Materials. E1367-92. 26pp.

ASTM. (1994a) *American Society for Testing and Materials: Standard guide for designing biological test with sediments*. Philadelphia: American Society for testing and Materials. E1525-94.

ASTM. (1994b) *American Society for Testing and Materials: Terminology*. Philadelphia: American Society for testing and Materials. E943-94.

Atkinson, C. A., Jolley, D. F. & Simpson, S. L. (2007) Effect of overlying water pH, dissolved oxygen, salinity and sediment disturbances on metal release and sequestration from metal contaminated marine sediments. *Chemosphere*, **69**, 1428–37.

Barber, T. R., Chappie, D. J., Duda, D. J., Fuchsman, P. C. & Finley, B. L. (1998) Using a spiked sediment bioassay to establish a no-effect concentration for dioxin exposure to the amphipod Ampelisca abdita. *Environmental Toxicology and Chemistry*, **17**, 420–24.

Beg, M. U., Al-Muzaini, S., Saeed, T., *et al.* (2001) Chemical contamination and toxicity of sediment from a coastal area receiving industrial effluents in Kuwait. *Archives of*

Environmental Contamination and Toxicology, **41**, 289–97.

Black, H. J., Dainat, M., Koster, M. & Meyer-Reil, L. A. (2002) A multiple corer for taking virtually undisturbed samples from shallow water sediments. Estuarine Coastal and Shelf *Science*, **54**, 45–50.

Boehm, P. D., Page, D. S., Brown, J. S., Neff, J. M. & Bence, A. E. (2005) Comparison of mussels and semi-permeable membrane devices as intertidal monitors of polycyclic aromatic hydrocarbons at oil spill sites. *Marine Pollution Bulletin*, **50**, 740–50.

Boisson, F., Hartl, M. G. J., Fowler, S. W. & Amiard-Triquet, C. (1998) Influence of chronic exposure to silver and mercury in the field on the bioaccumulation potential of the bivalve Macoma balthica. *Marine Environmental Research*, **45**, 325–40.

Boularbah, A., Schwartz, C., Bitton, G. & Morel, J. L. (2006) Heavy metal contamination from mining sites in South Morocco: 1. Use of a biotest to assess metal toxicity of tailings and soils. *Chemosphere*, **63**, 802–10.

Boxall, A. B. A., Brown, C. D. & Barrett, K. L. (2002) Higher-tier laboratory methods for assessing the aquatic toxicity of pesticides. *Pest Management Science*, **58**, CP1–646.

Brassard, P. (1991) Measurement of Eh and pH in aquatic sediments. In *Manual of Physico-Chemical Analysis of Aquatic Sediments*, A. Murdoch, J. J. E. Azcue, & Murdoch, P. (eds), 47–67. Boca Raton, FL: CRC Press.

Brown, J. A., Colling, A., Park, D., Phillips, J., Rothery, D. & Wright, J. (1992) *Seawater: Its Composition, Properties and Behaviour*. Oxford: Pergamon Press. 165pp.

Buchanan, J. B. (1971) Measurement of the physical and chemical environment. In *Methods for the Study of Marine Benthos*, N. A. Holme & A. D. McIntyre, (eds), International Biological Programme, 30–52. London: Blackwell Scientific Publications.

Bufflap, S. E. & Allen, H. E. (1995) Sediment pore water collection methods for trace metal analysis: A review. *Water Research*, **29**, 165–77.

Burgess, R. M., Cantwell, M. G., Pelletier, M. C., Ho, K. T., Serbst, J. R., Cook, H. F. & Kuhn, A. (2000) Development of a toxicity identification evaluation procedure for characterizing metal toxicity in marine sediments. *Environmental Toxicology and Chemistry*, **19**, 982–91.

Burgess, R. M., Pelletier, M. C., Ho, K. T., *et al.* (2003) Removal of ammonia toxicity in marine sediment TIEs: a comparison of Ulva lactuca, zeolite and aeration methods. *Marine Pollution Bulletin*, **46**, 607–18.

Burgess, R. M., Perron, M. M., Cantwell, M. G., *et al.* (2007) Marine sediment toxicity identification evaluation methods for the anionic metals arsenic and chromium. *Environmental Toxicology and Chemistry*, **26**, 61–67.

Burton, G. A. (1992) Freshwater benthic toxicity tests. In *Sediment Toxicity Assessment*, 213–40. Chelsea, MI: Lewis Publishers.

Byrne, P. A. & O'Halloran, J. (1999) Aspects of assaying sediment toxicity in Irish estuarine ecosystems. *Marine Pollution Bulletin*, **39**, 97–105.

Byrne, P. A. & O'Halloran, J. (2000) Acute and sublethal toxicity of estuarine sediments to the manila clam, Tapes semidecussatus. *Environmental Toxicology*, **15**, 456–68.

Carlton, R. G. & Wetzel, R. G. (1985) A box corer for studying metabolism of epipelic microorganisms in sediment under in situ conditions. *Limnology and Oceanography*, **30**, 422–26.

Carr, R. S. & Chapman, D. C. (1995) Comparison of methods for conducting marine and estuarine sediment porewater toxicity tests:extraction, storage, and handling techniques. *Archives of Environmental Contaminantion and Toxicology*, **28**, 69–77.

Carr, R. S., Williams, J. W. & Fragata, C. T. B. (1989) Development and evaluation of a novel marine sediment pore water toxicity test with the polychaete Dinophilus gyrociliatus. *Environmental Toxicology and Chemistry*, **8**, 533–43.

Carson, R. 1962. *Silent Spring*. Boston, MA: Houghton Mifflin. 304pp.

Casado-Martinez, M. C., Fernandez, N., Forja, J. M. & DelValls, T. A. (2007) Liquid versus solid phase bioassays for dredged material toxicity assessment. *Environment International*. **33**, 456–62.

CCME. (1995) *Protocol for the Derivation of Sediment Quality Guidelines for the Protection of Aquatic Life*. Canadian Council of Ministery of the Environment, Ottawa, Canada. CCME EPC-98E. 35pp.

Chambers, J. E., Boone, J. S., Carr, R. L., Chambers, H. W. & Straus, D. L. (2002) Biomarkers as predictors in health and ecological risk assessment. *Human and Ecological Risk Assessment*, **8**, 165–76.

Chandler, G. T., Green, A. S. (1996) A 14-day harpacticoid copepod reproduction bioassay for laboratory and field contaminated muddy sediments. In *Techniques in Aquatic Toxicology*, G. K. Ostrander (ed.), 23–39. CRC Lewis Publishers. Boca Raton.

Chapman, P. M. (1989) *Sediment quality criteria-developmental approaches*. In: *Oceans '89*, vol. **2**. IEEE. New York. Report, 412–14.

Chapman, P. M. (2002) Integrating toxicology and ecology: putting the "eco" into ecotoxicology. *Marine Pollution Bulletin*, **44**, 7–15.

Chapman, P. M., Wang, F. Y. (2001) Assessing sediment contamination in estuaries. *Environmental Toxicology and Chemistry*, **20**, 3–22.

Cheung, K. C., Wong, M. H. & Yung, Y. K. (2003) Toxicity assessment of sediments containing tributyltin around Hong Kong harbour. *Toxicology Letters*, **137**, 121–31.

Cheung, Y. H., Neller, A., Chu, K. H., *et al.* (1997) Assessment of sediment toxicity using different trophic organisms. *Archives of Environmental Contamination and Toxicology*, **32**, 260–67.

Chiron, S., Roy, S., Cottier, R. & Jeannot, R. (2000) Speciation of butyl- and phenyltin compounds in sediments using pressurized liquid extraction and liquid chromatography – inductively coupled plasma mass spectrometry. *Journal of Chromatography A*, **879**, 137.

Ciarelli, S., Vonck, W. & vanStralen, N. M. (1997) Reproducibility of spiked-sediment bioassays using the marine benthic amphipod, *Corophium volutator*. *Marine Environmental Research*, **43**, 329–43.

Cohn, S. A. & McGuire, J. R. (2000) Using diatom motility as an indicator of environmental stress: effects of toxic sediment elutriates. *Diatom Research*, **15**, 19–29.

Cook, N. H. & Wells, P. G. (1996) Toxicity of Halifax harbour sediments: an evaluation of the Microtox® solid-phase test. *Water Quality Research Journal of Canada*, **31**, 673–708.

Côte, C., Blaise, C., Michaud, J. R., Menard, L., Trottier, S., Gagne, F. & Lifshitz, R. (1998) Comparisons between microscale and whole-sediment assays for freshwater sediment toxicity assessment. *Environmental Toxicology and Water Quality*, **13**, 93–110.

Coughlan, B. M., Hartl, M. G. J., O'Reilly, S. J., *et al.* (2002) Detecting genotoxicity using the Comet assay following chronic exposure of Manila clam *Tapes semidecussatus* to polluted estuarine sediments. *Marine Pollution Bulletin*, **44**, 1359–65.

Dalmacija, B., Prica, M., Ivancev-Tumbas, I., *et al.* (2006) Pollution of the Begej Canal sediment-metals, radioactivity and toxicity assessment. *Environment International*, **32**, 606–15.

Davoren, M., Ni Shuilleabhain, S., O'Halloran, J., *et al.* (2005) A test battery approach for the ecotoxicological evaluation of estuarine sediments. *Ecotoxicology*, **14**, 741–55.

Day, K. E., Dutka, B. J., Kwan, K. K., Batista, N., Reynoldson, T. B. & Metcalfesmith, J. L. (1995) Correlations between solid phase microbial screening assays, whole sediment toxicity tests with macroinvertebrates and in situ benthic community structure. *Journal of Great Lakes Research*, **21**, 192–06.

Dellamatrice, P. M., Monteiro, R. T. R., Blaise, C., Slabbert, J. L., Gagne, F. & Alleau, S. (2006) Toxicity assessment of reference and natural freshwater sediments with the LuminoTox assay. *Environmental Toxicology*, **21**, 395–402.

Depledge, M. H., Amaral-Mendes, J. J., Daniel, B., *et al.* (1993) The conceptual basis of the biomarker approach. In *Biomarkers: Research and Application in the assessment of Environmental Health*, D. B. Peakall & L. R. Shugart, (eds), vol. **68**, NATO ASI Series H: Cell Biology, 15–29. Berlin: Springer.

Di Toro, D. M., Zarba, C. S. & Hansen, D. J., *et al.* (1991) Technical basis for establishing sediment quality criteria for nonionic organic-chemicals using equilibrium partitioning. *Environmental Toxicology and Chemistry*, **10**, 1541–83.

Doherty, F. G. (2001) A review of the Microtox® toxicity test system for assessing the toxicity of sediments and soils. *Water Quality Research Journal of Canada*, **36**, 475–18.

Duan, Y. H., Guttman, S. I., Oris, J. T., Huang, X. D. & Burton, G. A. (2000) Genotype and toxicity relationships among *Hyalella azteca*: II. Acute exposure to fluoranthene-contaminated sediment. *Environmental Toxicology and Chemistry*, **19**, 1422–26.

Eggleton, J. & Thomas, K. V. (2004) A review of factors affecting the release and bioavailability of contaminants during sediment disturbance events. *Environment International*, **30**, 973–80.

Elskens, M., Leermakers, M., Panutrakul, S., Monteny, F. & Baeyens, W. (1991) Microbial activity in sandy and muddy estuarine sediments. *Geo-Marine Letters*, **11**, 194–98.

Environment Canada. (1997) *Test for growth and survival in sediment using the freshwater amphipod* Hyalella azteca. Environment Canada. EPS 1/RM/33. Ottawa, ON, Canada. 123pp.

Faimali, M., Garaventa, F., Piazza, V., *et al.* (2006) Swimming speed alteration of larvae of *Balanus amphitrite* as a behavioural end-point for laboratory toxicological bioassays. *Marine Biology*, **149**, 87–96.

Fenchel, T. (1969) The ecology of marine microbenthos IV. Structure and function of the benthic ecosystem, its chemical and physical factors and microfauna communities with special reference to the ciliated protozoa. *Ophelia*, **6**, 1–182.

Fernandez, N., Bellas, J., Lorenzo, J. I. & Beiras, R. (2008) Complementary approaches to assess the environmental quality of estuarine sediments. *Water Air and Soil Pollution*, **189**, 163–77.

Flower, R. J., Monteith, D. T., Mackay, A. W., Chambers, J. M. & Appleby, P. G. (1995) The design and performance of a new box corer for collecting undisturbed samples of soft subaquatic sediments. *Journal of Paleolimnology*, **14**, 101–11.

Forrester, G. E., Fredericks, B. I., Gerdeman, D., *et al.* (2003) Growth of estuarine fish is associated with the combined concentration of sediment contaminants and shows no adaptation or acclimation to past conditions. *Marine Environmental Research*, **56**, 423–42.

Fuchsman, P. C., Sferra, J. C. & Barber, T. R. (2000) Three lines of evidence in a sediment toxicity evaluation for hexachlorobutadiene. *Environmental Toxicology and Chemistry*, **19**, 2328–337.

Fulton, M., Key, P., Wirth, E., *et al.* (2006) An evaluation of contaminated estuarine sites using sediment quality guidelines and ecological assessment methodologies. *Ecotoxicology*, **15**, 573–81.

Galloway, T. S., Brown, R. J., Browne, M. A., *et al.* (2004) A multibiomarker approach to environmental assessment. *Environmental Science & Technology*, **38**, 1723–31.

Geffard, A., Geffard, O., Amiard, J. C., His, E. & Amiard-Triquet, C. (2007) Bioaccumulation of metals in sediment elutriates and their effects on growth, condition index, and metallothionein contents in oyster larvae. *Archives of Environmental Contamination and Toxicology*, **53**, 57–65.

Gomez-Ariza, J. L., Giraldez, I., Morales, E., Ariese, F., Cofino, W. & Quevauviller, P. (1999) Stability and storage problems in organotin speciation in environmental samples. *Journal of Environmental Monitoring*, **1**, 197.

Graney, R. L., Giesy, J. P. & Clark, J. R. (1997) Field studies. In *Fundamentals of Aquatic Toxicology – Effects, Environmental Fate and Risk Assessment*, G. M. Rand (ed.), 257–305. London: Taylor & Francis.

Hagger, J. A., Jones, M. B., Leonard, D. R. P., Owen, R. & Galloway, T. S. (2006) Biomarkers and integrated environmental risk assessment: are there more questions than answers? *Integrated Environmental Assessment and Management*, **2**, 312–29.

Hallare, A. V., Kosmehl, T., Schulze, T., Hollert, H., Kohler, H. R. & Triebskorn, R. (2005) Assessing contamination levels of Laguna Lake sediments (Philippines) using a contact assay with zebrafish (*Danio rerio*) embryos. *Science of the Total Environment*, **347**, 254–71.

Harris, J. R. W. & Cleary, J. J. (1987) *Particle–water partitioning and organotin dispersal in an estuary*. In: *Organotin Symposium Oceans '86*, vol. **4** Institute of Electrical and Electronic Engineers. Washington, DC, Report, 1370–74.

Harris, J. R. W., Cleary, J. J. & Valkirs, A. O. (1996) Particle-water partitioning and the role of sediments as a sink and secondary source of TBT. In *Organotin. Environmental fate and effects*, M. A. Champ, & P. F. Seligman, (eds), 459–74. London: Chapman & Hall.

Hartl, M. G. J. (2002) Benthic fish as sentinel organisms of estuarine sediment toxicity. In *The Vienna School of Marine Biology: A Tribute to Jörg Ott*, M. Bright, P. C. Dworschak & M. Stachowitsch (eds), 89–100. Vienna: Facultas Universitätsverlag.

Hartl, M. G. J., Hutchinson, S. & Hawkins, L. E. (2000) Reduced blood osmolality in freshwater-adapted European 0-group flounder, Platichthys flesus (L.), exposed to environmental levels of sediment-associated tributyl- and triphenyltin. *Marine Pollution Bulletin*, **40**, 792–94.

Hartl, M. G. J., Kilemade, M., Coughlan, B. M., *et al.* (2006) A two-species biomarker model for the assessment of sediment toxicity in the marine and estuarine environment. *Journal of Environmental Science and Health A*, **41**, 939–53.

Hartl, M. G. J., Kilemade, M., Sheehan, D., O'Halloran, J., O'Brien, N. M. & van Pelt, F. N. A. M. (2007) Hepatic biomarkers of sediment-associated pollution in juvenile turbot, *Scophthalmus maximus* L. *Marine Environmental Research*, **64**, 191–208.

Hartl, M. G. J. & Ott, J. A. (1999) An in-situ study on the influence of ascidian suspension feeding on the subtidal nepheloid layer in the northern Adriatic Sea. *Marine Ecology-Pubblicazioni Della Stazione Zoologica Di Napoli I*, **20**, 359–72.

Hartl, M. G. J., van Pelt, F. N. A. M. & O'Halloran, J. (2005) Development and application of sediment toxicity tests for regulatory purposes. In *Water Encyclopedia: Water Quality and Resource Development*, J. H. Lehr & J. Keeler (eds), vol. **5**, 383–87. New York: John Wiley.

Hesslein, R. H. (1976) An in situ sampler for close interval pore water studies. *Limnol. Oceanography*, **21**, 912–14.

Ho, K. T., Burgess, R. M., Pelletier, M. C., *et al.* (2004) Use of powdered coconut charcoal as a toxicity identification and evaluation manipulation for organic toxicants in marine sediments. *Environmental Toxicology and Chemistry*, **23**, 2124–31.

Ho, K. T., Kuhn, A., Pelletier, M. C., Burgess, R. M. & Helmstetter, A. (1999) Use of *Ulva lactuca* to distinguish pH-dependent toxicants in marine waters and sediments. *Environmental Toxicology and Chemistry*, **18**, 207–12.

Horowitz, A. J. & Elrick, K. A. (1988) Interpretation of bed sediment trace metal data: Methods for dealing with the grain size effect. In *Chemical and Biological Characterisation of Sludges, Sediments, Dredge Spoils and Drilling Muds*, J. J. Lichtenberg, J. A. Winter, C. I. Weber, & Fradkin, L. (eds), STP 976, 114–28. Philadelphia: American Society for Testing and Materials.

Huang, J. H. (2004) Reducing blank values for trace analysis of ionic organotin compounds and their adsorption to different materials. *International Journal of Environmental Analytical Chemistry*, **84**, 255–65.

Humphries III, L. F. (2006) Effects of polycyclic aromatic hydrocarbon exposure on three life stages of freshwater mussels (Bivalvia: Unionidae). MSc thesis, North Carolina State University. 54pp.

Hutchins, C. M., Teasdale, P. R., Lee, J. & Simpson, S. L. (2007) The effect of manipulating sediment pH on the porewater chemistry of copper- and zinc-spiked sediments. *Chemosphere*, **69**, 1089–99.

Hyland, J. L., Snoots, T. R. & Balthis, W. L. (1998) Sediment quality of estuaries in the southeastern US. *Environmental Monitoring and Assessment*, **51**, 331–43.

Ingersoll, C. G. (1995) Sediment tests. In *Fundamentals of Aquatic Toxicology – Effects, Environmenatl Fate and Risk Assessment*, G. M. Rand (ed.), 231–56. London: Taylor & Francis.

Jimenez-Tenorio, N., Morales-Caselles, C., Kalman, J., *et al.* (2007) Determining sediment quality for regulatory proposes using fish chronic bioassays. *Environment International*, 33, 474–80.

Keddy, C. J., Greene, J. C. & Bonnell, M. A. (1995) Review of whole-organism bioassays – soil, freswater sediment, and freshwater assessment in Canada. *Ecotoxicology and Environmental Safety*, 30, 221–51.

Keely, J. E. & Engler, R. M. (1974) Dredged material research program. Miscellaneous paper. D-74-14.

Keiter, S., Rastall, A., Kosmehl, T., *et al.* (2006) Ecotoxicological assessment of sediment, suspended matter and water samples in the upper Danube River – a pilot study in search for the causes for the decline of fish catches. *Environmental Science and Pollution Research*, **13**, 308–19.

Kemble, N. E., Hardesty, D. G., Ingersoll, C. G., Johnson, B. T., Dwyer, F. T. & MacDonald, D. D. (2000) An evaluation of the toxicity of contaminated sediments from Waukegan Harbor, Illinois, following remediation. *Archives of Environmental Contamination and Toxicology*, **39**, 452–61.

Kilemade, M. F., Hartl, M. G. J., Sheehan, D., *et al.* (2004) Genotoxicity of field-collected inter-tidal sediments from Cork Harbor, Ireland, to juvenile turbot (*Scophthalmus maximus* L.) as measured by the Comet assay. *Environmental and Molecular Mutagenesis*, **44**, 56–64.

Knezovich, J. P., Harrison, F. L. & Wilhelm, R. G. (1987) The bioavailibility of sediment-sorbed organic chemicals: a review. *Water Air Soil Pollution*, **32**, 233.

Kolok, A. S., Huckins, J. N., Petty, J. D. & Oris, J. T. (1996) The role of water ventilation and sediment ingestion in the uptake of benzo[a]pyrene in gizzard shad (*Dorosoma cepedianum*). *Environmental Toxicology and Chemistry*, **15**, 1752–59.

Lamberson, J. O. & Swartz, R. C. (1992) *Spiked-sediment toxicity test approach*. Sediment Classification Methods Compendium. US EPA 823-R-92-006, Chapter 4.

Lee, K., Prince, R. C., Greer, C. W., et al. (2003) Composition and toxicity of residual Bunker C fuel oil in intertidal sediments after 30 years. *Spill Science & Technology Bulletin*, **8**, 187–99.

Lenihan, H. S., Kiest, K. A., Conlan, K. E., Slattery, P. N., Konar, B. H. & Oliver, J. S. (1995) Patterns of survival and behavior in Antarctic benthic invertebrates exposed to contaminated sediments – field and laboratory bioassay experiments. *Journal of Experimental Marine Biology and Ecology*, **192**, 233–55.

Leung, K. M. Y., Furness, R. W., Svavarsson, J., Lau, T. C. & Wu, R. S. S. (2008) Field validation, in Scotland and Iceland, of the artificial mussel for monitoring trace metals in temperate seas. *Marine Pollution Bulletin*, **57**, 790–800.

Lewis, M. A., Daniels, C. B. & Chancy, C. A. (2006) Microbial genotoxicity as an environmental indicator for near-coastal sediment pore waters. *Environmental Toxicology*, **21**, 193–204.

Libes, S. M. (1992) *An Introduction to Marine Biogeochemistry*. Chichester, UK: John Wiley. 734pp.

Looise, B. A. S., Holwerda, D. A. & Foekema, E. M. (1996) Induction of glutathione S-transferase in the freshwater bivalve Sphaerium corneum as a biomarker for short-term toxicity tests? *Comparative Biochemistry and Physiology C-Pharmacology Toxicology & Endocrinology*, **113**, 103–07.

Looser, P. W., Berg, M., Fent, K., Muhlemann, J. & Schwarzenbach, R. P. (2000) Phenyl- and butyltin analysis in small biological samples by cold methanolic digestion and GC/MS. *Analytical Chemistry*, **72**, 5136–41.

Losso, C., Novelli, A. A., Picone, M., *et al.* (2007) Potential role of sulfide and ammonia as confounding factors in elutriate toxicity bioassays with early life stages of sea urchins and bivalves. *Ecotoxicology and Environmental Safety*, **66**, 252–57.

Luoma, S. N. (1995) Prediction of metal toxicity in nature from bioassays: limitations and research needs. In *Metal Speciation and Bioavailability*, A. Tessier & D. R. Turner (eds), 609–59. Chichester, UK: John Wiley.

Luoma, S. N. & Ho, K. T. (1998) Appropriate uses of marine and estuarine sediment bioassays. In *Handbook of Ecotoxicology*, P. Calow (ed.), 193–26. Berlin: Blackwell Science.

Machan, R. & Ott, J. A. (1972) Problems and methods of continuous in situ measurements of redox potential in marine sediments. *Limnology and Oceanography*. **17**, 622–26.

Malins, D. C., Stegeman, J. J., Anderson, J. W., Johnson, P. M., Gold, J. & Anderson, K. M. (2004) Structural changes in gill DNA reveal the effects of contaminants on puget sound fish. *Environmental Health Perspectives*, **112**, 511–15.

Malueg, K. W., Schuytema, G. S., Franwczyk, D. F. & Gakstatter, J. H. (1984) Laboratory sediment toxicity tests, sediment chemistry and distribution of benthic macorinvertebrates in sediments from the Keweenaw Waterway, Michigan. *Environmental Toxicology and Chemistry*, **3**, 233–42.

Martin-Diaz, M. L., Blasco, J., Sales, D. & Del Valls, T. A. (2004) Biomarkers as tools to assess sediment quality. *Laboratory and field surveys. Trends in Analytical Chemistry*, **23**, 807–18.

Martin-Diaz, M. L., Blasco, J., Sales, D. & Delvalls, T. A. (2008) Field validation of a battery of biomarkers to assess sediment quality in Spanish ports. *Environmental Pollution*, **151**, 631–40.

Mato, Y., Isobe, T., Takada, H., Kanehiro, H., Ohtake, C. & Kaminuma, T. (2001) Plastic resin pellets as a transport medium for toxic chemicals in the marine environment. *Environmental Science & Technology*, **35**, 318–24.

Maycock, D. S., Prenner, M. M., Kheir, R., *et al.* (2003) Incorporation of in situ and biomarker assays in higher-tier assessment of the aquatic toxicity of insecticides. *Water Research*, **37**, 4180–90.

Mayer, L. M. (1976) Chemical water sampling in lakes and sediments with dialysis bags. *Limnology and Oceanography*, **21**, 909–12.

Mayer, L. M., Weston, D. P. & Bock, M. J. (2001) Benzo[a] pyrene and zinc solubilization by digestive fluids of benthic invertebrates – a cross-phyletic study. *Environmental Toxicology and Chemistry*, **20**, 1890–900.

Melo, S. L. R. & Nipper, M. (2007) Sediment toxicity tests using the burrowing amphipod *Tiburonella viscana* (Amphipoda : Platyischnopidae). *Ecotoxicology and Environmental Safety*, **66**, 412–20.

Morales-Caselles, C., Jimenez-Tenorio, N., de Canales, M. L. G., Sarasquete, C. & Del Valls, T. A. (2006) Ecotoxicity of sediments contaminated by the oil spill associated with the tanker "Prestige" using juveniles of the fish *Sparus aurata*. *Archives of Environmental Contamination and Toxicology*, **51**, 652–60.

Moreira, S. M., Lima, I., Ribeiro, R. & Guilhermino, L. (2006) Effects of estuarine sediment contamination on feeding and on key physiological functions of the polychaete *Hediste diversicolor*: laboratory and in situ assays. *Aquatic Toxicology*, **78**, 186–201.

Mucha, A. P., Leal, M. F. C., Bordallo, A. A. & Vasconcelos, M. T. S. D. (2003) Comparison of the response of three microalgae species exposed to elutriates of estuarine sediments based on growth and chemical speciation. *Environmental Toxicology and Chemistry*, **22**, 576–85.

Mueller, D. C., Bonner, J. S., McDonald, S. J., *et al.* (2003) The use of toxicity bioassays to monitor the recovery of oiled wetland sediments. *Environmental Toxicology and Chemistry*, **22**, 1945–55.

Müller, A., Heininger, P., Wessels, M., *et al.* (2002) Contaminant levels and ecotoxicological effects in sediments of the river Odra. *Acta Hydrochimica Et Hydrobiologica*, **30**, 244–55.

Nakajima, F., Saito, K., Isozaki, Y., *et al.* (2006) Transfer of hydrophobic contaminants in urban runoff particles to benthic organisms estimated by an in vitro bioaccessibility test. *Water Science and Technology*, **54**(6–7), 323–30.

Nendza, M. (2002) Inventory of marine biotest methods for the evaluation of dredged material and sediments. *Chemosphere*, **48**, 865–83.

Neuparth, T., Correia, A. D., Costa, F. O., Lima, G. & Costa, M. H. (2005) Multi-level assessment of chronic toxicity of estuarine sediments with the amphipod *Gammarus locusta*: I. Biochemical endpoints. *Marine Environmental Research*, **60**, 69–91.

Ni Shuilleabhain, S., Davoren, M., Hartl, M. G. J. & O'Halloran, J. (2003) In vitro ecotoxicological assessment of aquatic sediments. In: *In vitro Methods in Aquatic Toxicology*, C. Mothersill, C., Austin, B. (eds), 327–74. Berlin: Springer.

Northcott, G. L. & Jones, K. C. (2000) Developing a standard spiking procedure for the introduction of hydrophobic organic compounds into field-wet soil. *Environmental Toxicology and Chemistry*, **19**, 2409–17.

OSPAR. (1997) *Agreed Background/Reference Concentrations for Contaminants in Sea Water, Biota and Sediment*. OSPAR Convention. 1997-14. London, UK.

Ott, J. A. (1988) *Meereskunde*. Verlag Eugen Ulmer. Stuttgart. 386pp.

Ownby, D. R., Newman, M. C., Mulvey, M., Vogelbein, W. K., Unger, M. A. & Arzayus, L. F. (2002) Fish (*Fundulus heteroclitus*) populations with different exposure histories differ in tolerance of creosote-contaminated sediments. *Environmental Toxicology and Chemistry*, **21**, 1897–902.

Paine, R. T. (1966) Food web complexity and species diversity. *American Naturalist*, **100**, 65–75.

Papadopoulou, D. & Samara, C. (2002) Polycyclic aromatic hydrocarbon contamination and LUMIStox(R) solvent extract toxicity of marine sediments in the north Aegean sea, Greece. *Environmental Toxicology*, **17**, 556–66.

Papucci, C., Jennings, C. D. & Lavarello, O. (1986) A modified box corer and extruder for marine pollution studies. *Continental Shelf Research*, **6**, 671–75.

Penttinen, O. P., Kukkonen, J. & Pellinen, J. (1996) Preliminary study to compare body residues and sublethal energetic responses in benthic invertebrates exposed to sediment-bound 2,4,5-trichlorophenol. *Environmental Toxicology and Chemistry*, **15**, 160–66.

Pereira, C. D. S., Abessa, D. M. D. S., Bainy, A. C. D., *et al.* (2007) Integrated assessment of multilevel biomarker responses and chemical analysis in Mussels from Sao Sebastiao, Sao Paulo, Brazil. *Environmental Toxicology and Chemistry*, **26**, 462–69.

Pesch, G. G. & Stewart, N. E. (1980) Cadmium toxicity to three species of estuarine invertebrates. *Marine Environmental Research*, **3**, 145–56.

Phelps, H. J., Hardy, J. T., Pearson, W. H. & Apts, C. W. (1983) Clam burrowing behaviour: inhibition by copper-enriched sediment. *Marine Pollution Bulletin*, **14**, 452–55.

Power, E. A. & Chapman, P. M. (1992) Assessing sediment quality. In: *Sediment Toxicity Assessment*, G. A. Burton Jr (ed.), 1–18. Boca Raton, FL: CRC/Lewis.

Quevauviller, P. & Ariese, F. (2001) New sediment reference material for the quality control of butyltin and phenyltin analysis. *Trends in Analytical Chemistry*, **20**, 207–18.

Ragnarsdottir, K. V. (2000) Environmental fate and toxicology of organophosphate pesticides. *Journal of the Geological Society*, **157**, 859–76.

Ramsak, A., Stopar, K., Sepcic, K., Berden-Zrimec, M., Bajt, O. & Malej, A. (2007) Reflection of hydrocarbon pollution on hepatic EROD activity in the black goby (*Gobius niger*). *Environmental Toxicology and Pharmacology*, **24**, 304–10.

Rand, G. M., Wells, P. G. & McCarty, L. S. (1997) Introduction to aquatic toxicology. In *Fundamentals of Aquatic Toxicology – Effects, Environmenatl Fate and Risk Assessment*, G. M. Rand (ed.), 3–67. London: Taylor & Francis.

Regoli, F., Frenzilli, G., Bocchetti, R., *et al.* (2004) Time-course variations of oxyradical metabolism, DNA integrity and lysosomal stability in mussels, *Mytilus galloprovincialis*, during a field translocation experiment. *Aquatic Toxicology*, **68**, 167–78.

Reynoldson, T. B. & Rodriguez, P. (1999) Field methods and interpretation for sediment bioassessment. In *Manual of bioasessment of aquatic sediment quality*, A. Mudroch, J. M. Azcue, & P. Mudroch (eds), 135–76. Boca Raton, FL: Lewis Publishers.

Riba, I., Blasco, J., Jimenez-Tenorio, N., de Canales, M. L. G. & Del Valls, T. A. (2005) Heavy metal bioavailability and effects: II. Histopathology-bioaccumulation relationships caused by mining activities in the Gulf of Cadiz (SW, Spain). *Chemosphere*, **58**, 671–82.

Riedl, R. & Ott, J. A. (1982) Water movement through porous sediments. In *The Dynamic Environment of the Ocean Floor*, F. T. Manheim & K. A. Fanning, (eds), 29–56. Heath, MA: Lexington Books.

Robertson, E. L. & Liber, K. (2007) Bioassays with caged *Hyalella* azteca to determine in situ toxicity downstream of two Saskatchewan, Canada, uranium operations. *Environmental Toxicology and Chemistry*, **26**, 2345–55.

Rodriguez, P. & Reynoldson, T. B. (1999) Laboratory methods and criteria for sediment bioassessment. In *Manual of Bioassessment of Aquatic Sediment Quality*, A. Mudroch, J. M. Azcue, & P. Mudroch, P. (eds), 83–134. Boca Raton, FL: Lewis Publishers.

Roulier, J. L., Tusseau-Nuillemin, M. H., Coquery, M., Geffard, O. & Garric, J. (2008) Measurement of dynamic mobilization of trace metals in sediments using DGT and comparison with bioaccumulation in Chironomus riparius: first results of an experimental study. *Chemosphere*, **70**, 925–32.

Santschi, P. H., Guo, L. D. & Wen, L. S. (2001) Box coring artifacts in sediments affected by a waste water outfall. *Marine Pollution Bulletin*, **42**, 267–72.

Schlenk, D. (1999) Necessity of defining biomarkers for use in ecological risk assessments. *Marine Pollution Bulletin*, **39**, 48–53.

Schratzberger, M., Wall, C. M., Reynolds, W. J., Reed, J. & Waldock, M. J. (2002) Effects of paint-derived tributyltin on structure of estuarine nematode assemblages in experimental microcosms. *Journal of Experimental Marine Biology and Ecology*, **272**, 217–35.

Schreiber, E. A., Otter, R. R. & van den Hurk, P. (2006) A biomarker approach to measure biological effects of contaminant exposure in largemouth bass from Lake Conestee, South Carolina, USA. *Environmental Toxicology and Chemistry*, **25**, 1926–32.

Schwab, K. & Brack, W. (2007) Large volume TENAX (R) extraction of the bioaccessible fraction of sediment-associated organic compounds for a subsequent effect-directed analysis. *Journal of Soils and Sediments*, **7**, 178–86.

Shuba, P. J., Tatem, H. E. & Carroll, J. H. (1978) *Biological Assessment Methods to Predict the Impact of Open-Water Disposal of Dredged Material.* US Corps of Engineers. Technical Report D-78-5Q. Washington, DC.

Simpson, S. L., Batley, G. E., Chariton, A. A., *et al.* (2005) *Handbook of Sediment Quality Assessment. Bangor*, New South Wales: Centre for Environmental Contaminants Research. 126pp.

Simpson, S. L., Micevska, T., Adams, M. S., Stone, A. & Maher, W. A. (2007) Establishing cause–effect relationships in hydrocarbon-contaminated sediments using a sublethal response of the benthic marine alga, *Entomoneis cf punctulata. Environmental Toxicology and Chemistry*, **26**, 163–70.

Simpson, S. L., Rosner, J. & Ellis, J. (2000) Competitive displacement reactions of cadmium, copper, and zinc added to a polluted, sulfidic estuarine sediment. *Environmental Toxicology and Chemistry*, **19**, 1992–99.

Sormunen, A. J., Leppanen, M. T. & Kukkonen, J. V. K. (2008) Influence of sediment ingestion and exposure concentration on the bioavailable fraction of sediment-associated tetrachlorobiphenyl in oligochaetes. *Environmental Toxicology and Chemistry*, **27**, 854–63.

Spencer, K. L., Dewhurst, R. E. & Penna, P. (2006) Potential impacts of water injection dredging on water qualit and ecotoxicity in Limehouse Basin, River Thames, SE England, UK. *Chemosphere*, **63**, 509–21.

Stronkhorst, J., van Hattum, B. & Bowmer, T. (1999) Bioaccumulation and toxicity of tributyltin to a burrowing heart urchin and an amphipod in spiked, silty marine sediments. *Environmental Toxicology and Chemistry*, **18**, 2343–51.

Suedel, B. C., Rodgers, J. H. (1994) Development of formulated reference sediments for fresh-water and estuarine sediment testing. *Environmental Toxicology and Chemistry*, **13**, 1163–75.

Sundby, B. (1991) Geochemical aspects of metal bioavailability: an overview of sediment geochemistry. In *17th Annual Aquatic Toxicology Workshop*, vol. **1**, P. Chapman, F. Bishay, E. Power, *et al.* (eds), 323–30 Canadian Technological Report in Fisheries and Aquatic Science No. 1774. Vancouver, BC.

Swartz, R. C. (1987) Toxicological methods for determining the effect of contaminated sediment on marine organisms. In: *Fate and Effects of Sediment-bound Chemicals in Aquatic Systems*, K. L. Dickson, A. W. Maki, & W. A. Brung, (eds), 183–98. New York: Pergamon Press.

Thomson, E. A., Luoma, S. N., Johansson, C. E. & Cain, D. J. (1984) Comparison of sediments and organisms in identifying sources of biologically available trace metal contamination. *Water Research*, **18**, 755–65.

Truhaut, R. (1977) Eco-toxicology – objectives, principles and perspectives. *Ecotoxicology and Environmental Safety*, **1**, 151–73.

US Army Corps of Engineers. (1997) *Guidance for Low-Level Radioactive Waste and Mixed Waste Treatment and Handling*. Washington, DC: Department of the Army. EM 1110-1-4002. 126pp.

US Environmental Protection Agency. (1972) *Marine Protection, Research, and Sanctuaries Act of 1972*. US Environmental Protection Agency, Washington, DC.

US Environmental Protection Agency. (1977) Dredged material sample collection and preparation. Appendix B. In: *Ecological Evaluation of Proposed Discharge of Dredged Material into Ocean Waters*. Vicksburg, Mississippi: Environmental Protection Agency/Corps of Engineers Technical Committee for Dredged and Fill Material. Environmental Effects Laboratory, US Army Engineer Waterways Experiment Station.

US Environmental Protection Agency. (1986) *Quality Criteria for Water*. Washington, DC: US Environmental Protection Agency: EPA-4405/5-86-001.

US Environmental Protection Agency. (1992) *Sediment Classification Methods Compendium*. Washington, DC: Environmental Protection Agency. EPA 823-R-92-006. 229pp.

US Environmental Protection Agency. (1993) *Technical Basis for Deriving Sediment Quality Criteria for Nonionic Organic Contaminants for the Protection of Benthic Organisms by Using Equilibrium Partitioning*. EPA-822/R-93-011. Washington, DC. 62pp.

US Environmental Protection Agency. (2001a) *Methods for Assessing the Chronic Toxicity of Marine and Estuarine Sediment-associated Contaminants with the Amphipod Leptocheirus plumulosus*. First Edition. EPA600/R-01/020. 120pp.

US Environmental Protection Agency. (2001b). *Methods for the Collection, Storage and Manipulation Of Sediments for Chemical and Toxicological Analyses: Technical Manual*. EPA 823-B-01-002. Washington, DC: US Environmental Protection Agency, Office of Water. 208pp.

Vella, A. J. & Adami, J. P. T. (2001) Preliminary evidence for in vitro methylation of tributyltin in a marine sediment. *Applied Organometallic Chemistry*, **15**, 901–06.

Viguri, J. R., Irabien, M. J., Yusta, I., *et al.* (2007) Physico-chemical and toxicological characterization of the historic estuarine sediments: a multidisciplinary approach. *Environment International*, **33**, 436–44.

Watanabe, N., Takatsuki, H. & Sakai, S. (1997) Desorption of tributyltin, dibutyltin and zinc from resuspended sediment. *Applied Organometallic Chemistry*, **11**, 273–79.

Watson, P. G., Frickers, P. E. & Goodchild, C. M. (1985) Spatial and seasonal variations in the chemistry of sediment interstitial waters in the Tamar Estuary. *Estuarine, Coastal and Shelf Science*, **21**, 105–09.

Wauhob, T. J., Nipper, M. & Billiot, B. (2007) Seasonal variation in the toxicity of sediment-associated contaminants in Corpus Christi Bay, TX. *Marine Pollution Bulletin*, **54**, 1116–26.

Weaver, P. P. E. & Schultheiss, P. J. (1990) Current methods for obtaining, logging and splitting marine sediment cores. *Marine Geophysical Researches*, **12**, 85–100.

Webb, P. (1989) A comparison of core and grab sampling techniques. BSc thesis, University of Southampton, UK. 51pp.

Weber, J., Kreutzmann, J., Plantikow, A., *et al.* (2006) A novel particle contact assay with the yeast Saccharomyces cerevisiae for ecotoxicological assessment of freshwater sediments. *Journal of Soils and Sediments*, **6**, 84–91.

Werner, I., Kline, K. F. & Hollibaugh, J. T. (1998) Stress protein expression in *Ampelisca abdita* (Amphipoda) exposed to sediments from San Francisco Bay. *Marine Environmental Research*. **45**, 417–30.

Yang, X. & Baumann, P. C. (2006) Biliary PAH metabolites and the hepatosomatic index of brown bullheads from Lake Erie tributaries. *Ecological Indicators*, **6**, 567–74.

7 Tools for assessing contaminated sediments in freshwater, estuarine, and marine ecosystems

Donald D. MacDonald[1] & Christopher G. Ingersoll[2]

[1]*MacDonald Environmental Sciences Ltd., Canada*
[2]*United States Geological Survey, USA*

7.1 Introduction

Traditionally, concerns about the management of aquatic resources in aquatic ecosystems have focused primarily on water quality. As such, early water resource management efforts were often directed at assuring the potability of surface water or groundwater sources. Subsequently, the scope of these management initiatives expanded to include protection of instream (i.e., fish and aquatic life), agricultural, industrial, and recreational water uses. Although initiatives undertaken in the past 30 years have unquestionably improved water quality conditions, a growing body of evidence indicates that management efforts directed solely at the attainment of surface-water quality criteria may not provide an adequate basis for protecting the designated uses of aquatic ecosystems.

In recent years, concerns about the health and vitality of aquatic ecosystems have begun to re-emerge in North America. One of the principal reasons for this is that many toxic and bioaccumulative chemicals, which are found in only trace amounts in water, can accumulate to elevated levels in sediments. Some of these pollutants, such as organochlorine (OC) pesticides and polychlorinated biphenyls (PCBs), were released into the environment long ago. The use of many of these substances has been banned in North America for 30 years or more; nevertheless, these chemicals continue to persist in the environment. Other contaminants enter our waters every day from industrial and municipal discharges, urban and agricultural runoff, and atmospheric deposition from remote sources. Owing to their physical and chemical properties, many of these substances tend to accumulate in sediments. In addition to providing sinks for many chemicals, sediments can also serve as potential sources of pollutants to the water column when conditions change in the receiving water system (for example during periods of anoxia, after severe storms).

7.2 Sediment quality issues and concerns

Sediments represent essential elements of freshwater, estuarine, and marine ecosystems. Nevertheless, the available information on sediment quality conditions indicates that sediments throughout North America are contaminated by a wide range of toxic and bioaccumulative substances, including metals, polycyclic aromatic hydrocarbons (PAHs), PCBs, OC pesticides, pyrethroid pesticides, a variety of semi-volatile organic chemicals (SVOCs), and polychlorinated dibenzo-*p*-dioxins and furans (PCDDs and PCDFs) (International Joint Commission (IJC) 1988; US Environmental Protection Agency (USEPA) 1997, 2000a). Contaminated sediment has been identified as a source of ecological impacts throughout North America. In the Great Lakes basin, for example, sediment quality issues and concerns are apparent at 42 of the 43 areas of concern (AOCs) that have been identified by the International Joint Commission (IJC 1988). In British Columbia, such issues and concerns have been identified in the lower Fraser River basin, the lower Columbia River basin, and elsewhere in the province (Mah *et al.* 1989; MESL 1997; Macfarlane 1997). Such issues have also emerged in Florida, in some cases raising concerns about human health and aquatic-dependent wildlife (MacDonald 2000).

Sedimentology of Aqueous Systems, 1st edition.
Edited by Cristiano Poleto and Susanne Charlesworth.

Contaminated sediments represent an important environmental concern for several reasons. First, contaminated sediments have been demonstrated to be toxic to sediment-dwelling organisms and fish. As such, exposure to contaminated sediments can result in decreased survival, reduced growth, or impaired reproduction in benthic invertebrates and fish. Additionally, certain sediment-associated contaminants (termed bioaccumulative substances) are taken up by benthic organisms through a process called bioaccumulation. When larger animals feed on these contaminated prey species, the pollutants are taken into their bodies and are passed along to other animals in the food web in a process called biomagnification. As a result, benthic organisms, fish, birds, and mammals can be adversely affected by contaminated sediments. Contaminated sediments can also compromise human health owing to direct exposure when wading, swimming, or through the consumption of contaminated fish and shellfish. Human uses of aquatic ecosystems can also be compromised by the presence of contaminated sediments through reductions in the abundance of food or sportfish species or by the imposition of fish consumption advisories. As such, contaminated sediments in aquatic ecosystems can pose potential hazards to sediment-dwelling organisms (i.e., epibenthic and infaunal invertebrate species), aquatic-dependent wildlife species (i.e., fish, amphibians, reptiles, birds, and mammals), and human health.

Although contaminated sediment does not represent a specific use impairment, a variety of beneficial use impairments have been documented in association with contaminated sediments. For example, the imposition of fish consumption advisories (i.e., resulting from the bioaccumulation of sediment-associated contaminants) has adversely affected commercial, sport, and food fisheries in many areas. In addition, degradation of the benthic community (i.e., resulting from direct exposure to contaminated sediments) and other factors have contributed to the impairment of fish and wildlife populations. Furthermore, fish from areas with contaminated sediments have been observed to have higher incidences of tumors and other abnormalities than fish from reference areas (i.e., because of exposure to carcinogenic and teratogenic substances that accumulate in sediments). Contaminated sediments have also threatened the viability of many commercial ports through the imposition of restrictions on dredging of navigational channels and disposal of dredged materials (IJC 1997). A summary of use impairments and how they can be affected by contaminated sediments is presented in Table 7.1.

7.3 Indicators of sediment quality conditions

Owing to the potential effects of contaminated sediments on aquatic organisms, aquatic-dependent wildlife, and human health and well-being, regulatory agencies require information on sediment quality conditions. As comprehensive monitoring programs to assess sediment quality can be resource intensive, investigators often select one or more indicators of sediment quality conditions to obtain the requisite information in a cost-effective manner. Several factors need to be considered in the selection of indicators for assessing sediment quality conditions. First, the indicators that are selected must be related to the ecosystem goals and objectives established for the body of water under investigation (Environment Canada 1996). Second, a suite of indicators should be selected to reduce the potential for errors in decisions that are made based on the results of sediment quality monitoring programs (Environment Canada 1996). Third, the selection of indicators should be guided by selection criteria that reflect the stated purpose of the monitoring program (as described in Table 7.2).

Relative to sediment contamination, chemicals of potential concern (COPCs) can be classified into two general categories based on their potential effects on ecological receptors, including toxic substances and bioaccumulative substances. For toxic substances that partition into sediments, evaluation of direct effects on sediment-dwelling organisms is likely to represent the primary focus of sediment quality investigations. For bioaccumulative substances, sediment quality assessments are likely to focus on evaluating effects on aquatic-dependent wildlife (i.e., fish, amphibians, reptiles, birds, and mammals) and on human health. In this way, such investigations can provide the information needed to evaluate attainment of the sediment management objectives for the site and the objectives that have been recommended for soft substrate habitats in freshwater ecosystems.

Table 7.1 A summary of use impairments potentially associated with contaminated sediment.

Use impairment	How contaminated sediment may affect use impairment
Restrictions on fish and wildlife consumption	* Contaminant uptake through contact with sediment or through the food web
Degradation of fish and wildlife populations	* Contaminant degradation of habitat * Contaminant impacts through direct sediment contact * Food web uptake
Fish tumors or other deformities	* Contaminant transfer through contact with sediment or through the food web * Possible metabolism to carcinogenic or more carcinogenic compounds
Bird or animal deformities or reproduction problems	* Contaminant degradation of habitat * Contaminant impacts through direct sediment contact * Food web uptake
Degrdation of benthos	* Contact * Ingestion of toxic contaminants * Nutrient enrichment leading to a shift in species composition and structure owing to oxygen depletion
Restrictions on dredging activities	* Restrictions on disposal in open water owing to contaminants and nutrients and their potential impacts on biota
Eutrophication or undesirable algae	* Nutrient recycling from temporary sediment sink
Degradation of esthetics	* Resuspension of solids and increased turbidity * Odors associated with anoxia
Added costs to agriculture or industry	* Resuspended solids * Presence of toxic substances and nutrients
Degradation of phytoplankton or zooplankton populations	* Toxic contaminant release * Resuspension of solids and absorbed contaminants and subsequent ingestion
Loss of fish and wildlife habitat	* Toxicity to critical life-history stages * Degradation of spawning and nursery grounds owing to siltation

From (IJC 1997)

Table 7.2 Desirable characteristics of sediment quality indicators for different monitoring purposes.

	Purpose of monitoring program				
Characteristic of indicator	Assessment of status of sediment quality conditions	Assessment of trends in sediment quality conditions	Early warning of degraded sediment quality conditions	Diagnostic of courses of degraded conditions	Evaluation of linkages between sources and effects
Biologically relevant	3	3	2	2	2
Socially relevant	3	3	2	2	2
Sensitive	*	*	*	*	*
Broadly applicable	2	2	2	1	1
Diagnostic	1	1	1	3	1
Measurable	*	*	*	*	*
Interpretable	3	3	2	1	1
Cost effective	*	*	*	*	*
Integrative	2	2	1	1	2
Historical data	*	*	*	*	*
Anticipatory	1	1	3	1	2
Nondestructive	*	*	*	*	*
Continuity	2	3	1	1	1
Appropriate scale	*	*	*	*	*
Lack of redundancy	*	*	*	*	*
Timeliness	2	2	3	3	2

Table entries are on a scale of importance from 1 to 3, where 1 indicates lower importance and 3 indicates an essential attribute. Characteristics that are universally desirable and do not differ between purposes are marked with an asterisk.
From IJC (1991).

There is a wide range of indicators that can be used to evaluate sediment quality conditions. In the past, physical and chemical indicators have been primarily used to provide a means of assessing environmental quality conditions. More recently, significant effort has also been directed at the development of biological indicators of ecosystem integrity (which are often termed biocriteria) (OEPA 1988). These biological indicators can apply to one or more levels of organization and encompass many metrics, ranging from biochemical variables to community parameters (for example species richness). Ideally, environmental monitoring programs would include each of the physical, chemical, and biological variables that could, potentially, be affected by anthropogenic activities. However, limitations on human and financial resources preclude this possibility. For this reason, identifying the most relevant indicators for assessing sediment quality conditions is necessary.

The scoring system developed by the IJC (1991) provides a basis for evaluating candidate indicators relative to the intended purpose of the resultant monitoring data (Table 7.2). Application of the IJC (1991) criteria is dependent on identifying the most desirable characteristics of the indicators and subsequently evaluating the candidate indicators for these characteristics. Based on the information presented in Table 7.2, it is essential that indicators for any monitoring purpose be sensitive, measurable, cost-effective, supported by historical data, non-destructive, of appropriate scale, and non-redundant (i.e., these are the essential characteristics of indicators). For sediment quality evaluations that focus on status and trends assessment, indicators that are biologically relevant, socially relevant, interpretable, and provide continuity of measurements over time are likely to be the most relevant (i.e., these are the important characteristics of indicators for this monitoring application). Application of the IJC (1991) evaluation criteria facilitates the identification of the indicators that are the most relevant for assessing sediment quality conditions. MacDonald and Ingersoll (2000) and MacDonald *et al.* (2002a,b) evaluated a variety of candidate indicators and concluded that the following were particularly relevant for assessing sediment quality conditions in aquatic ecosystems.

Receptors of interest	Indicator of sediment quality conditions
Sediment-dwelling organisms	Chemistry of whole sediments
	Chemistry of pore water
	Toxicity of sediments to invertebrates
	Structure of benthic invertebrate communities
Wildlife resources	Toxicity of sediments to fish
	Health of fish
	Status of fish communities
	Chemistry of whole sediments
	Chemistry of fish and invertebrate tissues
Human health	Chemistry of whole sediments
	Chemistry of fish and invertebrate tissues
	Presence of fish and wildlife consumption advisories

Again, the selection of indicators must be guided by the sediment quality issues and concerns that are identified at the site under investigation. Where sediments are primarily contaminated by toxic substances, focusing sediment quality assessments on the receptors that are most likely to be directly affected by contaminated sediments is reasonable (i.e., sediment-dwelling organisms and fish). At sites contaminated by bioaccumulative substances, sediment quality assessments need to have a broader focus, potentially including sediment-dwelling organisms, wildlife resources, and human health. Importantly, the significance of the decisions (i.e., size of the site, potential clean-up costs) that may be made based on the results of the assessment should be a central consideration when developing a suite of indicators for assessing contaminated sediments.

The problem formulation process provides an effective framework for identifying the issues and concerns that should be addressed in sediment quality assessments (USEPA 1997, 1998). In addition to supporting identification of the indicators that should be incorporated into sediment quality monitoring programs, the problem-formulation process also enables investigators to select the variables that will be measured to provide the requisite information. These variables or metrics provide the data needed to evaluate the status of each of the selected indicators of sediment quality conditions.

7.3.1 Selection of metrics for whole-sediment chemistry and pore-water chemistry

Several types of information can be used to support the selection of appropriate metrics for sediment chemistry. First, current and historic land and water use activities in the vicinity of the site should be determined. Historical data should include information on the nature and location of industrial developments (and associated management practices that could lead to releases of chemical substances) and municipal infrastructure (combined sewer overflows, sewage treatment plants), on the nature and location of any spills that have occurred, and on the nature and general location of non-point pollution sources. In addition, information on the location, composition, and volumes of stormwater and effluent discharges is useful for identifying the chemicals that have been or may have been released into surface waters near the site. Evaluation of the environmental fate of these chemicals provides a basis for identifying the substances that are likely to partition into sediments. Finally, existing sediment chemistry data should be assembled and used to identify the chemicals that have been measured at elevated levels (i.e., through comparisons with sediment quality guidelines (SQGs)) in surficial (i.e., top 10 cm) and deeper sediments. Together, this information can be used to develop a list of COPCs for the site. This list of COPCs can then be used to establish the primary metrics for sediment chemistry at the site. Additional metrics, such as total organic carbon (TOC), grain size, acid volatile sulfides (AVS), ammonia, and hydrogen sulfide should also be included to support interpretation of the resultant data for the primary metrics. The final list of chemical analytes to be measured is also influenced by the equipment, technology, facilities, and funds that are available for the project.

The chemicals that are typically analyzed in whole-sediment samples collected near urbanized and industrial areas include trace metals, PAHs, PCBs, and various other organic constituents (for example PCDDs/PCDFs; chlorophenols, and phthalates). In areas that may be affected by inputs from agricultural activities, it may be appropriate to measure the concentrations of pesticides (such as OCs, carbamates, and organophosphates) in sediment samples. Chemical concentrations are generally reported on a dry weight basis, based on the results of total extraction of sediment samples. However, several other measures of sediment chemistry have also been used in various assessments. For example, the concentrations of non-ionic organic contaminants may be normalized to TOC concentrations in sediment (Swartz *et al.* 1987; Di Toro *et al.* 1991; USEPA 2003). In addition, AVS-normalization procedures may be used to interpret data on the levels of simultaneously extracted metals (SEMs) (Di Toro *et al.* 1992; Ankley *et al.* 1996; USEPA 2005). Furthermore, chemical concentrations can be normalized to percent fines. These normalization procedures are intended to define better the bioavailable fraction of the substance under consideration.

Pore water is the water that occupies the spaces between sediment particles. It can be isolated from the sediment matrix to conduct toxicity testing or to measure the concentrations of chemical substances. ASTM (2008a) and USEPA (2000a) describe procedures for isolating pore water from whole-sediment samples. Evaluation of the concentrations of COPCs in pore water is important because sediment-dwelling organisms are directly exposed to the substances that occur in this sediment phase. For this reason, pore water assessments can provide useful information on the potential effects of sediment-associated contaminants, particularly on infaunal species (i.e., those species that use habitats within the sediment matrix). Importantly, the toxicity of sediments to aquatic organisms has been correlated to the concentrations of COPCs in pore water (Di Toro *et al.* 1991; Ankley *et al.* 1996; USEPA 2003, 2005). COPCs in pore water also represent hazards to water-column species because these substances can be transported into overlying waters through chemical partitioning, diffusion, bioturbation, or resuspension processes. However, data on the concentrations of chemicals in pore water may not fully represent the total exposure of sediment-dwelling organisms to sediment-associated contaminants, particularly for compounds with higher octanol–water partition coefficients (K_{ow}) that bind strongly to organic carbon in the sediment (Harkey *et al.* 1994). For this reason, pore-water chemistry alone should not be used to evaluate total exposure to sediment-associated COPCs.

Selection of appropriate metrics for pore-water chemistry should be consistent with the process used to select the metrics for whole-sediment chemistry.

In addition to the substances that are expected to partition into sediments (owing to their physical–chemical properties), it may be appropriate to include additional COPCs that are likely to partition primarily into water. It is necessary to include several variables (e.g., pH, water temperature, water hardness, dissolved oxygen) that will provide ancillary information for interpreting the data on the primary chemical metrics.

Sediment chemistry data provide information that is directly relevant for determining if sediments within an assessment area are contaminated with toxic and/or bioaccumulative substances. However, information on the concentrations of contaminants in whole sediments (i.e., the metrics for sediment chemistry) does not, by itself, provide a basis for determining if the ecosystem goals and objectives are being achieved. For this reason, it is necessary to establish sediment quality targets for sediment chemistry that define the levels of each metric (i.e., the COPCs and mixtures of COPCs) that are likely to support the designated uses of the aquatic ecosystem (i.e., the benthic invertebrate community). These targets can be established by selecting appropriate SQGs for each COPC at the site. Such SQGs can be derived using information on contemporary background levels and/or on the concentrations associated with a pre-selected probability of observing adverse biological effects (see, for example, MacDonald *et al.* 2000b; Field *et al.* 2002). The recommended procedures for assessing sediment chemistry data are summarized in Fig. 7.1 and Table 7.3.

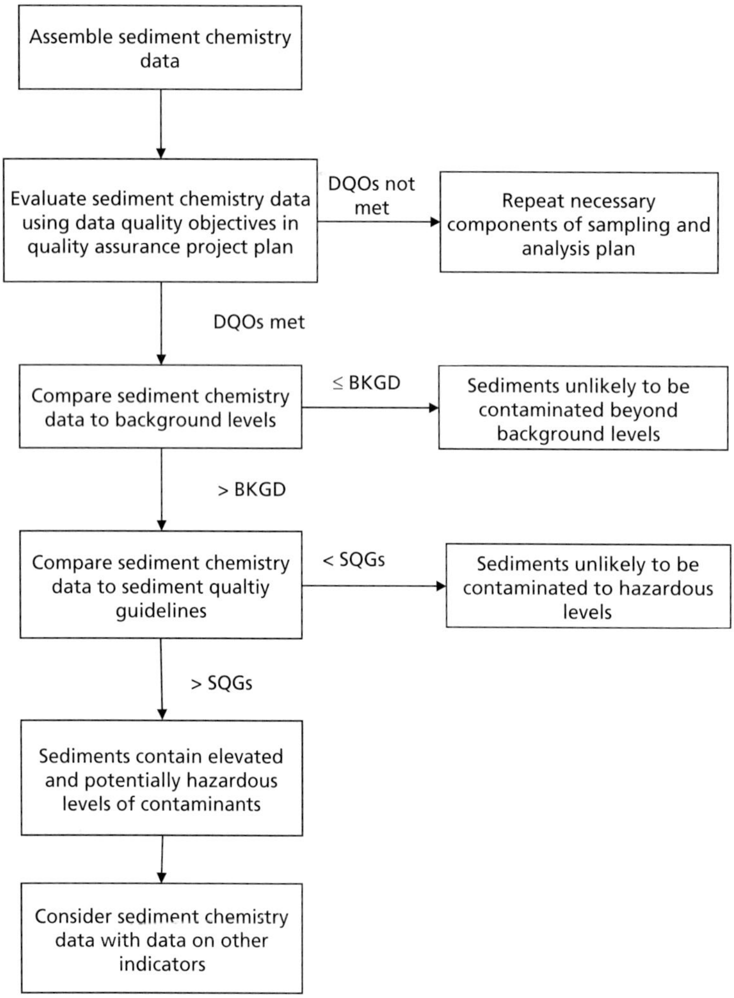

Fig. 7.1 Recommended procedure for assessing sediment chemistry data.

Table 7.3 Toxicity screening values (TSVs) for freshwater and marine or estuarine sediments.

Chemical class/analyte	CAS number	Freshwater sediment TSV (mg/kg DW)[1]	Marine or estuarine sediment TSV (mg/kg DW)[2]
Metals			
Arsenic	7440-38-2	7.15	7.95
Cadmium	7440-43-9	0.991	1.31
Chromium	7440-47-3	20.2	25
Copper	7440-50-8	25.2	28.7
Lead	7439-92-1	35.3	41.3
Mercury	7439-97-6	0.158	0.145
Nickel	7440-02-0	18.7	19.5
Zinc	7440-66-6	124	142
Carbamate pesticides			
Aldicarb	116-06-3	NBA	NBA
Carbaryl	63-25-2	NBA	NBA
Carbofuran	1563-66-2	0.002	NBA
Chlorinated benzenes			
1,2,3-Trichlorobenzene	87-61-6	NBA	NBA
1,2,4-Trichlorobenzene	120-82-1	8.16	9.2
1,2-Dichlorobenzene	95-50-1	0.173	0.218
1,3-Dichlorobenzene	541-73-1	1.61	1.7
1,4-Dichlorobenzene	106-46-7	0.247	0.341
Chlorobenzene	108-90-7	0.363	0.313
Hexachlorobenzene	118-74-1	0.0552	0.0337
PCNB (pentachloronitrobenzene)	82-68-8	NBA	NBA
Nitrogen/phosphorus/sulfur pesticides			
Azinphos methyl	86-50-0	0.00001	0.00003
Bromacil	314-40-9	NBA	NBA
Captan	133-06-2	NBA	NBA
Chlorothalonil	1897-45-6	NBA	NBA
Chlorpyrifos	2921-88-2	0.053	0.0072
Demeton-A/B	8065-48-3	NBA	NBA
Demeton-O	298-03-3	NBA	NBA
Demeton-S	126-75-0	NBA	NBA
Dimethoate	60-51-5	NBA	NBA
Ethyl parathion	56-38-2	0.000757	
Linuron	330-55-2	NBA	NBA
Malathion	121-75-5	0.000495	0.000495
Metribuzin	21087-64-9	NBA	NBA
Tebuthiuron	34014-18-1	NBA	NBA
Trifluralin	1582-09-8	NBA	NBA
Organometallic compounds			
Tributyltin chloride	1461-22-9	NBA	NBA
Persistent organochlorine pesticides			
4,4′-DDD	72-54-8	0.00509	0.00275
4,4′-DDE	72-55-9	0.00261	0.00228
4,4′-DDT	50-29-3	0.00266	0.00158
Aldrin	309-00-2	0.002	0.002
alpha-BHC	319-84-6	0.006	0.006
beta-BHC	319-85-7	0.005	0.005
Chlordane	57-74-9	0.00262	0.000716
delta-BHC	319-86-8	71.5	NBA
Dieldrin	60-57-1	0.00493	0.00339

Table 7.3 *Continued*

Chemical class/analyte	CAS number	Freshwater sediment TSV (mg/kg DW)[1]	Marine or estuarine sediment TSV (mg/kg DW)[2]
Endosulfan I	959-98-8	0.00297	0.0029
Endosulfan II	33213-65-9	0.00943	0.014
Endrin	72-20-8	0.0046	0.00389
Endrin aldehyde	7421-93-4	0.48	NBA
Endrin ketone	53494-70-5	NBA	NBA
gamma-BHC (Lindane)	58-89-9	0.00233	0.00136
Heptachlor	76-44-8	0.00537	0.068
Heptachlor epoxide	1024-57-3	0.00173	0.00395
Kepone	143-50-0	0.00331	NBA
Methoxychlor	72-43-5	0.0141	0.0142
Mirex	2385-85-5	0.007	0.007
Toxaphene	8001-35-2	0.00279	0.00684
Phenols			
2,3,4,6-Tetrachlorophenol	58-90-2	0.129	NBA
2,3,5,6-Tetrachlorophenol	935-95-5	NBA	NBA
2,3,5-Trichlorophenol	933-78-8	NBA	NBA
2,3,6-Trichlorophenol	933-75-5	NBA	NBA
2,4,5-Trichlorophenol	95-95-4	NBA	0.003
2,4-Dichlorophenol	120-83-2	0.0817	0.005
2,6-Dichlorophenol	87-65-0	NBA	NBA
2-Chlorophenol	95-57-8	0.0319	0.008
m-Chlorophenol	108-43-0	NBA	NBA
m-Cresol	108-39-4	0.0524	NBA
o-Cresol	95-48-7	0.0316	0.0275
p-Cresol	106-44-5	0.333	0.416
Pentachlorophenol	87-86-5	0.733	0.195
Phenol	108-95-2	0.0667	0.163
Phenoxyacetic acids			
Dicamba	1918-00-9	NBA	NBA
Dinoseb	88-85-7	0.0145	NBA
MCPA	94-74-6	NBA	NBA
Polychlorinated biphenyls			
PCB-1016	12674-11-2	0.00442	0.007
PCB-1221	11104-28-2	0.0988	0.0814
PCB-1232	11141-16-5	0.6	0.6
PCB-1242	53469-21-9	0.17	0.17
PCB-1248	12672-29-6	0.03	0.03
PCB-1254	11097-69-1	0.06	0.06
PCB-1260	11096-82-5	0.005	0.005
Total PCBs	1336-36-3	0.0404	0.0236
Polychlorinated dibenzo-*p*-dioxins and dibenzofurans			
2,3,7,8-Tetrachlorodibenzo-*p*-dioxin	1746-01-6	0.00000138	0.000003
Polycyclic aromatic compounds			
2-Methylnaphthalene	91-57-6	0.114	0.0728
Acenaphthene	83-32-9	0.0983	0.0356
Acenaphthylene	208-96-8	0.0783	0.044
Anthracene	120-12-7	0.151	0.142
Benzo(a)anthracene	56-55-3	0.132	0.226
Benzo(a)pyrene	50-32-8	0.205	0.342
Benzo(b)fluoranthene	205-99-2	4.74	2.64

Table 7.3 *Continued*

Chemical class/analyte	CAS number	Freshwater sediment TSV (mg/kg DW)[1]	Marine or estuarine sediment TSV (mg/kg DW)[2]
Benzo(g,h,i)perylene	191-24-2	0.252	0.327
Benzo(k)fluoranthene	207-08-9	0.139	0.24
Biphenyl	92-52-4	1.1	1.1
Chrysene	218-01-9	0.195	0.313
Dibenzo(a,h)anthracene	53-70-3	0.0596	0.0779
Fluoranthene	206-44-0	0.505	0.502
Fluorene	86-73-7	0.0841	0.0413
Indeno(1,2,3-cd)pyrene	193-39-5	0.193	0.257
Naphthalene	91-20-3	0.176	0.145
Phenanthrene	85-01-8	0.234	0.266
Pyrene	129-00-0	0.36	0.506
Semivolatile chlorinated organic compounds			
Hexachlorobutadiene	87-68-3	0.0205	0.025
Triazine herbicides			
Atrazine	1912-24-9	NBA	NBA
Simazine	122-34-9	NBA	NBA
Volatile chlorinated organic compounds			
1,1,1-Trichloroethane	71-55-6	0.126	0.0954
1,1,2,2-Tetrachloroethane	79-34-5	0.921	0.94
1,2-Dichloroethane	107-06-2	0.253	0.25
Carbon tetrachloride	56-23-5	0.56	0.408
Tetrachloroethene	127-18-4	0.397	0.312
Trichloroethene	79-01-6	0.738	0.974
Vinyl chloride	75-01-4	0.59	NBA
Volatile organic compounds			
Acetone	67-64-1	0.0141	0.0087
Benzene	71-43-2	0.117	0.11
Chloroform	67-66-3	0.388	0.022
Ethanol	64-17-5	NBA	NBA
Ethyl acetate	141-78-6	NBA	NBA
Ethylbenzene	100-41-4	0.471	0.318
Methanol	67-56-1	NBA	NBA
Methyl ethyl ketone	78-93-3	0.146	0.27
Methylene chloride	75-09-2	0.279	0.37
m-Xylene	108-38-3	0.025	0.025
o-Xylene	95-47-6	NBA	NBA
p-Dioxane	123-91-1	0.119	NBA
p-Xylene	106-42-3	NBA	NBA
Styrene	100-42-5	0.254	NBA
Toluene	108-88-3	0.581	0.479

CAS, chemical abstracts; NBA, no benchmark available; DW, dry weight.

[1] The toxicity threshold benchmark is the geometric mean of the Draft Freshwater Sediment Benchmarks by USEPA Region (USEPA compilation; February 12, 2004 draft; received from Marc Greenberg on September 16, 2004). Benchmarks that were expressed on an organic carbon (OC) normalized basis were converted to a dry weight basis at 1% OC before calculating the geometric mean.

[2] The toxicity threshold benchmark is the geometric mean of the Draft Marine Sediment Benchmarks by USEPA Region (USEPA compilation; July 9, 2003 draft; received from Marc Greenberg on September 16, 2004). Benchmarks that were expressed on an OC normalized basis were converted to a dry weight basis at 1% OC before calculating the geometric mean.

Effects-based SQGs represent tools that can be used to help establish sediment quality targets that correspond to the specific management goals that have been established for the site under consideration. A variety of numerical SQGs have been developed to support sediment quality assessments in North America. The approaches selected by individual jurisdictions depend on the receptors that are to be considered (for example, sediment-dwelling organisms, wildlife, or humans), the degree of protection that is to be afforded, the geographic area to which the values are intended to apply (for example, site-specific, regional, or national), and their intended uses (for example, screening tools, remediation objectives, identifying toxic and non-toxic samples, bioaccumulation assessment). Although such SQGs can be used in many applications, the USEPA generally advocates their use primarily in screening-level assessments of sediment quality conditions.

Guidelines for assessing sediment quality relative to the potential for adverse effects on sediment-dwelling organisms in freshwater systems have been derived using a combination of theoretical and empirical approaches, primarily including the equilibrium partitioning approach ((EqPA) which is used to develop equilibrium partitioning-derived sediment benchmarks (ESBs)) (Di Toro *et al.* 1991; USEPA 1997, 2003, 2005; NYSDEC 1999), screening-level concentration approach (SLCA) (Persaud *et al.* 1993), effects range approach (ERA) (Long *et al.* 1995; USEPA 1996), effects level approach (ELA) (Smith *et al.* 1996; USEPA 1996), the apparent effects threshold approach (AETA) (Cubbage *et al.* 1997), the consensus-based approach (Swartz 1999; MacDonald *et al.* 2000a,b, 2002a,b; USEPA 2000b; Ingersoll *et al.* 2001, 2002), and the logistic regression modeling approach (LRM) (Field *et al.* 1999, 2002). Application of these methods has resulted in the derivation of numerical SQGs for many COPCs in freshwater, estuarine, and marine sediments. Table 7.3 provides a summary of SQGs that can be applied in screening-level assessments of sediment quality conditions. Information on uses of such SQGs is available in Engler *et al.* (2005), Ingersoll *et al.* (2005), and Word *et al.* (2005).

In addition to causing direct effects on aquatic biota, sediment-associated COPCs can accumulate in the tissues of sediment-dwelling organisms. Because many benthic and epibenthic species represent important components of the food web, such contaminants can be transferred to higher trophic levels in the food web. In this way, contaminated sediments represent a potential hazard to the wildlife species that consume aquatic organisms. As such, sediment chemistry represents an important indicator for the potential for effects on aquatic-dependent wildlife species. Information on the applications of bioaccumulation-based SQGs is provided in Moore *et al.* (2005).

Residue-based SQGs provide practical tools for establishing targets for sediment chemistry relative to the potential for bioaccumulation (Cook *et al.* 1992). Residue-based SQGs define the maximum concentrations of individual chemicals or classes of chemicals in sediments that are predicted to result in tolerable levels of those substances in the tissues of aquatic organisms (i.e., below the levels associated with adverse effects in piscivorous wildlife). The first step in the development of residue-based SQGs involves the derivation or selection of an appropriate tissue residue guideline (TRG) for the substance or substances under consideration (e.g., the New York State Department of Environmental Conservation fish flesh criteria for piscivorous wildlife) (Newell *et al.* 1987). Subsequently, relations between concentrations of COPCs in sediments and COPC residues in aquatic biota need to be established. In general, the necessary biota–sediment accumulation factors (BSAFs) are determined from field studies, based on the results of bioaccumulation tests, and/or estimated using various modeling approaches. The SQGs are then derived by dividing the TRG by the BSAF (Cook *et al.* 1992; NYSDEC 1999). Because it is difficult to predict accurately relations between sediment chemistry and the concentrations of COPCs in the tissues of aquatic organisms, potential risks to piscivorous wildlife identified using the SQGs should be confirmed using site-specific tissue residue data and appropriate TRGs.

Contaminated sediment represents a significant environmental concern for the protection of human health. Humans can be directly exposed to contaminated sediments through primary contact recreation, including swimming and wading in affected waterbodies. In addition, indirect exposure to sediment-associated contaminants can occur when humans consume fish, shellfish, or wildlife tissues that have become contaminated owing to bioaccumulation in the food web (Crane 1996). Therefore, sediment chemistry represents an important indicator for assessing the potential effects of COPCs on human health. The bioaccumulation-based SQGs for the protection of human health that were developed by the New York State Department of Environmental Conservation (NYSDEC 1999) and the Washington State Department of Health (1995, 1996) provide a basis for establishing sediment quality targets relative to the protection of human health.

7.3.2 Selection of metrics for whole-sediment and pore-water toxicity testing

The objective of a sediment toxicity test is to determine whether contaminated sediments are harmful to benthic organisms (USEPA 2000a, ASTM 2008a). These tests can be used to measure the interactive toxic effects of complex chemical mixtures in sediment. Furthermore, knowledge of specific pathways of interactions among sediments and test organisms is not necessary to conduct the tests. Sediment tests can be used to: (1) determine the relation between toxic effects and bioavailability; (2) investigate interactions among chemicals; (3) compare the sensitivities of different organisms; (4) determine spatial and temporal distribution of contamination; (5) evaluate hazards of dredged material; (6) measure toxicity as part of product licensing or safety testing; (7) rank areas for clean up; and (8) estimate the effectiveness of remediation or management practices.

The results of sediment toxicity tests can be used to assess the bioavailability of contaminants in field-collected sediments. The responses of organisms exposed to field-collected sediments are often compared with the response of organisms exposed to a negative control material and/or to appropriately selected reference sediments. The results of toxicity tests on sediments spiked with one or more chemicals can also be used to help establish cause and effect relations between chemical concentrations and biological responses. The results of toxicity tests with test materials spiked into sediments at different concentrations are often reported in terms of a median lethal concentration (LC_{50}), a median inhibition concentration (IC_{50}), a no observed effect concentration (NOEC), or a lowest observed effect concentration (LOEC) (USEPA 2000a; ASTM 2008a).

The choice of a test organism has a major influence on the relevance, success, and interpretation of a test. As no one organism is best suited for all applications, considering the intended uses of the resultant data is important in the selection of toxicity tests. The following criteria were considered in the selection of the methods and species that were to be described in ASTM (2008a) and USEPA (2000a) (Tables 7.4 and 7.5). Ideally, a test organism should:

- have a toxicological database demonstrating relative sensitivity and discrimination to a range of COPCs in sediment;
- have a database for inter-laboratory comparisons of procedures (for example, round-robin studies);
- be in contact with sediment (e.g., water column versus sediment-dwelling organisms);
- be readily available through culture or from field collection;
- be easily maintained in the laboratory;
- be easily identified;
- be ecologically or economically important;
- have a broad geographical distribution, be indigenous to the site being evaluated (either present or historical), or have a niche similar to organisms of concern at the site (for example, similar feeding guild or behavior to the indigenous organisms);
- be tolerant of a broad range of sediment physico chemical characteristics (for example grain size); and,
- be compatible with selected exposure methods and endpoints; the method should also be peer reviewed and confirmed with responses with natural populations of benthic organisms.

Of these criteria, a database demonstrating relative sensitivity to contaminants, contact with sediment, ease of culture in the laboratory, inter-laboratory comparisons, tolerance of varying sediment physico-chemical characteristics, and confirmation with responses of natural benthos populations were the primary criteria used for selecting the amphipod *Hyalella azteca* and the midge *Chironomus dilutus* for describing test methods for freshwater sediments, as outlined by ASTM (2008a) and USEPA (2000a) (Table 7.4). Procedures for conducting sediment tests with oligochaetes, mayflies, and other amphipods or midges are also outlined in ASTM (2008a) and in Environment Canada (1997). However, USEPA (2000a) chose to not develop methods for

Table 7.4 Rating of selection criteria for freshwater sediment toxicity testing organisms (USEPA 2000a; ASTM 2008a).

Criterion	*Hyalella azteca*	*Diporeia* spp.	*Chironomus dilutus*	*Chironomus riparius*	*Lumbriculus variegatus*	*Tubifex tubifex*	*Hexagenia* spp.	Mollusks	*Daphnia* spp. and *Ceriodaphnia* spp.
Relative sensitivity toxicity database	+	–	+	–	+	–	–	–	–
Round-robin studies conducted	+	–	+	–	–	–	–	–	–
Contact with sediment	+	+	+	+	+	+	+	+	–
Laboratory culture	+	–	+	+	+	+	–	–	+
Taxonomic identification	+	+/-	+/–	+/–	+	+	+	+	+
Ecological importance	+	+	+	+	+	+	+	+	+
Geographical distribution	+	+/–	+	+	+	+	+	+	+/–
Sediment physicochemical tolerance	+	+	+/–	+	+	+	–	+	NA
Response confirmed with benthos populations	+	+	+	+	+	+	+	–	+
Peer reviewed	+	+	+	+	+	+	+	–	+/–
Endpoints monitored	S,G,M	S,B,A	S,G,E	S,G,E	B,S	S,R	S,G	B	S,G,R
Overall Assessment	10+	5+	8+	7+	9+	8+	5+	5+	4+

+ or – rating indicates a positive or negative attribute; NA, not applicable; S, survival; G, growth; M, maturation; E, emergence; B, bioaccumulation; R, reproduction.

Table 7.5 Rating of selection criteria for estuarine or marine amphipod sediment toxicity testing (ASTM 2008c).

Criterion	Species			
	Ampelisca abdita[a]	*Eohaustorius estuarius*	*Leptocheirus plumulosus*	*Rhepoxynius abronius*
Relative sensitivity toxicity database	+/–	+	+	+
Round-robin studies conducted	+	+	+	+
Contact with sediment	+	+	+	+
Laboratory culture	+/–	–	+	–
Taxonomic identification	+	+	+	+
Ecological importance	+	+	+	+
Geographical distribution	Atlantic coast, Pacific coast, and Gulf of Mexico	Pacific coast	Atlantic coast	Pacific coast
Sediment physico-chemical tolerance	+	+	+	+
Response confirmed with benthos	+	–	–	+
Populations peer reviewed	+	+	+	+
Endpoints monitored	Survival	Survival, reburial	Survival	Survival, reburial

+, postive attribute; –, negative attribute. [a]*Ampelisca abdita* is a tube-dwelling species, which could reduce exposure to sediment-associated COPCs.

conducting sediment toxicity tests with these additional organisms because they did not meet all the required selection criteria listed in Table 7.4. For both of the selected species (*H. azteca* and *C. dilutus*), survival is the principal endpoint measured in 10- to 14-day acute toxicity tests (although growth is also commonly measured), whereas survival, growth, emergence (midges only) and/or reproduction are the principal endpoints measured in longer-term exposures.

The USEPA (2000b) evaluated relative endpoint and organism sensitivity in a database developed from 92 published reports that included a total of 1657 field-collected samples with high-quality matching sediment toxicity and chemistry data. The database comprised primarily 10- to 14-day or 28- to 42-day toxicity tests with the amphipod *H. azteca* (designated as the HA10 or HA28 tests) and 10- to 14-day toxicity tests with the midges *C. dilutus* or *C. riparius* (designated as the CS10 test). Endpoints reported in these tests were primarily survival or growth. For each test and endpoint, the incidence of effects above and below various mean probable effects concentration (PEC) quotients (mean quotients of 0.1, 0.5, 1.0, and 5.0) was determined. In general, the incidence of sediment toxicity increased consistently and markedly with increasing levels of sediment contamination. See MacDonald *et al.* (2000b) for additional information on the calculation of mean PEC quotients.

A higher incidence of toxicity with increasing mean PEC quotients was observed in the HA28 test compared with the short term HA10 or CS10 tests and may be due to the duration of the exposure or the sensitivity of the growth endpoint in the longer HA28 test. A 50% incidence of toxicity in the HA28 test corresponds to a mean PEC quotient of 0.63 when survival or growth were used to classify a sample as toxic (Fig. 7.2) (USEPA 2000b). By comparison, a 50% incidence of toxicity is expected at a mean PEC quotient of 3.2 when survival alone was used to classify a sample as toxic in the HA28 test. In the CS10 test, a 50% incidence of toxicity is expected at a mean PEC quotient of 9.0 when survival alone was used to classify a sample as toxic, or at a mean PEC quotient of 3.5 when survival or growth were used to classify a sample as toxic (Fig. 7.2). In contrast, similar mean PEC quotients resulted in a 50% incidence of toxicity in the HA10 test when survival alone (mean PEC quotient of 4.5) or when survival or growth (mean PEC quotient of 3.4) were used to classify a sample as toxic. The results of these analyses indicate that both the duration of the exposure and the endpoints measured can influence whether a sample is found to be toxic or not. The longer-term tests in which growth and survival are measured tended to be more sensitive

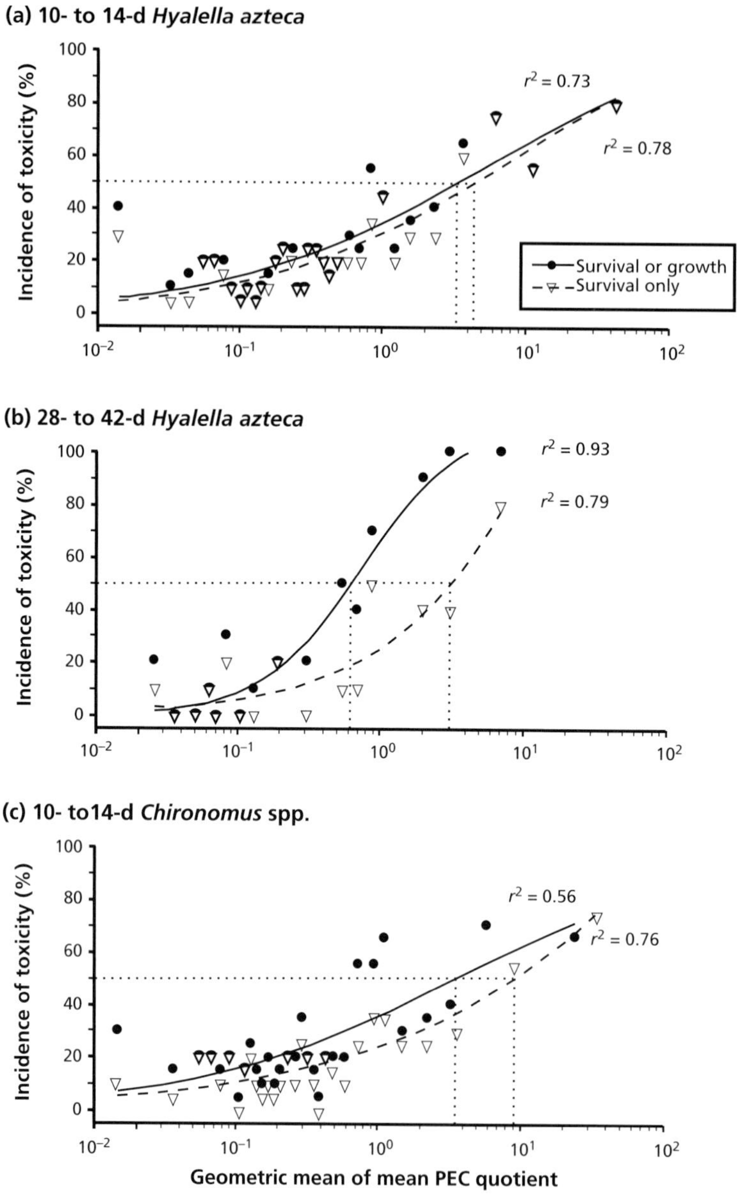

Fig. 7.2 Relation between mean probable effect concentration quotient (PEC quotient) toxicity in and the incidence of freshwater toxicity tests.
From USEPA (2000b).

than shorter-term tests, with an acute to chronic ratio on the order of six indicated for *H. azteca*. Based on these analyses, if only one of these tests were performed, it would be desirable to conduct chronic (i.e., 28- to 42-day) sediment toxicity tests with *H. azteca* measuring survival and growth (as length) instead of 10- to 14-day tests with *H. azteca*, *C. dilutus*, or *C. riparius*.

Relative species sensitivity frequently varies among chemicals; consequently, both ASTM (2008a) and USEPA (2000a) recommend the use of a battery of tests to assess sediment quality, including organisms representing different trophic levels. However, testing multiple species with every sediment sample can be very costly. An alternative approach could be to perform a preliminary evaluation on a few samples from a site using a battery of tests (i.e., see procedures for various species outlined in ASTM 2008a). This preliminary evaluation could be used to identify sensitive species or endpoints to include in a more comprehensive assessment at the site. The preliminary evaluation should include samples representing a gradient of contamination at the site of interest. This approach was taken by Kemble *et al.* (1994) in an assessment of the toxicity of metal-contaminated sediments in the Clark Fork River in Montana. A battery of acute and chronic whole-sediment and pore-water tests were conducted with samples collected from this site. The results of this investigation indicated that a 28-day whole-sediment toxicity test with *H. azteca* measuring survival and growth (as length) was the most sensitive metric across a gradient of metal-contaminated stations at the site. The results of chronic toxicity tests with *H. azteca* were also predictive of effects observed on benthic community structure at the site (Canfield *et al.* 1994). Therefore, Kemble *et al.* (1994) recommended that future evaluations of sediment toxicity at the site should use chronic tests with *H. azteca* rather than testing a suite of toxicity tests.

A diverse array of whole-sediment and pore-water toxicity tests are available to evaluate contaminated sediments at marine and estuarine sites. It is generally recognized that 10-day whole-sediment toxicity tests with marine and estuarine amphipods represent an essential element of the suite of toxicity tests that should be used to assess marine and estuarine sites. Although *Eohaustorius estuarius* and *Rhepoxynius abronius* are the most highly recommended species for conducting such toxicity tests, toxicity testing can be conducted using other amphipod species, considering additional endpoints (i.e., survival, growth, emergence, reburial, and reproduction) and exposure durations (i.e., up to 28-day tests for *Leptocheirus plumulosus*). It should be recognized that *Ampelisca* is a tube-dwelling species and, hence, may receive less exposure to COPCs than other amphipod species (potentially making it less sensitive to sediment-associated COPCs). Toxicity testing with amphipods is recommended because they tend to be sensitive species and their responses are often correlated with responses of the benthic community in the field. At least one study documented effects on the benthic community at mean SQG quotients substantially lower than those associated with toxicity to marine amphipods in 10-day exposures, however (Hyland *et al.* 1999).

Toxicity testing with other species, evaluating non-lethal endpoints over longer durations of exposure, can provide relevant information of assessing contaminated sediments. For example, 20- to 28-day whole-sediment toxicity tests with polychaetes (e.g., *Neanthes arenoceodentata*; endpoints survival and growth) can provide useful information for assessing risks to benthic invertebrates associated with exposure to contaminated sediments. In addition, 48- to 96-hour sediment–water interface toxicity tests with echinoderm (e.g., *Arbacia punctulata*) or bivalve mollusk larvae (e.g., *Mytilus edulis*; endpoint development) represent emerging toxicity tests that could provide broader taxonomic coverage and reduce uncertainties associated with the traditional use of these species and life stages (i.e., in pore-water exposures). The need for standardization of the sediment–water interface toxicity testing protocols has been identified as one of the current limitations associated with applying these tests on a routine basis. Therefore, the relevance of such toxicity tests should be evaluated on a case-by-case basis to determine if one or more of these ancillary tests should be used to assess contaminated sediments at a site.

Certain other toxicity tests may be relevant for assessing marine and estuarine sediments. However, it is now generally agreed that elutriate toxicity tests should not be included in the core suite of tests that are applied at marine and estuarine sites (except in dredged material disposal analysis applications). In addition, some challenges associated with the use of

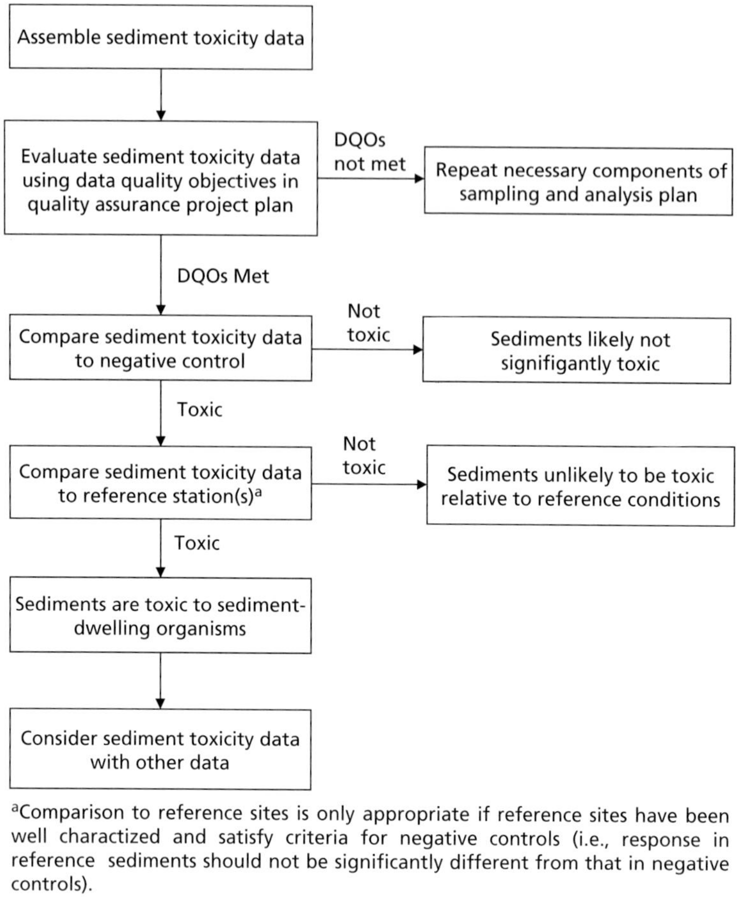

Fig. 7.3 Recommended procedure for assessing sediment toxicity data.

pore-water toxicity tests have been identified, including responsiveness to hydrogen sulfide and ammonia, and depletion of hydrophobic organics during the course of the test. However, a recent evaluation of data from multiple studies showed that ammonia and hydrogen sulfide were rarely confounding factors for pore-water toxicity tests with the sea urchin *Arbacia punctulata* (Carr *et al.* 2006). Elutriate toxicity tests are considered more relevant for assessing the effects of open-water disposal of dredged materials than evaluating the toxicity of in-place sediments. Neither solid-phase nor aqueous-phase toxicity tests with the bacterium *Vibrio fisheri* (i.e., Microtox®) are currently recommended for assessing contaminated sediments at marine or estuarine sites, as these tests provide an indication of exposure to contaminants rather than specific measures of effects on benthic organisms.

The recommended procedures for assessing sediment toxicity data are presented in Fig. 7.3. Importantly, evaluation of the usability of the data represents the first step in this process. Comparison of the results to negative control data and the reference envelope provides a basis for designating samples as toxic or non-toxic (MacDonald *et al.* 2002c). Such designations of toxicity are useful for evaluating sediment quality conditions on a sample-by-sample basis and for deriving concentration–response relations for the site as a whole (Fig. 7.4) (MacDonald *et al.* 2008).

7.3.3 Selection of metrics for benthic invertebrate community assessment

Benthic communities are assemblages of organisms that live in or on the bottom sediment. Because most

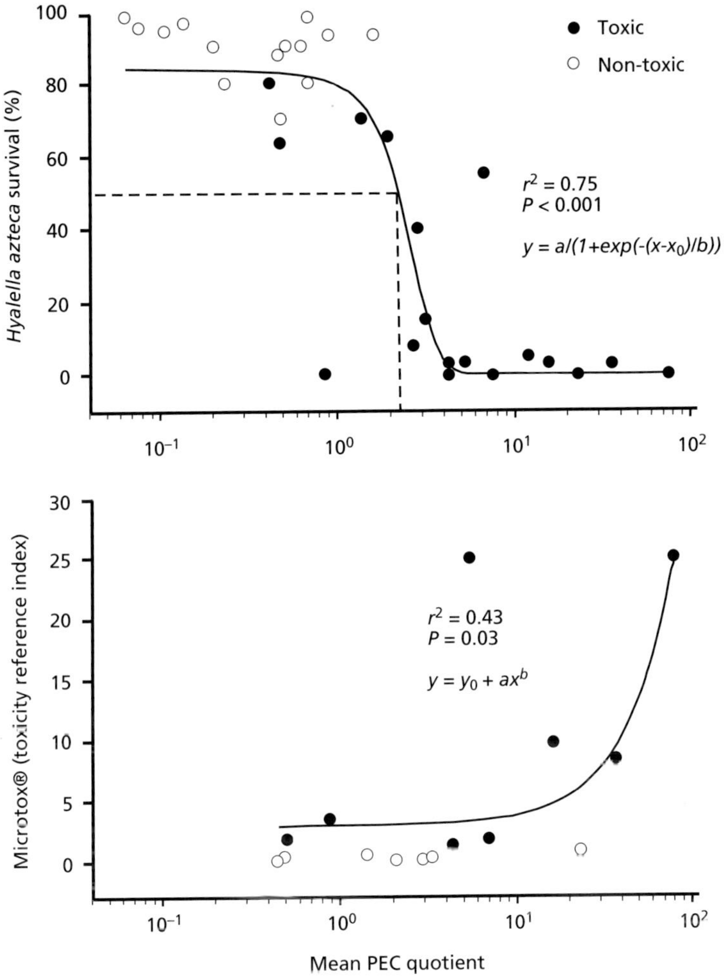

Fig. 7.4 Relation between the mean PEC quotient and the response of *Hyalella azteca* in the 10-day tests (as percentage survival) or the response in the Microtox® solid-phase sediment toxicity test (as the EC_{50} expressed as a toxicity reference index). Sediment samples were collected from the Grand Calumet River and Indiana Harbor Canal located in northwestern Indiana.
From Ingersoll *et al.* (2002).

benthic macroinvertebrates are relatively sedentary and are closely associated with the sedimentary environment, they tend to be sensitive to both short-term and long-term changes in habitat, sediment, and water quality conditions (Davis & Lathrop 1992). Therefore, data on the distribution and abundance of these species can provide important information on the health of the aquatic ecosystem. As such, benthic invertebrate community structure (BICS) represents a candidate indicator of sediment quality conditions.

Assessments of BICS have been used to describe reference conditions, to establish baseline conditions, and to evaluate the effects of natural and anthropogenic disturbances (Striplin *et al.* 1992). In terms of evaluating sediment quality, such assessments are focused on establishing relations between various community structure metrics (for example species richness, total abundance, relative abundance of various taxonomic groups, macroinvertebrate index of biotic integrity (mIBI)) and measures of sediment quality (for example chemical concentrations and organic content). Data from benthic community assessments have the potential to provide relevant information for identifying impacted sites and, with appropriate supporting data, the factors that are contributing to any adverse effects that are observed (USEPA 1992a,b, 1994).

The IJC (1988) suggested that benthic community surveys should be the first assessment tool used to evaluate areas of the Great Lakes with suspected sediment contaminant problems. If no effects are demonstrated in an initial survey, IJC (1988) recommended no further assessment. However, the absence of benthic organisms in sediment does not necessarily indicate that contaminated sediment caused the observed response. Benthic invertebrate distributions may exhibit high spatial or temporal variability. Furthermore, short-term exposure to chemical (for example ammonia, dissolved oxygen) or physical (for example temperature, abrasion) factors can influence benthic invertebrate distribution and abundance, even in the absence of measurable levels of COPCs in sediment. Most importantly, evaluations of BICS only infrequently have sufficient statistical power to detect effects associated with exposure to contaminated sediments (i.e., owing to high variability in the selected metrics). Therefore, information on BICS alone is not always indicative of ambient sediment quality conditions and is certainly not diagnostic of sediment contamination or sediment toxicity (USEPA 1992a,b, 1994).

One objective of a BICS assessment is to determine whether sediment-associated COPCs may be contributing to a change in the distribution of benthic organisms in the field. These assessments can be used to measure interactive toxic effects of complex chemical mixtures in sediment. Furthermore, knowledge of specific pathways of interactions among sediments and test organisms is not necessary for assessments of the benthic community. Assessments of the benthic invertebrate community can be used to:

- determine the relation between toxic effects and bioavailability;
- investigate interactions among chemicals;
- compare the sensitivities of different organisms;
- determine spatial and temporal distribution of contamination;
- rank areas for clean up; and
- evaluate the effectiveness of remediation or management practices.

The results of benthic community assessments can also be used to assess the bioavailability of COPCs in field-collected sediments. The responses of organisms collected from test sites are often compared with the responses of organisms collected from reference sites. Reynoldson *et al.* (1995, 1997) and MacDonald & Ingersoll (2000) describe procedures for assessing BICS (Fig. 7.5). Similarly, procedures for evaluating fish health data are shown in Fig. 7.6.

Although the BICS assessment should provide useful information for evaluating the status of benthic invertebrate communities, all of these potential applications of BICS assessments are limited by uncertainties in the relation between exposure to COPCs and effects on the benthic community. This uncertainty arises because the sampling methods used in BICS evaluations only rarely provide matching data on whole-sediment and/or pore-water chemistry. In addition, variability in BICS metrics among selected reference sites (i.e., owing to factors other than sediment contamination) makes it difficult to discriminate COPC-related effects on the benthic community. As a result, BICS data are often of limited value for evaluating the effects of contaminated sediments in freshwater, estuarine, and marine ecosystems.

7.3.4 Selection of metrics for whole-sediment bioaccumulation assessment

Contaminated sediments represent important sources of the substances that accumulate in aquatic food webs (Ingersoll *et al.* 1997). Because these contaminants can adversely affect aquatic-dependent wildlife species and/or human health, tissue chemistry represents an important indicator in sediment quality assessments (USEPA 2000a; ASTM 2008b). In general, the concentrations of bioaccumulative COPCs in the tissues of sediment-dwelling organisms represent the primary metrics for tissue chemistry. As wildlife species typically consume the entire prey organism, whole-body COPC levels are the most relevant for assessing risks to aquatic-dependent wildlife. In contrast, the levels of COPCs in edible tissue represent the most important metrics for human health assessments. Assessments that are directed at evaluating COPC residues in the tissues of benthic macroinvertebrates should focus on the bioaccumulative COPCs that are known or suspected to occur in sediments at the site under investigation. Typically, the COPCs that are considered

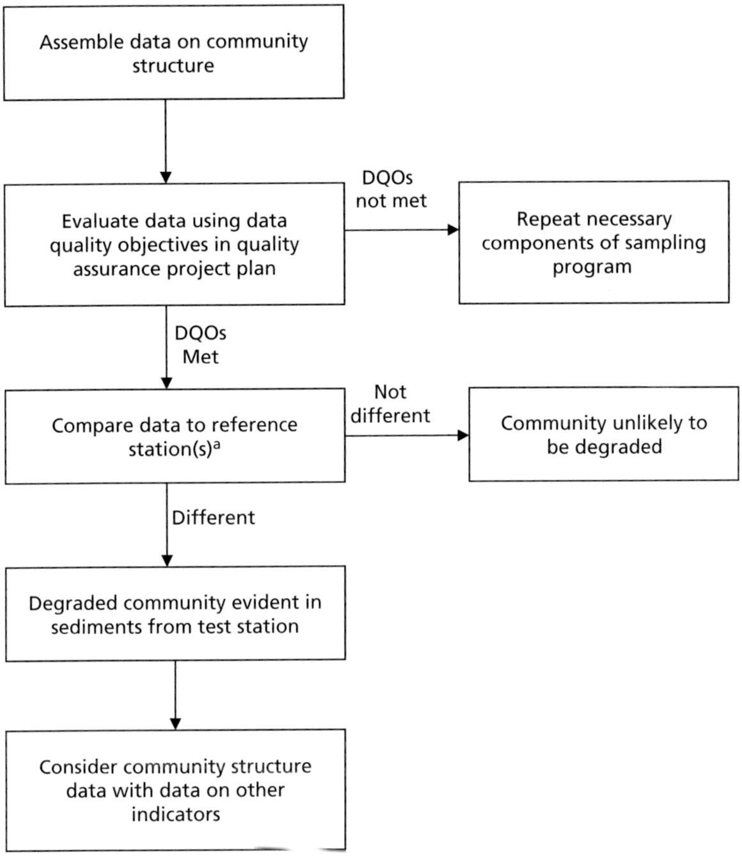

[a]Comparison with reference sites is only appropriate if reference sites have been well characterized and satisfy criteria for negative controls (that is, response in reference sediments should not be significantly different from that in negative controls).

Fig. 7.5 Recommended procedure for assessing benthic invertebrate or fish community structure.

in such assessments include metals, methyl mercury, PAHs, PCBs, OC pesticides, chlorophenols, and/or PCDDs/PCDFs. However, this list should be refined based on the land and water use activities that have been documented near the site.

The selection of species for inclusion in assessments of bioaccumulation requires an understanding of the predator–prey relations in the ecosystem under investigation. For example, the levels of COPCs in benthic macroinvertebrates are likely to be relevant when evaluating risks associated with dietary uptake of COPCs by bottom-feeding fish or sediment-probing birds. Conversely, emergent insects may be the primary focus of an investigation if swallows or bats represent the primary receptor of concern. In cases where fish-eating birds and mammals represent the wildlife species of special concern, fish would be the primary species targeted in sampling and analytical programs. In this way, sampling programs can be tailored to answer the key risk questions that are being posed by the investigators. Bioaccumulation is not an appropriate assessment approach for COPCs that are rapidly metabolized or otherwise not accumulated in the tissues of the organism(s) being evaluated.

Ingersoll *et al.* (1997) identified four general approaches for conducting bioaccumulation assessments, including the following:

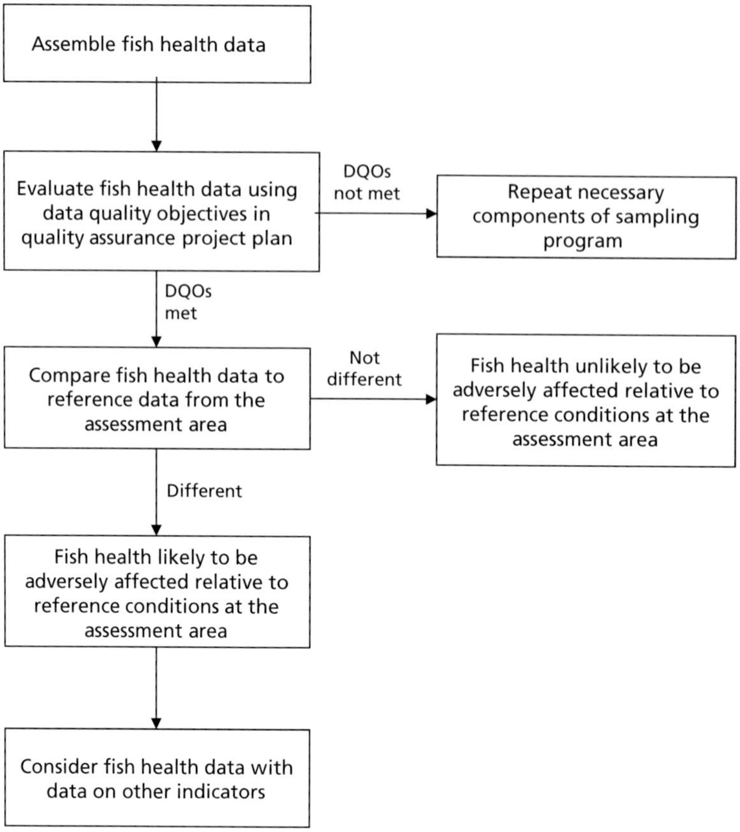

Fig. 7.6 Recommended procedure for evaluating fish health data.

- a laboratory approach, which involves exposing organisms to sediment under controlled conditions;
- a field approach, which involves collecting organisms from a study area;
- assessment of food-web transfer; and
- models to predict bioaccumulation processes.

In the laboratory approach, individuals of a single species are exposed under controlled laboratory conditions to sediments collected from the study area being assessed (USEPA 2000a; ASTM 2008b). After an established period of exposure, the tissues of the organisms are analyzed for the COPCs. Bioaccumulation has occurred if the final concentrations in tissues exceed concentrations that were present before the exposure was started. This requires that individuals representative of initial conditions also be analyzed. This approach has been routinely applied in the assessment of contaminated sediments (USEPA 2000a; ASTM 2008b).

In the field approach, concentrations of COPCs in tissues are determined by collecting one or more species exposed to sediments at the study area being assessed. In addition, organisms representing various trophic levels may be collected and analyzed to determine tissue residue levels. These concentrations are compared with those that have been measured in the tissues of organisms collected from appropriately selected reference area(s). Two methods have been used to determine bioaccumulation in the field:

- organisms resident at the area are collected *in situ* for analysis; or
- organisms are transplanted from another location (presumably with a history of little contaminant exposure) to the area of concern then re-collected,

and tissues are analyzed after an established period of exposure.

In some cases, semipermeable membrane devices (SPMDs) are deployed in the field for specified periods to simulate exposures of aquatic organisms to COPCs (Williamson *et al.* 2002).

Models that describe bioaccumulation are relatively well developed both for organic and inorganic contaminants (Thomann 1989; Luoma & Fisher 1997; ASTM 2008b). Toxicokinetic models have a long history, as do simpler models of bioaccumulation processes. Site-specific models predict bioaccumulation based on laboratory-determined characterization of biological processes in the species of interest and field-determined chemical measurements at the area of concern. Some uncertainties remain unresolved in most models and consensus does not exist about the appropriate model to apply for some (if not all) COPCs (Luoma & Fisher 1997).

Equilibrium models are commonly used in assessments of bioaccumulation and are available for both organic and inorganic COPCs (Di Toro *et al.* 1991; Ankley *et al.* 1996). The models assume that the concentrations of COPCs among all compartments of the environment are controlled by thermodynamics and at least approach equilibrium conditions. If thermodynamic equilibrium exists and if one route of uptake is known or can be predicted, overall bioaccumulation is inferred. Recent applications use an extension of the equilibrium models, termed kinetic or pathway models (ASTM 2008b). These models incorporate geochemical principles and address uncertainties in the assumptions of equilibrium. Kinetic models assume that routes of bioaccumulation are additive and must be determined independently. Kinetic models and equilibrium models may yield similar results if COPC distributions and concentrations in an environment are at equilibrium (although not always), but can yield very different results where environmental compartments are not at equilibrium (for example if biological processes control concentrations, speciation, or phase partitioning of COPCs) (Ingersoll *et al.* 1997).

Tissue residue guidelines for the protection of piscivorous wildlife species and/or human health represent candidate sediment-quality targets that are used to interpret the results of bioaccumulation assessments (Fig. 7.7). However, a variety of risk-based procedures have also been developed to evaluate the results of such assessments (i.e., by calculating average daily doses of COPCs for specific receptor groups and comparing them with no or lowest observed effect doses). These tools can also be used to back-calculate to the concentrations of COPCs in sediment that will protect human health and ecological receptors.

7.4 Integration of information on multiple indicators of sediment quality conditions

Sediment quality assessments are typically conducted to determine if sediments have become contaminated as a result of land- or water-use activities. When such contamination is indicated, the results of sediment quality assessments need to provide the information required to evaluate the nature, severity, and areal extent of sediment contamination. In turn, this information can be used to identify actual and probable use impairments in the assessment area. As indicated previously, investigators can select a variety of indicators for evaluating sediment quality conditions. Data on such indicators can provide useful information for assessing effects on aquatic life, wildlife, or human health.

Although individual indicators of sediment quality each have an inherent level of uncertainty associated with their application, the uncertainty associated with an overall assessment of sediment contamination can be reduced by integrating information from each of these individual indicators. For example, sediment chemistry, sediment toxicity, and benthic community data can be used together in a sediment quality triad assessment to establish a weight of evidence linking contaminated sediments to adverse effects on sediment-dwelling organisms (Table 7.6). The integration of multiple tools using a weight-of-evidence approach has the potential to reduce substantially uncertainty associated with risk assessments of contaminated sediment and, thereby, improve management decisions (Long & Chapman 1985; Chapman 1992; Canfield *et al.* 1996; Ingersoll *et al.* 1997; Wenning & Ingersoll 2002).

The first step in the evaluation of sediment quality data should be to determine if individual indicators exceed the established targets. For example, the following questions should be addressed:

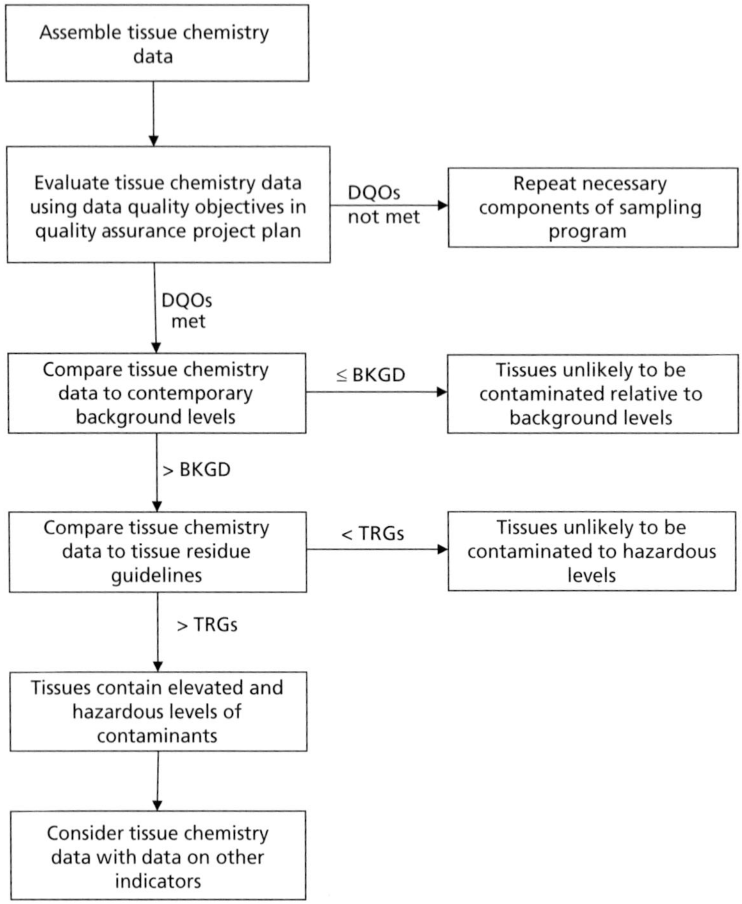

Fig. 7.7 Recommended procedure for assessing tissue chemistry data.

- Do the concentrations of COPCs in sediments exceed applicable SQGs (Fig. 7.1)?
- Are sediments toxic relative to control and/or reference treatments (Fig. 7.3)?
- Are communities of invertebrates or fish in the field degraded relative to reference conditions (Fig. 7.5)?
- Is the health of fish compromised relative to reference conditions (Fig. 7.6)?
- Do the concentrations of COPCs in tissues exceed TRGs (Fig. 7.7)?

The answers to these questions will help to establish if metrics associated with each of these individual indicators are adversely affected at the test stations relative to the reference stations. However, it is also important to determine the relations among individual indicators measured at the assessment area. These relations can be evaluated most directly by using scatter plots of the data to determine if there is correspondence between pairs of indicators and associated metrics measured on splits of individual samples collected from stations in the assessment area (for example sediment toxicity versus sediment chemistry). Alternatively, the scatter plots can be used to evaluate broader trends across geographic reaches within the assessment area (for example fish community status, or fish health versus sediment chemistry). Comparisons of fish community status or tissue chemistry of fish are often made across multiple stations sampled for sediment chemistry to account for the movements of fish within the assessment area.

Statistical regression analyses can be used to determine if there are significant relations between pairs

Table 7.6 Contingency table for assessing impacts of contaminated sediments on aquatic life based on three separate indicators of sediment quality.

Possible outcome	Sediment chemistry	Toxicity test	Benthic community	Possible conclusions
1	+	+	+	Impact highly likely: contaminant-induced degradation of sediment-dwelling organisms evident.
2	–	–	–	Impact highly unlikely: contaminant-induced degradation of sediment dwelling organisms not evident.
3	+	–	–	Impact unlikely: contaminants unavailable to sediment-dwelling organisms.
4	–	+	–	Impacts possible: unmeasured contaminants or conditions exist that have the potential to cause degradation.
5	–	–	+	Impacts unlikely: no degradation of sediment-dwelling organisms in the field apparent relative to sediment contamination; physical factors may be influencing benthic community.
6	+	+	–	Impact likely: toxic chemicals probably stressing the system.
7	–	+	+	Impact likely: unmeasured toxic chemicals are probably contributing to the toxicity.
8	+	–	+	Impact likely: sediment-dwelling organisms degraded by toxic chemicals, but toxicity tests not sensitive to chemicals present.

+, Indicator classified as affected; as determined based on comparison to the established target.
–, Indicator not classified as affected; as determined based on comparison to the established target.
Adapted from Chapman (1992) and Canfield *et al.*(1996).

of indicators and associated metrics. For example, Fig. 7.4 illustrates the relation between sediment chemistry (as a function of mean PEC quotients) and sediment toxicity (as a function of toxicity to *H. azteca* in 10-day sediment tests). Similarly, relations between metrics for a particular indicator can also be evaluated using scatter plots. Figure 7.8 illustrates the relation between two metrics for sediment chemistry: SEM normalized to AVS (i.e., SEM–AVS) and toxic units of metals measured in pore water from these same samples. The results of these types of analysis can be used to establish concordance among various indicators (i.e., high chemistry and toxic, low chemistry and not toxic). Additionally, these analyses can help to establish the rate of false positives (i.e., high chemistry and not toxic) or false negatives (i.e., low chemistry and toxic) among various indicators.

An expanded version of the sediment-quality triad approach has been developed to incorporate measures of bioaccumulation with the traditional measures of sediment quality (MacDonald 1998). Specifically, integration of data from sediment chem-

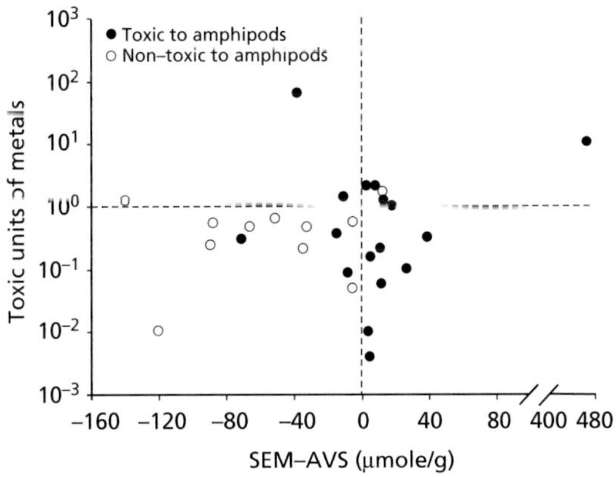

Fig. 7.8 Relation between the molar concentration of simultaneously extracted metals to acid volatile sulfide (SEM–AVS) and toxic units of metals in the sediment samples. Toxicity of samples was determined using 10-day whole-sediment tests with *Hyalella azteca*.
From Ingersoll *et al.* (2002).

istry, sediment toxicity, community status, and/or tissue chemistry provides important information for assessing sediment quality conditions. The contingency table presented in Table 7.7 provides a means of interpreting the data generated from multiple indicators of sediment quality using a weight-of-evidence approach. The results of these analyses can be used to estimate the likelihood of impacts of sediment contamination on aquatic life (sediment-dwelling organisms), wildlife (vertebrates), or human health.

Table 7.7 Contingency table for assessing impacts of contaminated sediments on aquatic life based on four separate indicators of sediment quality.

Possible outcome	Sediment chemistry	Toxicity test	Benthic community	Tissue chemistry	Possible conclusions
1	+	+	+	+	Contaminant-induced impacts on sediment-dwelling organisms and higher trophic levels are likely to be observed; elevated levels of sediment-associated contaminants are likely contributing to sediment toxicity and benthic community impairment; and, bioaccumulation of sediment-associated contaminants has the potential to adversely affect aquatic-dependent wildlife and/or human health.
2	–	–	–	+	Contaminant-induced impacts on higher trophic levels are likely to be observed; adverse effects on sediment-dwelling organisms are unlikely to be observed; and, bioaccumulation of sediment-associated contaminants has the potential to adversely affect aquatic-dependent wildlife and/or human health.
3	+	–	–	+	Contaminant-induced impacts on higher trophic levels are likely to be observed; the bioavailability of sediment-associated contaminants is likely to be limited; and, bioaccumulation of sediment-associated contaminants has the potential to adversely affect aquatic-dependent wildlife and/or human health.
4	–	+	–	+	Contaminant-induced impacts on higher trophic levels are likely to be observed; unmeasured factors (e.g., physical factors or contaminants) are likely to be contributing to sediment toxicity; and, bioaccumulation of sediment-associated contaminants has the potential to adversely affect aquatic-dependent wildlife and/or human health.
5	–	–	+	+	Contaminant-induced impacts on sediment-dwelling organisms and higher trophic levels are likely to be observed; adverse effects on sediment-dwelling organisms are likely due to physical factors and/or unmeasured chemicals are stressing benthos and toxicity tests are not sensitive enough to detect effects; and, bioaccumulation of sediment-associated contaminants has the potential to adversely affect aquatic-dependent wildlife and/or human health.
6	+	+	–	+	Contaminant-induced impacts on sediment-dwelling organisms and higher trophic levels are likely to be observed; high variability in the benthic community metrics may be masking contaminant-related effects; and, bioaccumulation of sediment-associated contaminants has the potential to adversely affect aquatic-dependent wildlife and/or human health.
7	–	+	+	+	Contaminant-induced impacts on sediment-dwelling organisms and higher trophic levels are likely to be observed; unmeasured contaminants are likely contributing to sediment toxicity and benthic impairment; and, bioaccumulation of sediment-associated contaminants has the potential to adversely affect aquatic-dependent wildlife and/or human health.

Table 7.7 *Continued*

Possible outcome	Sediment chemistry	Toxicity test	Benthic community	Tissue chemistry	Possible conclusions
8	+	–	+	+	Contaminant-induced impacts on sediment-dwelling organisms and higher trophic levels are likely to be observed; toxicity tests are not sensitive enough to detect adverse effects; and, bioaccumulation of sediment-associated contaminants has the potential to adversely affect aquatic-dependent wildlife and/or human health.
9	+	+	+	–	Contaminant-induced impacts on sediment-dwelling organisms are likely to be observed; elevated levels of sediment-associated contaminants are likely contributing to sediment toxicity and benthic community impairment; and, bioaccumulation of sediment-associated contaminants is unlikely to be adversely affect aquatic-dependent wildlife and/or human health.
10	–	–	–	–	Contaminant-induced impacts are unlikely to be observed; sediment-associated contaminants are unlikely to adversely affect sediment-dwelling organisms; and, bioaccumulation of sediment-associated contaminants is unlikely to adversely affect aquatic-dependent wildlife and/or human health.
11	+	–	–	–	Contaminant-induced impacts are unlikely to be observed; the bioavailability of sediment-associated contaminants is likely to be limited; and, bioaccumulation of sediment-associated contaminants is unlikely to adversely affect aquatic-dependent wildlife and/or human health.
12	–	+	–	–	Contaminant-induced impacts are unlikely to be observed, based on the COPCs that were evaluated; unmeasured factors (e.g., physical factors or contaminants) are likely to be contributing to sediment toxicity; and, bioaccumulation of sediment-associated contaminants is unlikely to adversely affect aquatic-dependent wildlife and/or human health.
13	–	–	+	–	Contaminant induced impacts on sediment-dwelling organisms are unlikely to be observed, based on the COPCs that were evaluated; adverse effects on sediment-dwelling organisms are likely due to physical factors and/or unmeasured chemicals are stressing benthos and toxicity tests are not sensitive enough to detect effects; and, bioaccumulation of sediment-associated contaminants is unlikely to adversely affect aquatic-dependent wildlife and/or human health.
14	+	+	–	–	Contaminant-induced impacts on sediment-dwelling organisms are likely to be observed; high variability in the benthic community metrics may be masking contaminant-related effects; and, bioaccumulation of sediment-associated contaminants is unlikely to adversely affect aquatic-dependent wildlife and/or human health.
15	–	+	+	–	Contaminant-induced impacts on sediment-dwelling organisms are likely to be observed, based on the COPCs that were evaluated; unmeasured contaminants are likely contributing to sediment toxicity and benthic impairment; and, bioaccumulation of sediment-associated contaminants is unlikely to adversely affect aquatic-dependent wildlife and/or human health.
16	+	–	+	–	Contaminant-induced impacts on sediment-dwelling organisms are likely to be observed; toxicity tests are not sensitive enough to detect adverse effects; and, bioaccumulation of sediment-associated contaminants is unlikely to adversely affect aquatic-dependent wildlife and/or human health.

+, Indicator classified as affected; as determined based on comparison to the established target.
–, Indicator not classified as affected; as determined based on comparison to the established target.

7.5 Summary and conclusions

Contaminated sediments have the potential to affect adversely sediment-dwelling organisms, wildlife, and/or human health. Whenever practicable, multiple lines of evidence (i.e., data on multiple indicators of sediment quality conditions) should be used to assess the quality of freshwater, estuarine, and marine sediments. Procedures for determining if individual lines of evidence indicate that the beneficial uses of sediments are being impaired have been described in this chapter. The contingency tables presented in this chapter provide a basis for integrating the information on multiple indicators of sediment quality conditions and, in so doing, supporting informed decisions about the management of contaminated sediments.

Importantly, the weight of evidence generated should be proportional to the weight of the decision in the management of contaminated sediments. At small and uncomplicated sites, the costs associated with detailed site investigations are likely to exceed the costs associated with the removal and disposal of contaminated sediments. In these cases, SQGs represent cost-effective tools for establishing clean-up targets and developing remedial action plans (Wenning & Ingersoll 2002). At larger, more complicated sites, it is prudent to conduct further investigations when preliminary screening indicates that contaminated sediments are present. In such cases, the application of toxicity testing, bioaccumulation assessments, and other tools provides a means of confirming the severity and extent of degraded sediment quality conditions (Wenning & Ingersoll 2002). Application of toxicity-identification evaluation procedures and/or sediment spiking studies provides a basis of confirming the identity of the substances that are causing or substantially contributing to sediment toxicity (Ingersoll *et al.* 1997).

References

Ankley, G. T., Di Toro, D. M., Hansen, D. J. & Berry, W. J. (1996) Technical basis and proposal for deriving sediment quality criteria for metals. *Environmental Toxicology and Chemistry*, **15**, 2056–66.

ASTM (American Society for Testing and Materials). (2008a) Standard test methods for measuring the toxicity of sediment-associated contaminants with freshwater invertebrates. E1706-05E1. In *ASTM. 2008. Annual Book of Standards. Volume 11.05*. West Conshohocken, Pennsylvania: ASTM.

ASTM (American Society for Testing and Materials). (2008b) Standard guide for determination of the bioaccumulation of sediment-associated contaminants by benthic invertebrates. E1688-00(2007). In *ASTM. 2008. Annual Book of Standards. Volume 11.05*. West Conshohocken, Pennsylvania: ASTM.

ASTM (American Society for Testing and Materials). (2008c). Standard test method for measuring the toxicity of sediment-associated contaminants with estuarine and marine invertebrates. E1367-03(2008). In *ASTM. 2008. Annual Book of Standards. Volume 11.05*. West Conshohocken, Pennsylvania: ASTM.

Canfield, T. J., Kemble, N. E., Brumbaugh, W. G., Dwyer, F. J., Ingersoll, C. G. & Fairchild, J. F. (1994) Use of benthic invertebrate community structure and the sediment quality triad to evaluate metal-contaminated sediment in the upper Clark Fork River, MT. *Environmental Toxicology and Chemistry*, **13**, 1999–2012.

Canfield, T. J., Dwyer, F. J., Fairchild, J. F., *et al.* (1996) Assessing contamination in Great Lakes sediments using benthic invertebrate communities and the sediment quality triad approach. *Journal of Great Lakes Research*, **22**, 565–83.

Carr, R. S., Biedenbach, J.M. & Nipper, M. (2006) Influence of potentially confounding factors on sea urchin porewater toxicity tests. *Archives of Environmental Contamination and Toxicology* **51**, 573–79.

Chapman, P. M. (1992) Sediment quality triad approach. In *Sediment Classification Methods Compendium*. EPA 823-R-92-006. Office of Water. United States Environmental Protection Agency. Washington, District of Columbia.

Cook, P. M., Carlson, A. R. & Lee, H. (1992) Tissue residue approach. In *Sediment Classification Methods Compendium*. EPA 823-R-92-006. Washington, District of Columbia: Office of Water, United States Environmental Protection Agency.

Crane, J. L. (1996) Carcinogenic human health risks associated with consuming contaminated fish from five Great Lakes areas of concern. *Journal of Great Lakes Research*, **22**, 653–68.

Cubbage, J., Batts, D. & Briedenbach, S. (1997) *Creation and analysis of freshwater sediment quality values in Washington State*. Olympia, Washington: Environmental Investigations and Laboratory Services Program, Washington Department of Ecology.

Davis, W. S. & Lathrop, J. E. (1992) Freshwater benthic macroinvertebrate community structure and function. In *Sediment Classification Methods Compendium*. EPA 823-R-92-006. Washington, District of Columbia: Office of Water, United States Environmental Protection Agency.

Di Toro, D. M., Zarba, C. S., Hansen, D. J., *et al.* (1991) Technical basis for establishing sediment quality criteria for non-ionic organic chemicals using equilibrium partitioning. *Environmental Toxicology and Chemistry*, **10**(12), 1541–83.

Di Toro, D. M., Mahony, J. D., Hansen, D. J., Scott, K. J., Carlson, A. R. & Ankley, G. T. (1992) Acid volatile sulfide predicts the acute toxicity of cadmium and nickel in sediments. *Environmental Science and Technology*, **26**(1), 96–101.

Engler, R. M., Long, E. R., Swartz, R. C., *et al.* (2005) Chronology of sediment quality assessment methods in the United States. In *Use of Sediment Quality Guidelines and Related Tools for the Assessment of Contaminated Sediments*, R. J. Wenning, G. Batley, C. G. Ingersoll & D. W. Moore (eds), 311–43. Pensacola, Florida: SETAC Press.

Environment Canada. (1996) *The Ecosystem Approach: Getting Beyond the Rhetoric*. Prepared by the Task Force on Ecosystem Approach and Ecosystem Science at Environment Canada. Prepared for the Environment Canada Long Term Strategic Plan for Ecosystem Initiatives. Ottawa, Ontario.

Environment Canada. (1997) *Biological test method: Test for growth and survival in sediment using the freshwater amphipod* Hyalella azteca. Technical Report EPS 1/RM/33. Ottawa, Ontario.

Field, L. J., MacDonald, D. D., Norton, S. B., Severn, C. G., Ingersoll, C. G. (1999) Evaluating sediment chemistry and toxicity data using logistic regression modeling. *Environmental Toxicology and Chemistry*, **18**(6), 1311–22.

Field, L. J., MacDonald, D. D., Norton, S. B., *et al.* (2002) Predicting amphipod toxicity from sediment chemistry using logistic regression models. *Environmental Toxicology and Chemistry*, **21**, 1993–2005.

Harkey, G. A., Landrum, P. F. & Klaine, S. J. (1994) Comparison of whole-sediment, elutriate and pore water exposures for use in assessing sediment-associated organic contaminants in bioassays. *Environmental Toxicology and Chemistry*, **13**, 1315–29.

Hyland, J. L., Van Dolah, R. F. & Snoots, T.R. (1999) Predicting stress in benthic communities of southeastern U.S. estuaries in relation to chemical contamination of sediments. *Environmental Toxicology and Chemistry* **18**(11), 2557–64.

IJC (International Joint Commission). (1988) *Procedures for the assessment of contaminated sediment problems in the Great Lakes*. Prepared by the Sediment Subcommittee and its Assessment Work Group. Windsor, Ontario: Great Lakes Regional Office. 140pp.

IJC (International Joint Commission). (1991) *A proposed framework for developing indicators of ecosystem health for the Great Lakes region*. Council of Great Lakes Research Managers. Windsor, Ontario. 47pp.

IJC (International Joint Commission). (1997) *Overcoming obstacles to sediment remediation in the Great Lakes Basin*. White Paper by the Sediment Priority Action Committee. Windsor, Ontario: Great Lakes Water Quality Board.

Ingersoll, C. G., Dillon, T., Biddinger, R. G. (eds). (1997) Methodological uncertainty in sediment ecological risk assessment. In *Ecological Risk Assessments of Contaminated Sediment*. Pensacola, Florida: SETAC Press. 389pp.

Ingersoll, C. G., MacDonald, D. D., Wang, N., *et al.* (2001) Predictions of sediment toxicity using consensus-based freshwater sediment quality guidelines. *Archives of Environmental Contamination and Toxicology*, **41**, 8–21.

Ingersoll, C. G., MacDonald, D. D., Brumbaugh, W. G., *et al.* (2002) Toxicity assessment of sediments from the Grand Calumet River and Indiana Harbor Canal in northwestern Indiana. *Archives of Environmental Contamination and Toxicology*, **43**, 156–67.

Ingersoll, C. G., Bay, S. M., Crane, J. L., *et al.* (2005) Ability of sediment quality guidelines to estimate effects of sediment-associated contaminants in laboratory toxicity tests or in benthic community assessments. In *Use of Sediment Quality Guidelines and Related Tools for the Assessment of Contaminated Sediments*, R. J. Wenning, G. Batley, C. G. Ingersoll & D. W. Moore (eds), 497–556. Pensacola, Florida: SETAC Press.

Kemble, N. E., Brumbaugh, W. G., Brunson, E. L., *et al.* (1994) Toxicity of metal-contaminated sediments from the upper Clark Fork River, MT to aquatic invertebrates in laboratory exposures. *Environmental Toxicology and Chemistry*, **13**, 1985–97.

Long, E. R., Chapman, P. (1985) A sediment quality triad: measurements of sediment contamination, toxicity, and infaunal community composition in Puget Sound. *Marine Pollution Bulletin*, **16**, 405–15.

Long, E. R., MacDonald, D. D., Smith, S. L. & Calder, F. D. (1995) Incidence of adverse biological effects within ranges of chemical concentrations in marine and estuarine sediments. *Environmental Management*, **19**(1), 81–97.

Luoma, S. N. & Fisher, N. (1997) Uncertainties in assessing contaminant exposure from sediments. In *Ecological Risk Assessment of Contaminated Sediment*, C. G. Ingersoll, T. Dillon & G. R. Biddinger (eds), 211–37. Pensacola, Florida: SETAC Press.

MacDonald, D. D. (1998) *An ecosystem-based framework for assessing sediment quality in the Great Lakes Basin*. A short course on collection, analysis, and interpretation of sediment quality data: Applications of sediment quality guidelines (SQGs) and various companion tools. Chicago, Illinois: Great Lakes National Program Office, United States Environmental Protection Agency.

MacDonald, D. D. (2000) *Interests and needs related to the development of freshwater sediment quality guidelines for the State of Florida*. Workshop summary report. Prepared for Florida Department of Environmental Protection. Tallahassee, Florida.

MacDonald, D. D. & Ingersoll, C. G. (2000) *An assessment of sediment injury in the grand Calumet River, Indiana Harbor Canal, Indiana Harbor, and the nearshore areas of Lake Michigan*. Volume I. Prepared for the United States Fish and Wildlife Service. Bloomington, Indiana. 238pp.

MacDonald, D. D., Di Pinto, L. M., Field, J., Ingersoll, C. G., Long, E. R. & Swartz, R. C. (2000a) Development and evaluation of consensus-based sediment effect concentrations for polychlorinated biphenyls (PCBs). *Environmental Toxicology and Chemistry*, **19**, 1403–13.

MacDonald, D. D., Ingersoll, C. G. & Berger, T. A. (2000b) Development and evaluation of consensus-based sediment quality guidelines for freshwater ecosystems. *Archives of Environmental Contamination and Toxicology*, **39**, 20–31.

MacDonald, D. D., Ingersoll, C. G., Smorong, D. E., *et al.* (2002a). Assessment of injury to fish and wildlife resources in the Grand Calumet River and Indiana Harbor area of concern. *Archives of Environmental Contamination and Toxicology*, **43**, 130–40.

MacDonald, D. D., Ingersoll, C. G., Smorong, D. E., *et al.* (2002b) An assessment of injury to sediments and sediment-dwelling organisms in the Grand Calumet River and Indiana Harbor area of concern. *Archives of Environmental Contamination and Toxicology*, **43**, 141–55.

MacDonald, D. D., Ingersoll, C. G., Moore, D. R. J., *et al.* (2002c). *Calcasieu Estuary remedial investigation/feasability study (RI/FS): Baseline ecological risk assessment (BERA)*. Technical report plus appendices. Contract No. 68-W5-0022. Prepared for CDM Federal Programs Corporation and United States Environmental Protection Agency. Dallas, Texas.

MacDonald, D. D., Smorong, D. E., Ingersoll, C. G., *et al.* (2008) *Evaluation of the matching sediment chemistry and sediment toxicity in the Tri-State Mining District (TSMD), Missouri, Oklahoma, and Kansas*. Preliminary Draft. Prepared for United States Environmental Protection Agency Region 6, and Region 7 and United States Fish and Wildlife Service, Columbia, Missouri. Prepared by MacDonald Environmental Sciences Ltd. Nanaimo, British Columbia. United States Geological Survey. Columbia, Missouri and CH2M Hill. Dallas, Texas.

Macfarlane, M. (1997) *Investigation and remediation of sediments*. File 26050-01/General. British Columbia Environment and Resource Management. British Columbia Ministry of Environment, Lands, and Parks. Victoria, British Columbia.

Mah, F. T. S., MacDonald, D. D., Sheehan, S. W., Tuominen, T. N. & Valiela, D. (1989) *Dioxins and furans in sediments and fish from the vicinity of ten inland pulp mills in British Columbia*. Vancouver, British Columbia: Water Quality Branch, Environment Canada. 77pp.

MESL (MacDonald Environmental Sciences Ltd). (1997) *Lower Columbia River from Birchbank to the International Boundary: Water quality and quantity assessment and objectives technical report*. Prepared for Environment Canada, Vancouver, British Columbia and the British Columbia Ministry of Environment, Lands and Parks, Victoria, British Columbia.

Moore D. W., Baudo, R., Conder, J. M., *et al.* (2005) Bioaccumulation in the assessment of sediment quality: Uncertainty and potential application. In *Use of Sediment Quality Guidelines and Related Tools for the Assessment of Contaminated Sediments*, R. J. Wenning, G. Batley, C. G. Ingersoll & D. W. Moore (eds), 429–95. Pensacola, Florida: SETAC Press.

Newell, A. J., Johnson, D. W. & Allen, L. K. (1987) *Niagara River Biota Contamination Project: Fish Flesh Criteria for Piscivorous Wildlife*. Technical Report 87-3. Albany, New York: New York State Department of Environmental Conservation. 182pp.

NYSDEC (New York State Department of Environmental Conservation). (1999) *Technical guidance for screening contaminated sediments*. Albany, New York: Division of Fish, Wildlife and Marine Resources. 39pp.

OEPA (Ohio Environmental Protection Agency). (1988) Biological criteria for the protection of aquatic life. Volume 2. *Users Manual for Biological Field Assessment of Ohio Surface Waters*. Columbus, Ohio: Ecological Assessment Section, Division of Water Quality Planning and Assessment.

Persaud, D., Jaagumagi, R. & Hayton, A. (1993) *Guidelines for the Protection and Management of Aquatic Sediment Quality in Ontario*. Toronto, Ontario: Water Resources Branch, Ontario Ministry of the Environment. 27pp.

Reynoldson T. B., Day, K. E., Bailey, R. C. & Norris, R. H. (1995) Methods for establishing biologically based sediment guidelines for freshwater quality management using benthic assessment of sediment. *Australian Journal of Ecology*, **20**, 198–219.

Reynoldson T. B., Norris, R. H., Resh, V. H., Day, K. E. & Rosenberg, D. M. (1997) The reference condition: A comparison of multimetric and multivariate approaches to assess water-quality impairment using benthic macroinvertebrates. *Journal of the North American Benthological Society*, **16**, 833–52.

Smith, S. L., MacDonald, D. D., Keenleyside, K. A., Ingersoll, C. G. & Field, J. (1996) A preliminary evaluation of sediment quality assessment values for freshwater ecosystems. *Journal of Great Lakes Research*, **22**, 624–38.

Striplin, B., Baun, G. & Bilyard, G. (1992) Marine benthic community structure assessment. In *Sediment Classification Methods Compendium*. EPA 823-R-92-006. Washington, District of Columbia: Office of Water, United States Environmental Protection Agency.

Swartz, R. C. (1999) Consensus sediment quality guidelines for PAH mixtures. *Environmental Toxicology and Chemistry*, **18**, 780–87.

Swartz, R. C., Schults, D. W., DeWitt, T. H., Ditsworth, G. R. & Lamberson, J. O. (1987) *Toxicity of fluoranthene in sediment to marine amphipods: A test of the equilibrium partitioning approach to sediment quality criteria*. 89th Annual Meeting, Society for Environmental Toxicology and Chemistry, Pensacola, Florida. November (1987) 12pp.

Thomann, R. V. (1989) Bioaccumulation model of organic chemical distributions in aquatic food chains. *Environmental Science and Technology*, **23**, 699–715.

USEPA (United States Environmental Protection Agency). (1992a) Freshwater benthic macroinvertebrate community structure and function. In *Sediment Classification Methods Compendium*, Chapter 8. EPA 813/R-92/006. Washington, District of Columbia.

USEPA (United States Environmental Protection Agency). (1992b) Marine benthic community structure assessment. In *Sediment Classification Methods Compendium*, Chapter 9. EPA 813/R-92/006. Washington, District of Columbia.

USEPA (United States Environmental Protection Agency). (1994) *Assessment and Remediation of Contaminated Sediments (ARCS) Program*. Chicago, Illinois: Great Lakes National Program Office. EPA 905/B-94/002.

USEPA (United States Environmental Protection Agency). (1996) *Assessment and Remediation of Contaminated Sediments (ARCS) Program*. Calculation and evaluation of sediment effect concentrations for the amphipod *Hyalella azteca* and the midge *Chironomus riparius*. EPA 905/R-96/008. Chicago, Illinois: Great Lakes National Program Office.

USEPA (United States Environmental Protection Agency). (1997) *The Incidence and Severity of Sediment Contamination in Surface Waters of the United States*. Volume 1. *National Sediment Quality Survey*. EPA/823/R-97/006. Washington, District of Columbia.

USEPA (United States Environmental Protection Agency). (1998) *Guidelines for Ecological Risk Assessment*. EPA/630/R-95/002F. Risk Assessment Forum. Washington, District of Columbia: National Center for Environmental Assessment.

USEPA (United States Environmental Protection Agency). (2000a) *Methods for Measuring the Toxicity and Bioaccumulation of Sediment-Associated Contaminants With Freshwater Invertebrates*, 2nd edition. EPA 600/R-99/064. Washington, District of Columbia: Office of Research and Development.

USEPA (United States Environmental Protection Agency). (2000b) *Prediction of Sediment Toxicity Using Consensus-Based Freshwater Sediment Quality Guidelines*. EPA 905/R-00/007. Chicago, Illinois: Great Lakes National Program Office.

USEPA (United States Environmental Protection Agency). (2003) *Procedures for the Derivation of Equilibrium Partitioning Sediment Benchmarks (ESBs) for the Protection of Benthic Organisms: PAH Mixtures*. EPA-600-R-02-013. Washington, District of Columbia: Office of Research and Development.

USEPA (United States Environmental Protection Agency). (2005) *Procedures for the Derivation of Equilibrium Partitioning Sediment Benchmarks (ESBs) for the Protection of Benthic Organisms: Metal Mixtures (Cadmium, Copper, Lead, Nickel, Silver, And Zinc)*. EPA-600-R-02-11. Washington, District of Columbia: Office of Research and Development.

Washington State Department of Health. (1995) *Development of Sediment Quality Criteria for the Protection of Human Health*. Tier I report. Prepared by Environmental Health Programs. Olympia, Washington.

Washington State Department of Health. (1996) *Development of Sediment Quality Criteria for the Protection of Human Health*. Tier II report. Olympia, Washington.

Wenning R. J. & Ingersoll, C. G. (2002) Use of sediment quality guidelines (SQGs) and related tools for the assessment of contaminated sediments. Summary from a SETAC Pellston Workshop. Pensacola, Florida: SETAC Press.

Williamson, K. S., Petty, J. D., Huckins, J. N., Lebo, J. A. & Kaiser, E. M. (2002) HPLC-PFD determination of priority pollutant PAHs in water, sediment, and semipermeable membrane devices. *Chemosphere*, **49**, 703–15.

Word, J. Q., Albrecht, B. B., Anghera, M. L., *et al.* (2005) Predictive ability of sediment quality guidelines. In *Use of Sediment Quality Guidelines and Related Tools for the Assessment of Contaminated Sediments*, R. J. Wenning, G. Batley, C. G. Ingersoll & D. W. Moore (eds), 121–61. Pensacola, Florida: SETAC Press.

Index

Page numbers in italic, e.g. *34*, indicate figures or tables